Ethics in Neurobiological Research with Human Subjects
The Baltimore Conference on Ethics

Edited by

Adil E. Shamoo

School of Medicine
University of Maryland
Baltimore

Gordon and Breach Publishers

Australia Canada China France Germany India Japan
Luxembourg Malaysia The Netherlands Russia Singapore
Switzerland Thailand United Kingdom

Amsteldijk 166
1st Floor
1079 LH Amsterdam
The Netherlands

The Baltimore Conference on Ethics was sponsored by Friends Medical Science Research Center, Inc.

Eleven of the chapters appearing in this book were published originally in the journal *Accountability in Research: Policies and Quality Assurance*, volume 4, numbers 3–4. They are marked with asterisks(*) in the Table of Contents.

British Library Cataloguing in Publication Data

Ethics in neurobiological research with human subjects :
    the Baltimore conference on ethics
    1. Neurobiology — Research — Moral and ethical aspects
    2. Human experimentation in medicine — Moral and ethical
    aspects
    I. Shamoo, Adil E.
    174.2'8

    ISBN 2-88449-161-9

CONTENTS

---

* Appeared in the journal *Accountability in Research: Policies and Quality Assurance*.

Ethics in Neurobiological Research with Human Subjects

INTRODUCTION

Excerpts of opening session remarks and later comments by Dr. Adil E. Shamoo, Chairman, First National Conference on Ethics in Neurobiological Research with Human Subjects, Baltimore, Maryland, January 7–9, 1995.

In the past forty years, there has been a significant increase in research conducted on neurobiological disorders in terms of basic scientific knowledge, as well as the development of new treatment therapies. This expansion in research has led to significant advances in the treatment of schizophrenia, manic-depression, and other neurobiological disorders. The lives of thousands of patients have been much improved. This has led to new-found optimism among patients, family members, researchers, and health-care providers that exciting breakthroughs are forthcoming in the development of more effective drugs for treatment of neurobiological disorders.

Development of drug treatments for neurobiological disorders eventually requires use of the newly developed drugs by patients. Successful introduction of neuroleptics in the 1950s led to major advances in the treatment of schizophrenia and other neurobiological disorders. However, these medications had wide-ranging and very serious side effects, such as tardive dyskensia and varied patient responses to medications.

In recent years, there has been growing concern about the potential vulnerability of human subjects with neurobiological disorders who participate in research. This concern stems from the fact that patients with neurobiological disorders often suffer from impaired cognition. Stories reported in the media and at congressional hearings have sharpened the debate, and advocacy groups have voiced their uncertainty on these issues.

This serious public policy issue requires careful attention and constructive cooperation in developing ethical, functional standards to guide progress toward newer and more effective therapies for patients. These efforts must succeed; the alternative risks Draconian rules, in response to public outcry, not serving the best interest of the patient.

The purpose of this conference is to bring together ethicists, psychiatrists, researchers, family members, consumers, and representatives of government, industry, and academia to discuss all issues openly and honestly. Conference speakers will address history and ethics of neurobiological research with human subjects, current practices, informed consent, government oversight/institutional review boards, and patient and family perspective.

I have organized over a dozen national and international conferences throughout my career. The organizational process of this conference has been a most difficult and humbling experience. But I truly believe that my efforts and those of the organizing committee have been worthwhile and that the myths and fears of some of the participants will be proven unfounded. It is unfortunate that the American

College of Neuropsychopharmacology (ACNP), through its President David Kupfer, initially called for ACNP members scheduled as speakers to boycott the meeting. I am pleased that all speakers and participants are present.

The national dialogue on the ethics of psychiatric research came about through the hard work of family members and patient advocates, followed by a report on the issue to the National Alliance for the Mentally Ill (NAMI) in 1993 by this author, and then by the Shamoo and Irving paper in 1993 (chapter 1). A special issue of the *Journal of the California Alliance for the Mentally Ill*, "Ethics in Neurobiological Research with Human Subjects," (1994) preceded this conference. In all these discussions, family members who brought this issue to the forefront have invited and sought participation of the psychiatric research community and listened to their views. For example, in the journal issue, 13 of 27 articles were authored by psychiatric researchers. In this conference, 10 of 40 scheduled speakers are psychiatric researchers. In contrast, the psychiatric research community invited only 1 speaker (the author) of 7 on one side of this issue to the American College of Neuropsychopharmacology meeting in December 1994 in San Juan, Puerto Rico. Also, the one-day planning meeting of the Institute of Medicine (IOM)/National Academy of Science in February 1995 in Washington, DC on this issue — held as a result of our conference — consisted mostly of caregivers, funders, agency officials, and staff.

It is probably because I am an immigrant from a tyrannical regime that I cherish daily the freedom of discussion and dissent. I have a deep personal commitment to open healthy discussion of all sides of all issues. I totally reject silencing or censoring any group. I subscribe to John Stuart Mill's 1859 statement:

All silencing of discussion is an assumption of infalibility. Its condemnation may be allowed to rest on this common argument, not the worse for being common.

My experience in this regard was rewarded in 1994 when I co-led a four-year coalition effort in Maryland to pass the first and most comprehensive law in the country on mental health parity.

* * * *

I would like to thank the members of the conference organizing committee for their help, support, and kind understanding of this difficult area: Jay Katz, MD, Yale Law School, New Haven, Connecticut; Robert Destro, JD, Catholic University, Columbus School of Law, Washington, DC; Robert J. Levine, MD, Yale University, School of Medicine; John Kane, MD, professor of psychiatry, Albert Einstein College of Medicine, New York; Steven Sharfstein, MD, president and CEO, Sheppard and Enoch Pratt Hospital, and Janet K. Brown, JD, Baltimore, Maryland. I gratefully acknowledge Dr. Sharfstein's considerable efforts to convince all that each group is attempting to serve the public good, and Dr. Katz for his moral support before, during, and after the conference. A "giant" in the area of ethics whose presence lent considerable credence to the issues, Dr. Katz's closing remarks were profound and significant contributions.

I express my gratitude to Friends Medical Science Research Center, Inc. and its president Reverend Robert Mordhorst for their financial and moral support throughout the conference. I thank, in particular, two staff members from Friends without whom this conference would not have come to fruition — Kathy Bogan and Eleanor Bruns. Also, I acknowledge the assistance of my secretary Lenora Reese and graduate student Cheryl Dunigan. Finally, thanks to all the speakers and participants who made this conference and the proceedings a scholarly product that will endure for some time to come.

Our hopes were to reach a consensus on how to conduct research using human subjects with neurobiological disorders in an ethical manner. Initially, the conference was to develop — through a drafting commitee — a set of broad policy recommendations entitled "Baltimore Declaration on Neurobiological Research with Human Subjects." Unfortunately, we did not achieve this goal. The Baltimore declaration was scrapped in order to alleviate the fears of some psychiatric research participants. It was fortunate, however, that after several years of debate and immediately after the conference, NAMI adopted a set of policy recommendations (chapter 32). Now, we must ensure that these recommendations and the conference proceedings are widely disseminated and discussed. We hope that institutions, agencies, and individuals will incorporate part or all of these recommendations and information into their regulations, guidelines, and — more importantly — practices.

Anyone wishing to contact Dr. Shamoo may do so at: Department of Biochemistry and Molecular Biology, School of Medicine, University of Maryland–Baltimore, 108 N. Greene St., Baltimore, Maryland 21201-1503, USA; telephone: 410-706-3327; fax: 410-706-3189; e-mail: ashamoo@umabnet.ab.umd.edu

CONFERENCE SUMMARY

These conference proceedings are a collection of manuscripts submitted within a few months after the conference. The papers therefore represent a more expanded and detailed analysis of the subject than the original presentations. Also, the papers may reflect revisions in response to some of the dialogue at the conference.

Background papers (chapters 1–7) were distributed to attendees to provide a detailed historical and philosophical underpinning of the problems.

Chapter 1 by Drs. Adil E. Shamoo and Dianne N. Irving first published in 1993 and is reprinted here. It reiterates the importance of research with persons suffering from mental illness in order to make important advances in ameliorating the symptoms of these dreadful diseases. However, the paper makes an equally strong argument that those with mental illness are vulnerable; and thus a great deal of care must be taken in order to justify legally and ethically any research — most especially high-risk research. The chapter reviews the historical foundation of ethical principles and how persons with mental illness should have been declared a vulnerable group as was recommended by the 1979 National Commission for the Protection of Human Subjects of Biomedical and Behavioral Research. It takes the Office for Protection from Research Risks (OPRR) to task for not enforcing current regulations where it clearly states that vulnerable status should be accorded to any group not specifically mentioned in the regulation. The paper ends with specific recommendations: among them that OPRR should enforce the current regulations toward persons with mental illness when they are used as human subjects in research.

Robert J. Levine (chapter 2) articulates the reasons underlying the fact that federal regulations concerning biomedical and behavioral research — involving persons who are institutionalized as mentally infirm — were never officially issued. While the President's Commission on this subject in the early 1980s was very clear on the need for this protection, the author clearly places the blame for this current lacuna on the Department of Health, Education, and Welfare (DHEW) regulations that were drafted and deviated markedly from the commission's recommendations. DHEW writers included in their proposals several provisions not part of the commission's report. The regulations governing the use of children in research, by virtue of timing and congruency, were approved and promulgated. Among the issues which resulted in the perception of a lack of consensus were *mandated* as opposed to the *discretionary* appointment of consent auditors, a procedure which would greatly amplify the bureaucratic aspects of any proposal project.

Another contentious issue was the DHEW focus on the seriously impaired institutionalized population. It would be helpful here to more clearly define the concept and role of consent auditor for the interested but not *au courant* reader. The role of the advocate proposed by DHEW as a guardian angel in the review/consent process is equally confusing. The most difficult question to resolve in a just manner is to ascertain the potential subjects' capacity for self-determination and the extent to which it exists on a temporary or permanent basis.

Dr. Levine reviews the original recommendations of the 1979 National Commission in a comprehensive manner and permits the reader to judge the extent of the deliberations encompassed by this group. The overview also strongly invoked a National Ethical Advisory Board, as an upper appeal panel to be resorted to in specific circumstances. It would appear that the substantial body of work accomplished by the original National Commission, possibly revisited and updated, could serve as a federal framework for the special vulnerable populations both within and outside the institutional environment.

The concept that such vulnerable populations may accrue benefit from participation in clinical trials is novel to the last 15 years. It is exemplified by the response of the AIDS-related community to the testing of AZT medications. The author correctly asserts that the pendulum swing to this point of view that research activity is predominantly therapeutic and beneficial should not preclude ongoing and rigid scrutiny of the ethical considerations.

George J. Annas and Leonard H. Glantz examine the legal issues surrounding informed consent to research on institutionalized mentally disturbed persons in chapter 3. A thorough exploration of all aspects of this subject is essential prior to the adoption of regulations designed to protect the rights and welfare of this group. Historically, such populations have been developed as research subjects, frequently in areas of investigation unrelated to their group characteristics. Their desirability in this role arises from the fact that they are a captive cohort with a fairly homogeneous lifestyle and are associated with reduced fiscal costs to the project.

The authors argue that competent adult mental patients have the same constitutional standing to refuse treatment or medications as free-living persons. They also state that there is no obvious reason why their constitutional rights should differ from those of medical patients. Since World War II and the developement of the Nuremberg Code, both ethical and legal principles demand that experimental interventions be voluntary, informed, and understanding — i.e. competent. There are legal precedents set forth for court adjudication of enforced treatment, namely: the patient as a threat to others, his capacity to decide on other medical considerations of alternative therapy, and the risk of serious effects. A later ruling defined the parameters very narrowly and required consent of the court or an appointed guardian in non-emergency cases. Therefore in therapeutic treatment, information consent or consent by proxy is acceptable. Only in the case of controlling disruptive behavior as a societal good in the penal system has involuntary consent obtained judicial approval.

The most egregious recent case of research experimentation in a New York facility for the mentally retarded is described. An outbreak of hepatitis occasioned an attempt to test a vaccine on incoming children by pressuring the parent guardians. The legal decree absolutely forbade medical experimentation. It is concluded that given the history of experimentation on such populations, even in recent times, research should not be conducted with this group unless it is scientifically important and can only be done with this particular group of persons.

Chapter 4 by Robert Destro is presented in a legalistic format and style very different from the other chapters. It is filled with detailed and thorough footnotes. The author does an excellent job of explaining the evolution of the complex web

of divergent and convergent interests that currently constitutes the research regulatory system. He correctly points out that promulgated rules governing the
conduct of research are largely post-hoc, in that they have been designed to circumvent real problems which have come to light. It is of considerable interest, in
this particular political climate, that the growing number of state governments that
have passed laws concerning research on human subjects have elaborate statute
provisions more comprehensive, in general, than existing federal guidelines. The
Federal Code of Regulations, for instance, does not explicitly mandate any special
protective measures for members of vulnerable groups. The welfare of research
subjects is, in law, assigned as the responsibility of the investigators associated
with the project. It is significant that to the average lay person, the term "informed
consent" has come to mean a signed document as opposed to a process of mutual
understanding that develops from effective dialogue and detailed communication.
Destro probes further into the dilemma surrounding the application of informa-
tional disclosure and provision of safeguards to non-competent persons. Such
vulnerable populations may not be capable of truly participating in the process of
informed consent because of a host of factors.

The determination of competence and capacity is adequately protected within
the courts and legal settings in all states. This is not necessarily so in the health care
milieu and both Wisconsin and Virginia have enacted laws providing for more detailed oversight and guardianship in proposed research projects. Therefore, it appears that the cutting edge of developing policies and procedures in this regard is
to be found within the legal framework of individual state initiatives.

While clear and unambiguous rules and guidelines constitute an impressive
start in delineating these issues, the personal morality and ethical orientation of
the investigators is much more difficult to legislate. This last issue, in fact, needs
to be addressed, and appropriate behaviors of professionals need to be inculcated
in their training. Although the author does not allude to this facet, ethics training
is increasingly mandated as a specific educational component within both the
medical and scientific professions.

Sue K. Hoppe presents a well-written historical explanation of the origins and current status of institutional review boards in chapter 5. The research study on schizophrenia at UCLA has been a major impetus, goaded by the press, to an evaluation of
the operation of such boards. This study was funded by the National Institute for
Mental Health (NIMH) and therefore fell under federal regulatory purview. Since
1966, all institutions conducting such research have been required to maintain a
committee on human research, charged with reviewing the ethics of the proposed
work. In recent years, in practice, this has become defacto assurance of prior review
of submitted proposals by the institution itself. In 1974, the first federal regulations
on IRBs were issued and a National Commission was created culminating in the
Belmont Report — a summary of how such evaluations should be carried out. By
1981, further revisions to the basic document were promulgated, with more delineated procedural directives, e.g. size and composition of the board, definition of
special populations at risk and so on. Surprisingly, no special consideration was
accorded those patients institutionalized as mentally infirm because of conflicting
points of view concerning the adequacy of existing regulations. At present, the IRBs

function as parochial structures within the academic research sphere, lending flexibility and local control as needed. As the author points out, however, the growing fluidity of academic/commercial research ventures may render this model less appropriate now, as compared to 25 years ago. There does not appear in this recounting how IRBs are perceived by different constituencies nor does there appear to be any federal oversight of their operation. Does the OPRR at the National Institute of Health (NIH), for instance, receive regular reports from the institutions, such that a summary profile could be developed?

Hoppe endorses a national ethics advisory board that could more effectively attempt to balance the thorny question of societal good versus individual protection from harm than is possible for a local group. It is apparent that a primary issue most IRBs wrestle with is the dual criteria of protection of right vis-à-vis risk benefit projections. it is also apparent, by now, that many review boards seriously evaluate the quality of scientific knowledge to be gained from the project. This represents an unanticipated development from the initial emphasis on "do no harm" via informed consent. The nature of this evaluative process has shifted to focus on the promotion of subject welfare and personal empowerment, a change attributable, according to the author, to both experience as well as societal modifications. This chapter is a very informative summary of accumulated experience with local review boards and the complex interwoven pattern of national oversight.

The major thesis of chapter 6 by Rami Kaminski is a new and potentially powerful idea. It concerns the welfare of institutionalized mental patients who have been characterized as refractory or non-responders to established regimens of therapy. Dr. Kaminski is a psychopharmacology consultant to three large state hospitals and has developed a wide range of experience with this neglected problem. He has observed that in non-academic research milieu, prescribing psychiatrists adhere to very rigid medication schedules. When the outcome is unsatisfactory and not the expected amelioration, augmentation strategies involving an armamentarium of drugs are often employed. The patient then becomes warehoused as refractory to any other treatment and the professional health team is profoundly disillusioned and discouraged. Dr. Kaminski believes that two major factors are responsible for this series of events. First, in these settings a thorough medical workup of the patient involving a search for possible neural, metabolic, and neuroendocrine disorders is rarely conducted. Second, medication is prescribed on the basis of symptomatology expressed and rarely is an attempt made to construct a differential diagnosis. A different approach, allowing careful experimentation with dosing schedules and a knowledge of clinical pharmacology, would perhaps be beneficial in more effectively tailoring the treatment to the unique characteristics of the non-responder. He makes a very good point that symptoms and diagnoses become very confused in institutional care.

Dr. Kaminski inaugurated a pilot study in which his academic institution adopted a large state hospital. Each week, a team of consultants conducted actual case conferences with all the members of the treatment team present. Coordinators at the hospital facilitated the treatment plan and the follow-up collection of data. The outcome has been marked improvement in the treatment culture at the hospital. Many patients, long considered hopeless, underwent marked-to-significant betterment in their lives and, most importantly, there was a dramatic change in the morale and enthusiasm of health caregivers. The perception that they could, with mentoring help, alter a situation they had learned to accept, most likely had a positive impact on each and every patient.

This simple idea of cross-linkages between academic investigators and those responsible for care of the most difficult patients deserves national attention by medical and scientific professionals. A properly constructed data collection and monitoring system, together with its dissemination, could conceivably result in significant information retrieval and a modification of treatment modalities. At present, no such systematic research project is under way.

The last background paper, authored by Drs. Irving and Shamoo (chapter 7), reviews the literature on washout/relapse on patients with schizophrenia. This paper refers to Shamoo and Keay's paper that appeared later in the *Cambridge Quarterly of Health Care Ethics*. The manuscript reviews 30 years of washout/relapse research on patients with schizophrenia. The most important aspects of the findings are that nearly 38% of patients participating in the U.S. studies underwent relapse and about 10% dropped out, without any information provided with respect to their current location. The most troubling experiments were those using a drug challange protocol in order to induce symptoms of the illness. The paper reviews the ethical considerations of such experiments, describes the ensuing controversy, and raises a host of ethical questions that ought to be asked about this kind of research.

The first conference paper, by Leonard Rubenstein (chapter 8), concerns the differing philosophies and points of view between the research community and the public at large, with respect to accountability standards and practice in research. Many of the conference presentations focused on the underlying principles and pragmatic or legal elaboration of informed consent. In the context of a particular case history (*Orlikow v. United States*) the author articulates the personal views of contemporary researchers. These can be stated as a substantial belief that ethical considerations are impractical and obtrusive. They also fail to see any significant distinction between their dual roles as physicians and research investigators. Inadequate attention to informed consent may in fact be "normal" operating procedure but very little data exists to verify these perceptions. Problems relating to consent continue to arise despite even more elaborate regulations and safeguards. Dr. Rubenstein puts forward the view that an unexamined explanation for this state of affairs is a lack of definitive clarification as to whose sense of ethics and whose standards of accountability apply. The model from which guidance in the matter is derived is highly critical. In the allied field of medical malpractice the standards involved are those of the community of practicing physicians.

In civil rights law, standards of accountability are predicated on those of society at large. The legislation alluded to above involved the development of mind control or brainwashing techniques, accompanied by technological invasiveness, by a highly respected, competent, and caring psychiatrist who became president of the APA. He used private patients, utterly unaware of what was being perpetrated upon them, and the results were widely reported and published and later adopted and fostered by the CIA. The federal government admitted the total lack of safeguards for the subjects but argued that the conduct should be judged by the ethical standards of the time. The opposing professional witnesses despite the promulgation of the Nuremberg Code after World War II, disagreed on the fundamental point as to whether or not accountability and informed consent were intrinsic facets of the research milieu in the 1950s and 1960s. If the ethos of medical researchers is judged by public standards and expectations, there is clearly a vast gap, even currently, between the latter and the actual mindset of many investigators that characterizes the medical research community.

The growing apparatus of regulation, including IRBs and detailed oversight, would appear to indicate the dominance of societal rather than the researcher-associated set of values. However, because of the intimate, often intense, relationships between potential subjects and health-care providers and their intramural institutional settings, the persistence of research community standards of accountability can be expected to endure for some considerable amount of time.

The last two decades have seen the emergence of consent/benefit analysis in balancing competing demands in many regulatory areas. While computations or assessments on this score are notoriously fraught with difficulties in any sphere, the author correctly concludes that further progress must openly acknowledge the existence and extent of different ethical approaches and consensus in the divergent views of society at large and those operating within the research community.

In chapter 9, Dr. Shamoo deals with the historical and global concepts of human rights applicable to persons with mental disabilities or illness. In the late 1940s, the United Nations first announced the Universal Declaration of Human Rights, declaring that these rights prevail, irrespective of race, sex, national origin, and, implicitly, of illness of any type. Nearly five decades later, the UN issued another declaration — Principles for the Protection of Persons with Mental Illness and for the Improvement of Mental Health Care. Dr. Shamoo contends that this specification of a certain class of human beings as being in need of special protection is stigmatizing in and of itself. He maintains that the initial proclamation of this world group should inherently encompass people with mental illness and that the proscribed exclusions in the original statement should have been expanded to include the terms "illness" and "disability."

In pragmatic terms, however, the 1991 declaration serves to emphasize that insufficient attention worldwide has been or is being paid to the serious difficulties encountered by people with mental problems. Modern emergence of ideas concerning the awareness of individual human rights was largely associated with the need for protection from abusive political structures and systems. Early philosophers considered that inherent individual rights could neither be taken away nor voluntarily surrendered.

The U.S. Declaration of Independence was one of the earliest documents that sought to make these rights explicit. It implies that all human beings possess these rights, derived from their Creator. The author asserts that unfortunately the framers of this document *did* intend to discriminate on the basis of race. The civil rights law, at least legally, remedied this in the 1960s.

It does appear that certain writers in the field sustain paternalistic notions regarding the category of individuals under discussion, even classifying incompetent humans alongside animals. Apart from the philosophical issue of how and when persons with mental illness possess moral competency, the vast contribution of many individuals to the advancement of humankind cannot be attributed to any specific range of mental attributes. The most intrinsic and irreducible arguments in this debate ought to center on the ethics of "sanctity of life" rather than relating to "quality of life" ethics.

Dr. Shamoo strongly believes that the rights of the mentally ill are firmly ensconced in our nation's Constitution. Open affirmation of this fact would dismiss the current ambivalence toward this group. The difficult task of translating rhetoric into enforceable practice still remains. The Constitution Equal Protection clause has not prevented medical insurance companies from denying coverage to these patients, despite the 1991 Americans with Disabilities Act.

Utilizing the mentally ill as subjects in research projects is the major theme of this conference. The dilemma posed here is the junctional conflict between two bedrock values within society, namely: freedom of scientific inquiry and the preservation of individual inviolability. Relegating the resolution of such oppositional forces to the individual research investigator constitutes an awesome burden. Among the outstanding questions to be addressed are:

1. Should participation be directly linked to the likelihood of positive benefit for the subject?

2. Should placebo protocols continue to be used in these studies?

3. Should the research investigator be the main purveyor of medical care to the subject?

The answer to the last question would appear to be a clear and definitive negative. Responses to 1. and 2. are still highly controversial and revolve around the merits attached to the use of an individual to promote greater future good to the many. The idea lies at the heart of efforts to "improve" the informed consent process.

Dr. Shamoo concludes that Nelson Mandela's pithy summary of equality, "The Oneness of the Human Race," is a phrase that ought to be adopted and expressed in both word and deed in the canon of behavior toward mentally ill persons by both the lay and professional communities.

Timothy Keay, MD (chapter 10) surveys a few historical cases of ethical lapses in the medical profession, and reviews the historical and philosophical underpinnings of informed consent. Dr. Keay cites recent data pertaining to his work with Dr. Shamoo on washout/relapse studies where some appalling practices such as washout and relapse are currently the norms. The author suggests improvement in informed consent content and process. Furthermore, he contends there should be a balance between proceeding with certain research and the cost of not doing such research.

Chapter 11 by Gregory Aller and Robert Aller is a pivotal chapter in the book, being an account rendered by Gregory Aller, a research subject in the UCLA schizophrenia project, and his father Robert Aller. They recount, alternatively, their individual experiences during and after participation in these investigative studies.

This project, the conduct and outcome of which was a major catalyst for this conference, was funded by NIMH in the early 1980s. Research was supported in order to follow the course of schizophrenia in more than 100 recent onset patient-subjects. The studies were designed in 3 phases. First, patients were treated with a fixed dose of antipsychotic medicaton for one year. Next, those deemed eligible entered a double-blind protocol consisting of random alternation of 12 weeks of placebo and 12 weeks of medication; and, finally, they entered a neuroleptic withdrawal phase in which medication would be discontinued for 1–1½ years in those who were clinically stable during the crossover phase.

Termination within the study could occur if the patient withdrew permission or if clinical relapse or psychotic exacerbation occurred. At this point in time, 4 of the involved families believe that there were significant breaches of ethical and legal adherence to federal regulations governing such research. Many research papers have been published arising from this project. The OPRR determined in 1993 that investigators had published clinical data obtained outside parameters of IRB approved research protocols.

The main issues raised by the families were: a. lack of proper informed consent; b. failure to meet patient needs; and c. failure of the research group or institution to respond to patient and family complaints. The underlying key questions are whether the subject-patients were deliberately exposed to severe exacerbation and whether relapse was an intrinsic component of the research design. Dr. Jay Katz, author of chapter 33, the final chapter in this book, holds the opinion that expectation of relapse was an integral part of the research design, intentionally induced by the investigators. This reading of case is substantiated by the investigators' own publications in the clinical literature. It is also clear that these goals were obscured and not disclosed in the informed consent process.

Both Gregory Aller's parents are alumni of UCLA and retain an abiding respect for the institution. They felt especially fortunate that their son could receive free care for his devastating illness at a premier treatment center, while contributing to advancement of knowledge concerning the disease.

In Gregory's account, he experienced the best year of his adult life when placed on the standardized medication. He enrolled in college, earned a 3.8 GPA, and made Dean's List while working 15 hours a week. The subjects as a group were cossetted and entertained with activities. The transfer to the next phase was ushered in with many positive reinforcements and no explanations of possible crossover/withdrawal from attending physicians. The principal investigator assured the family that Gregory would have the medication reinstated if and when his parents reported on the return of symptoms. In the withdrawal phase, his symptoms returned with severe exacerbation. He threatened and/or tried to kill both his parents. His physician saw him for only 5 minutes twice a month. Despite pleas from his parents and sister, orally and in writing, for resumption of medication, Gregory became more dangerously psychotic and eventually indigent and homeless. Medication, when finally accepted by the patient, proved to be needed at a higher dose than during phase 1.

Six months later, the Allers determined to present their concerns, which included fear of murder or suicide among participants, to UCLA. Their requests for information were met with extensive delays, and ultimately the surrender of altered medical records. They subsequently filed a formal complaint with NIH which resulted in a 3-year investigation focusing primarily on the informed consent process. This process, in retrospect, was woefully lacking in many respects. There was no designation of alternative treatment to be provided. Critical information, already known to the investigators at the time of grant submission, was withheld — e.g. that relapse upon withdrawal is significantly more intense than relapse during medication and can engender permanent damage in the form of loss of intellectual and social functioning. Both state and federal laws in force at the time mandated a full disclosure of risks.

In 1994, OPRR elicited compliance with the Code of Federal Regulations but UCLA continued to deny malfeasance. The experiences of two other patients in this program are also detailed here. The lack of medical care provided, coinciding and colliding with the results of the research protocol, is very striking. Indeed, Gregory Aller was comparatively fortunate in his family's educational background, resources, and commitment to his welfare. Abandonment and cynical disavowing by clinicians is now a shared experience of the families protesting the UCLA research conduct.

In summary, it is very clear, in this particular case, that the welfare of the subject-patients was grossly subordinated to the achievement of carefully suppressed research aims. These omissions were both material and deliberate and this conduct had complete institutional approval from the university.

It is hoped that publication of this paper in particular, as disturbing as it is, will lead to full and open discussions of this subject in broader settings.

The Allers have a dispute with the University of California currently in litigation, and the UCLA administration faxed a statement (chapter 12) to the conference which was read after the Allers' presentation. The statement, written in legal terms, claims that there have been allegations and misrepresentation of facts regarding the Clinical Research Center. The center maintains it is absolutely false that: 1. it deliberately withheld medication from subjects when needed; 2. one of the subjects committed suicide; and 3. it failed to adhere to ethical standards. Furthermore, the statement quotes exonerating segments of the OPRR investigation report on various allegations, and states that the center is forging alliances with all those involved in the research.

Chapter 13 is authored by Vera Hassner Sharav, chair of the Quality of Treatment Committee of the New York State Alliance for the Mentally Ill. In the past, this organization, in their desperate search for more effective therapies, wholeheartedly lent their full approval to all forms of neurobiological research. Their trust in the integrity and goodwill of research scientists has been gravely imperiled by the increasing number of reports of unethical conduct in certain studies. This growing concern has been translated into a conviction that the psychiatric research arena needs to be thoroughly explored and that intensive and independent vigilant oversight be brought to bear on all such research investigations. According to this author, the virulent objection of at least one professional society in the mental health field to the holding of this particular conference on neurobiological ethics and the general lack of commentary by the medical profession on the abuses which have occurred constitute a "fraternity of silence." She believes that a strong coalition of concerned professionals and lay people can be effective in developing and implementing proper safeguards and codes of conduct.

In particular, this chapter takes to task the philosophic position articulated by Dr. Frederick Goodwin, former director of NIMH. Dr. Goodwin has stated that the installation of protection for human subjects may damage the physician/patient relationship and lead to stigma. He has also advocated using institutionalized persons not currently protected by federal law for the pursuit of research questions irrelevant to their own psychiatric diagnosis. The author claims that cruel and unusual experimentation currently is being performed by psychiatrists operating within accredited institutions in New York State and Cincinnati, Ohio. Since the medical institution defends its members and their current mores and practices, Hassner Sharav concludes that the welfare of mentally disabled patients is crucially undermined by personal and professional capitalistic interests and also by lack of an independent patient advocacy system within IRBs. Her proposal for an Independent Research Intermediary is analogous to the NIMH Intramural Durable Power of Attorney (DPA). We are left wondering why the DPA philosophy and policy does not extend to extramural research projects funded by NIMH.

Janice Becker is from Baltimore, Maryland, where her younger daughter Laura was a research subject at the Maryland Psychiatric Research Center (MPRC) for over five years. Ms. Becker writes a heart-wrenching account (chapter 14) of a mother in pain watching her daughter tormented by illness and lack of attention. Little understanding and compassion were forthcoming when she brought her concerns to the attention of some of the MPRC staff. Mr. and Mrs. Becker recount the evasive answers they received to their questions during their daughter's stay. They were never told that Laura was in a drug-free interval: "Laura was kept in a drug-free interval for over a year. During this time, she was very psychotic. Laura suffered and we suffered with her."

Ms. Becker describes some of the difficulties her daughter experienced through those years; she feels betrayed and that she betrayed her daughter by not removing her from MPRC. She questions whether her daughter's suffering was scientifically justified.

Peggy Straw — one of the founders of NAMI and founder and past president of the New Hampshire AMI — reflects a point of view in chapter 15 that is rarely articulated in print. The author is a representative of the Alliance of the Mentally Ill and a member of the IRB for the Division of Mental Health in New Hampshire, and she discusses her experience in that capacity. Instructions to this board were to examine the impact of the proposed research projects on the human subjects and the legal aspects thereof. Proposals under consideration had already been approved by the academic institution where the research was to be conducted and any recommendations to the investigators would therefore be of an advisory nature.

The point is well made here that the experimental subjects are permanently dependent on their family support structure and such relationships should be preserved; this is a factor which is often ignored or overlooked by medical professionals in these studies. It is true that many physicians are drawn to their particular specialty because of personal experience. Their influence as "sensitized" experts is particularly helpful on such review committees.

In summary, this is a thoughtful commentary by a concerned and knowledgeable representative of the subject population and puts forward several cogent recommendations pertinent to the composition and operation of institutional review boards.

Chapter 16 by Frederick J. Frese, III presents a very unique point of view. Dr. Frese has been engaged at the professional level in clinical research in psychology and has suffered from schizophrenia for the past three decades, requiring numerous hospitalizations.

He believes the reassessment of mores, rules, and regulations in the conduct of psychiatric research needs to be continuous and ongoing. Psychiatric diagnostic manuals are, in fact, revised every seven years, as professional understanding and knowledge of mental illness alters. Dr. Frese wishes very much to participate as a volunteer in research, with the provision that he be allowed to choose the majority of the members of the relevant institutional review boards. He cites three specific individuals with professional credentials who have also suffered from the same disease.

Dr. Frese makes a strong point that close and caring relatives of subjects would function equally well on his behalf in this setting. Such a system offers protection to the research volunteer during periods of mental incapacitation. The need for research, he acknowledges, is overwhelming. His suggestion that a trusted guardian be the ultimate decision-maker is a sound one. His final recommendation that a recovered patient be involved in any review of the informed consent mechanism in order to ensure a humane and valid experience is novel and worthy of serious consideration.

The text of a talk delivered by Dr. Shamoo to the annual meeting of the American College of Neuropsycholopharmacologists (ACNP), December 13, 1994, is chapter 17 of this book. This speech primarily reviewed media attention to this problem, demonstrating how it can distort but emphasizing that this is the price of an open and democratic society. Dr. Shamoo explores the historical and philosophical aspects of the use of the mentally ill in research and discusses the washout/relapse data that he developed. The central theme of his presentation is the ethical dilemma arising from whether research has direct, indirect, or futuristic implications and provides no benefit to the subject. Dr. Shamoo states that high-risk research — e.g. washout/relapse — should proceed only if there are unique and compelling reasons why such a research protocol should be pursued.

Lana Skirboll, David Shore, MD, Andrea Baruchin, and Paul Scrovatka are the authors of chapter 18, which combines the content of two separate presentations during the conference by Drs. Skirboll and Shore. It deals with continuing efforts of the National Institute of Mental Health to pursue new initiatives regarding its responsibility to safeguard the rights and safety of participants in clinical research protocols. Formal mandates include compliance with federal human subject protection regulations elaborated in 45 CFR 46 and guidelines issued by the NIH Office for Protection from Research Risks (OPRR).

In 1993, the NIMH and the National Alliance for the Mentally Ill (NAMI) co-sponsored two meetings which explored ethical problems in psychiatric research involving human subjects. Attendees included NIMH staff, clinical investigators, members of institutional review boards, and family and consumer representatives. Conference chair Dr. Shamoo participated as a NAMI board member and a member of the NAMI science review committee. A primary objective of these sessions was the improvement of current informed consent processes. Agreement was reached on the need to focus on increasing awareness of the contemporary issues by *specific educational outreach measures* to each of the following groups:

- Members of the clinical research community;
- Members of institutional review boards;
- Potential participants in research projects.

All researchers who work with human subjects must comply with federal regulations now in place to protect said subjects. By virtue of mandated training in this area, clinical investigators should be familiar with these concerns as they embark on their careers. In 1993, NIMH alerted all program directors of training grants of the need to exert surveillance in this matter. While the principles of conduct remain unchanged, there is no specific iteration of what is unacceptable. The ethical standards by which research protocols are judged are considerably more restrictive than those of 10 or 15 years ago. Primary responsibility for an informed consent process resides with the Principal Investigator (PI) who in turn is monitored and advised by the local IRB. Prior to an NIH review, this group evaluates the proposal for submission of an application along with the human subject protection provisions. Any problems identified must be redressed before funding. To aid individual researchers, the OPRR mails out findings and interpretations of the relevant regulations. This initiative was prompted by the investigation of a funded project on schizophrenia.

In addition, the NIMH, in conjunction with the American Psychiatric Association, is developing an investigators' resource manual that will articulate the "pros and cons" of possible solutions to the concerns of all involved parties. Issues covered will encompass research design; methods of assessing quality of health care; and effective strategies for timely and effective informed consent.

There are approximately 4,000 IRBs operating at this time. By federal law, this responsibility is assigned at the local community level and their composition is dictated fairly specifically. A series of OPRR-sponsored regional workshops for IRB members and researchers, together with panels of expert psychiatric clinical scientists, has been initiated. It is intended to develop an NIMH package of materials for IRB members who have limited familiarity with modern psychiatric and psychological research. This resource should enable IRBs to execute their duties more appropriately.

In order for the funding sponsor (NIMH) to fulfill its overseeing function and to provide regular updates more effectively, in-house training has been instituted for NIMH staff. In addition, staff are strongly encouraged to participate in scientific and consumer meetings and to contribute to the literature in this field. There is an increasing acknowledgment that research subjects are not "guinea pigs" but should be active, involved participants in the pursuit of new knowledge. Since many of these issues have risen to public prominence through the efforts of consumers and their families, the NIMH knowledge-exchange office together with NAMI is developing a consumer guide to clinical research. This publication will provide a "road map" to aid potential research subjects and their families or designees in making crucial decisions. Further, NIMH sponsored a symposium at the 1995 NAMI annual meeting.

It is very clear from this chapter that the sensitivity of NIMH to the need for outreach, education, coalition, and consensus building has greatly escalated in recent years. Since regulatory authority over IRBs resides with OPRR, this role is consonant with the institute view that additional layers of new regulations recognizing mentally ill persons formally as needing protection would be both paternalistic and stigmatizing.

John Kane, MD and Michael Borenstein deal with the concept and design of controlled clinical trials as they pertain to research studies of therapeutic efficacy in mental health research (chapter 19). The construction of double-blind, random-assignment methodology with sophisticated statistical analysis is now a conviction of the medical health field. The issue addressed here is whether or not placebo controls are to be included in these trials.

It has been argued that it is both unscientific and unethical to use such controls routinely when a treatment is available. In this case the appropriate question is whether the proposed treatment is more effective than the old method? It is widely recognized, however, that the "placebo" concept covers a wide range of factors not easily subject to assessment in any other fashion.

The authors state that there are no universally accepted standards of efficacy and safety. Currently available neuroleptic drugs induce serious adverse reactions. The cumulative incidence of tardive dyskinesia, after 5 years of treatment of young adults, is reported to be in excess of 25%.

A major goal of drug research and development is the characterization of new and better formulas without these propensities. It is argued here that the use of placebo controls in such trials is necessary and that active (positive) controls will not suffice. They point out that there is an overlapping range of response between placebo and drug effect in many clinical trials. Since the reasons for this diversity are unknown, a true placebo is required. In specific trials and under particular circumstances where sampling error and sample size pose no problem, the placebo washout control can be dispensed with.

It is arguable that studies that attempt to show the superiority of a new treatment (using a placebo control) require fewer participants than those designed to demonstrate equivalence (using an active control).

The authors allude to the Gilbert Study, reviewing relapse incidence in 66 studies where the mean cumulative relapse rate was 53% in patients withdrawn from medication, and 16% in those on maintenance over a 9.7-month period. The variability was enormous and the relapse criteria were markedly varied. A good case is propounded here that in complex psychiatric illnesses such as schizophrenia, which are only partially treatable with toxic treatments, the issue of placebo controls should not be automatically rejected on the basis that proven therapies exist.

Chapter 20 by William T. Carpenter, Jr., MD seeks to challenge the charges that have been leveled at the conduct of modern schizophrenic research. It is the author's contention that proposals of the ethics and patient advocacy communities with respect to assertions of physical harm to research subjects lack documentary proof. Dr. Carpenter considers that many of the new proposals relating to the governance of such research do not take into account the clinical realities of the disease process and that they offer a serious impediment to research progress. He depicts the course of ordinary clinical care and profiles the clinical patterns following the first diagnosis. The majority of patients display a relapsing form of the illness; a significant minority will not relapse but the negative effects of medication render them seriously non-compliant over time. Undoubtedly from a clinical perspective, the vindication of valid indicators of relapse potential would be a most valuable tool and the author declares such research methodology not unethical, within the psychosocial context. With risk minimization and with the informed and voluntary consent form suitable modified, he concludes along with the Office of Protection from Research Risk that the California study did not pose harm to patients.

The author takes Drs. Shamoo, Keay, and Irving to task for their criticism of outpatient studies. The question posed in these studies is whether the routine excessive medication pattern currently deployed could be altered. Dr. Carpenter believes that these determinations of minimum effective dose have resulted in therapeutic benefit, while reducing side effects and encouraging compliance. He believes that the motivation of the investigators involved has been misrepresented. Dr. Carpenter cites a recent publication detailing a meta-analysis of 4,365 patients in drug withdrawal which found no lasting damage to those involved. The enriched clinical care in research settings may contribute to these findings and the author believes that outcomes are certainly no more deleterious than in standard clinical care. However, the majority of these patients underwent relapse. Therefore, the author makes the implicit assumption that relapse causes no harm in any form. But he puts forward no data to support his claim that relapses cause no serious physical harm to the brain. It is important to note that harm to the dignity of the patient is at least of equal importance.

Patients receiving clinical care in the research clinic of the Maryland Psychiatric research Center undergo very intense monitoring and psychosocial and psychoeducational services when off medication, resulting in fewer hospitalizations. The provision of such follow-up services may be quite the exception. However, studies of pharmacological challenges pose a different problem. Attempts to down-regulate dopamine receptors *in-vivo* would appear to be especially hazardous and this particular hypothesis could be pursued in alternative research models. It is, however, the firm belief of this clinical investigator that care in the research setting is vastly superior to that elsewhere and that significant benefits to the patient offset the risk elements associated with research procedures.

In summary, Dr. Carpenter believes that much of the current debate concerning these issues has been handled in a non-professional and non-collegial manner by the ethics community spokesperson. This concern is markedly heightened by the crisis in medical research funding, especially in the arena of mental health.

Taking Dr. Shamoo to task, Dr. Carpenter quotes from Dr. Shamoo's chapter 17. Dr. Shamoo contends that this paper is consistent with his advocacy for appropriate research and that washout/relapse are high-risk studies that have to be conducted only for unique and compelling reasons.

Trey Sunderland, MD and Ruth Dukoff, MD delineate the systematic approach that has been developed within the intramural research program at the NIMH to ensure adequate safeguards in the development of the cognitively impaired in research (chapter 21). They first discuss five major underlying issues, both clinical and ethical. There is an inherent degree of confusion in documenting cognitive impairment since it may be transient, progressive, or permanent. Dysfunctions articulated are descriptive terms that may be correlated with, inclusive of or exclusive of, specific medical diagnoses. With respect to legal and financial areas, the courts constitute the final decision-making path. There are no quantitative standards within the medical research community that assess the extent of impairment vis-à-vis informed consent.

It is interesting that, in the experience of this author (Dr. Shamoo), it is the community of neuropsychologists who have developed the most comprehensive test profiles for this purpose.

It is the role of IRBs to make specific recommendations for particular research projects under consideration. At this point in time, they could function in an evolutionary mode, absent federal law standard guidelines or experience. The NIH determined to establish its own guidelines for the intramural performance of research. Since 1985, the concept of Durable Power of Attorney (DPA) has been used at NIH and the experience of the past decade is recounted here. The potential research subject must understand the function of a DPA, and be able to discuss his wishes with his own designee. One important facet of this instrument is the selection of a DPA in anticipation of need. A series of eight case studies presented here aid in enabling an IRB to choose this route. The two factors that need to be balanced are the competency level of the individual and the degree of risk inherent in the research protocol. Increasing levels of risk and decreasing degrees of competency dictate more rigorous review of the particular case and possible exclusion. NIH requires consultation with a bioethicist when any change in these two factors occurs.

Drs. Sunderland and Dukoff recount the experience of the Geriatric Psychiatric Research Group at NIMH in utilizing this system. All 82 patients were assigned a DPA at enrollment. Because of this fact, many of these individuals were able to continue in the program despite progressive dementia. The patient assent is continuously solicited even though formal consent authority may reside with the DPA.

The authors argue that given the simplicity and effective operation of the DPA process with Alzheimer's disease, it deserves to be applied to other research projects. When used properly, the DPA model extends the choice of an individual to participate in research and gain a measure of self-esteem beyond the time that cognitive dysfunction deprives him of his capacity of informed consent.

Dr. David C. Thomasma's paper, chapter 22, is searching for a model in which to protect the vulnerable from harm — i.e. harm to dignity as well as physical harm — while leaving a window for valid and ethical research. This paper, like other papers in these proceedings, reviews the historical, philosophical, and ethical background considerations for the protection of human subjects — especially the vulnerable population with mental illness Dr. Thomasma ends his article with the statement:

Thus, the Principle of Dominion would state that for every intervention into a natural process, special care must be taken not to objectify or manipulate for our own ends, however noble, the person for whom the intervention is contemplated. Since that individual cannot make judgments for him- or herself, analyses of the consequences on his or her life, and on the lives of others, must be an intrinsic part of the decision to intervene. Under this principle it would be impossible to justify non-therapeutic research on neurologically impaired individual. Some direct benefit must be contemplated and designed into the protocol.

Chapter 23 by Dr. Paul S. Appelbaum is a continuation of a series of lengthy studies of competence of patients with mental illness. These competence tests are now called the MacArthur Competence Tests. In this chapter the author reviews what he and his colleagues have accomplished in determining a subject's competence to consent to research, such as the ability to communicate a choice; understand relevant information; appreciate the nature of the situation and its consequences; and manipulate information rationally. In order to make his point, Dr. Appelbaum contends that demanding total comprehension and appreciation by patients with mental illness is not appropriate since most "normal" subjects would fall short of this goal. The author is clearly thorough in his recitation of all the numerous variable in determining competence of patients. However, he recognizes the difficulties in dealing with this area and is therefore looking for new apprcaches by surrogate decision-makers on behalf of patients.

Eric Rosenthal (chapter 24) describes the UN International Covenant on Civil and Political Rights (ICCPR) that has been ratified by 115 nations including the United States. The ICCPR codifies the consent issue into human rights law. Article 7 of ICCPR provides clear and enforceable protections for human subjects in research. It states that: "no one shall be subjected without his free consent to medical or scientific experimentation." The U.S. research community is either unaware of Article 7 or tends to ignore it. In March 1995, in a UN committee hearing on U.S. compliance with the ICCPR, the committee recommended that federal and state laws sould comply with ICCPR in not allowing non-therapeutic research on minors and mentally ill patients on the basis of surrogate consent. The Mental Disability Rights International and its director, Dr. Rosenthal, were the first to bring this important issue to the attention of a U.S. audience.

Drs. J. Thomas Puglisi and Gary B. Ellis present the official position of OPRR in charge of overseeing the application of the Department of Health and Human Services (HHS) in chapter 25. Regulations in 45 CFR 46 require that all HHS-supported research involving human subjects be reviewed and approved by local institutional review boards (IRBs). This chapter reviews legal requirements with regard to informed consent and the IRB process, and considers the additional protections available for vulnerable subjects. The authors suggest strengthening the IRB and approval process in order to protect the vulnerable group of patients with mental illness by involving subjects' representatives as members of IRBs. Drs. Puglisi and Ellis recommend that informed consent documents clearly delineate treatment modes from research modes. They believe that when research warrants, an independent Data and Safety Monitoring Board should be established. It appears that the authors and their agency have been moved lately by the degree of concern presented by the subjects themselves and their families.

In chapter 26 by Dr. Robert Temple, the author argues strongly for placebo control trials in certain cases. He bases his argument on the fact that active control equivalence studies are uninformative because they do not discriminate between new and active control treatments. Dr. Temple supports his argument on two fronts. The first is based on data from Food and Drug Administration (FDA) files that show the new drug does not differ from active control treatments. The author understands that when standard therapy is known to prevent real, harmful, and irreversible damage including death, then this therapy should not be denied to any patient. But short of this calamity, he contends placebo controlled trials are necessary. His second argument is the ethical contention (contrary to the authors of chapter 27) that the 1964 Declaration of Helsinki did not mean the elimination of all placebo controlled trials when it used the phrase "best proven therapeutic method." If this interpretation is to hold true, it will lead to stoppage of research progress and continued suffering of future patients. However, what Dr. Temple should note is that in certain types of serious mental illness, the withholding of standard therapy from patients could lead to irreversible harm to patients.

Chapter 27 by Drs. Kenneth J. Rothman and Karen B. Michels was first published in 1994 in *The New England Journal of Medicine* and is reproduced here with permission. Dr. Rothman's presentation at the conference was derived from the original paper, but he has expanded upon it here and answered numerous questions. The authors claim that all placebo control experiments — citing six major studies, including one with persons with mental illness, are unethical. Drs. Rothman and Michels base their thesis on two fundamental premises. First: there is no scientific rationale for a placebo because it is more prudent to compare the new drug with the old treatment rather than with nothing. Second: they adhere to the provision of the Declaration of Helsinki which asserts that patients "should be assured of the best proven diagnostic and therapeutic method." The authors end their article with the following recommendation:

The burden of justification, however, should fall not on critics but on those responsible for the research, including investigators, regulatory agencies, research sponsors, institutional review boards, and journal editors. All these parties should adhere to the precept that patients ought not to face unnecessary pain or disease on account of a medical experiment, and they should question the ethical legitimacy of using placebos in any experiment.

Jean Campbell probes in some detail the historical and continuing polarity in outlook and perception between providers and consumers in the mental health field (chapter 28). The growing avalanche of consumerism in American affairs has been driven by the consumer rights protectionist movement and by market forces that seek to attract customers — even in the health care field — by determining and seeking to maximize consumer satisfaction. While there is general acceptance and adherence to these principles in the delivery of care to consumers of physical medical services, the application of these concepts in the psychiatric milieu has been tardy. The traditional paternalistic role of medical providers survives in that the mental health professional/researcher sees the role as directive and prescriptive; the role of the subject/patient is to be compliant and cooperative.

The creation and energy of the National Alliance for the Mentally Ill (NAMI) has been largely responsible for the partial success which has been achieved on these fronts. There has been a rapid expansion in the establishment of Offices of Consumer Affairs in state mental health agencies, and both state and nationwide consumer conferences have been held within the past few years.

Dr. Campbell correctly points out, however, that cognizance of these developments by clinical researchers and the implicit reforms needed bewilder clinical scientists who were trained and educated in the era when expertise of authority was the only operative criterion. She rightly asserts that dialogue and collaborative interaction are the only routes to prevent static legal consent documents and process from being honored only in the breach. In addition, she suggests adopting a tactic derived from the civil and equal rights movements: deploying a different vocabulary that is person-oriented rather than disease-oriented. A dialectic construction that puts persons first would be more effective in challenging social domination assumptions than confrontation and challenge.

It is also necessary to strive for excellence in consumer protection rather than minimal adherence to the specific regulations. Protocols vastly superior to the standard ones have been developed by a few concerned investigators, some of whom are subjects/patients themselves. It is evident from a survey of research projects that considerable abuse occurs and that greater risks exist in research participation than are currently recognized or acknowledged. There are many ethical issues and gray areas that often lie outside the familiar boundaries of experience of IRB members. Since most current policies and protocols in mental health research were derived without input from the consumer sector, it is critical to devise mechanisms for feedback of that nature. Ironically, the existence and operation of IRBs, when viewed as an appropriate protective shield, serve to counter creative new developments.

A new model that seeks to empower all is the creation of a multiparty structure that has been developed in the rehabilitation community. The Constituency Oriented Research and Dissemination Initiative (CORD) links together the producers, users, and disseminators of research outcomes to achieve a common objective — improving the lives of people with disabilities.

The author concludes by pointing out that the most difficult impediment to progress may lie in the necessary consciousness-raising of expert experimenters whose perceptions may differ vastly from those who are the target of their services. Nonetheless, the ethical conduct of future research activities will only be accomplished in collaboration with human subjects with neurological disorders.

Comments by Laurie M. Flynn, which comprise chapter 29, were delivered after the conference dinner as a preface to the remarks of Dr. Frederick Goodwin. Ms. Flynn, executive director of NAMI, emphasized the advocacy role of NAMI in vigorously supporting funding for basic and clinical research on the cause and treatment of serious mental illnesses such as schizophrenia. She stated that no research is more important than protecting the human rights of those who are vulnerable. Ms. Flynn projected a cautious tone and a balanced approach in strengthening the present system of informed consent/IRBs without damaging the research enterprise in the process. She agreed specifically with several speakers at the conference who indicated concern regarding placebo controls. Ms. Flynn ended her comments by saying: "Our goal is to alleviate human suffering while respecting human rights. This conference has been an extraordinary opportunity to begin that critical work."

Chapter 30 by Drs. Frederick Goodwin and Suzanne Hadley is written from the perspective of a senior eminent clinician and clinical investigator (F.G.). Dr. Goodwin, former director of the National Institute of Mental Health, focuses principally on the need for trust willingly proferred among patients, their family advocates, and the research community. The authors emphasize that mental illness is more burdensome and tragic than other chronic and debilitating medical diseases. The authors are concerned not only with the stigmatization associated with mental illness but also with the legitimization of the health caregivers and professionals who serve patients. Drs. Goodwin and Hadley fear that a legal and adversarial framework for dealing with these issues dispels trust and demoralizes investigators in this field. They believe that concrete research data are imperative to making a successful argument for equal treatment in fiscal reimbursement.

There is increasing evidence of partnerships among the major funding agency NIMH, the National Alliance for the Mentally Ill, and The National Depressive and Manic-Depressive Association. One outcome of this coalition building is the NIMH Knowledge Exchange Program, initiated by the author. The major objective of this program is the free transfer of information and perceptions between the patient community and the policy planners in research. Representatives of NAMI review the NIMH research portfolio and provide suggestions on its content and balance. NIMH is the first institute to formalize the inclusion of consumers in its decision-making about the protection of human subjects.

Drs. Goodwin and Hadley point out that most human endeavors of serious consequence cannot be considered risk-free. To assume that psychiatric research be conducted with risk standards different from other spheres of biomedical research activity would be a further source of stigmatization. He strongly posits that direct clinical research of more than minimal risk should, in the long run, contribute to the health of a particular class of patients in the future. It is his opinion that overreliance on the formal consent process may in fact impede the development of a true moral compass in research investigators. Many studies have shown that the past 20 years have given rise to a general erosion in the trust of the general public in institutions and professions. With a threefold increase in the number of lawyers per capita in the United States within the past 25 years, many serious scholars harbor grave concerns about the extension and even invasion of the principles and machinery of the law into societal areas for which a legal solution is not feasible. All concerned in the topic of this conference, including clinicians, consumers, and researchers, should develop a confluent approach to existing problems rather than employing the legalistic adversarial model.

It is probably correct to assert that personal motives are inextricably a part of the altruism in every person involved in this process and these need to be openly acknowledged. Unfortunately, workable guidelines often emerge from supervisory bureaucracies cast in stone. The authors believe that the flexibility and independence of local IRBs are crucial to the impact consumers can have in the local setting. They conclude by proposing that more research is needed on several topics: abusive clinical research, outcome research, consumers' attitudes, and experiences as research volunteers. It is possible that this conference could provide a blueprint for the necessary research agenda.

Herb Pardes, MD writes in chapter 31 that patients, their families, and health care providers should form a natural and national alliance, oriented towards the betterment of mental health. Parents should not be forced to accept a certain current level of treatment because clinical research is impeded. There is, however, a degree of tension between individualized care and the use of patients in research protocols for the greater good of humanity. Dr. Pardes cites the Hellmans, who as ethicists argue that human beings are inherently bearers of dignity and that such a right cannot be signed away. It is accepted, at this point in time, that sound data and the clinical judgments derived therefrom can only be developed in random clinical trials. The results of such painstaking systematic research can, in certain cases, be at variance with a particular physician's course of action. *The New England Journal of Medicine* recently published a second paper, alongside the Hellman article, that argues for the research approach. Dr. Pardes believes that the debate on this issue may be arcane because he foresees the growing restriction of health care resources, coupled with diminished support for research, as the most serious threat to the improvement of medical treatment in this field.

The author conducted a question-and-answer format for the latter part of his presentation. In response to a question concerning the evolution of managed care, he stated his belief that medicine for profit, via a corporate structure, is unlikely to surrender such funds for research or teaching. What is happening in the health care field, at the moment, is a very large and dramatic social experiment, of enormous impact, in which cost-containment is the driving force. The director of development of the NAMI network deplored the fact that speaking time and even exhibit space are denied to that organization at professional medical meetings and asked that Dr. Pardes urge his colleagues to be more forthcoming.

Mental health has generally been considered to be a "soft science." In the current competitive managed-care milieu, the availability of improved care will rest even more heavily on scientifically proven data. Jean Campbell, in the audience, commented strongly that individual researchers in the field who have been diagnosed with a psychiatric disorder are markedly stigmatized by their colleagues. Dr. Pardes agreed that this issue is a more widespread problem than is currently recognized. Dr. Robert J. Levine described the intense pressure within academic medical institutions to focus completely on income generation, to the detriment of optimal practice. Dr. Pardes concluded by emphasizing that the alliance of patients, families, and scientists is crucial in combatting the severe modifications which are occurring in the health care delivery system.

Chapter 32 authored by Dr. Shamoo, J. Rock Johnson, Ron Honberg, and Laurie Flynn reviews NAMI's four-year process to adopt "Standards for Protection of

Individuals with Severe Mental Illnesses Who Participate as Human Subjects in Research." NAMI's ten-point ethics standards emphasize the need for patient and family member involvement in the research process, informed consent, and IRB reviews and approval. Most important of NAMI's standards is item four, which recommends that research participants be evaluated for competency by an independent psychiatrist before they are admitted as research subjects.

The final paper of the conference, chapter 33, is authored by Dr. Jay Katz. This paper offers a final commentary on the diversity of viewpoints and perspectives which were delivered at this conference. This forum was designed to deliberately invite controversy rather than suppress it . The success of this pioneer meeting can, however, only be judged on the basis of creative integrative solutions to key problems that may be generated in the near future as a consequence of the opportunity offered in Baltimore.

The major point emerging from these deliberations was that near-universal agreement was reached on the necessity for continuing research in human mental illness. The majority of presentations focused on the myriad dilemmas surrounding how such research should be conducted with reasonable and appropriate safeguards. Dr. Katz proposes that physical harm to subjects in human research is of limited concern compared to the dignitary harm inflicted on subjects whose consent and participation are manipulated in the project. The mindset of physician investigators coupled with the overseeing function of IRBs assures that physical harm is limited to a minimum and is probably no more than would be experienced in the standard therapeutic environment with the pharmacological regimens available. In research, patient needs have to be subjugated to the objectives of research objectives and protocol.

One of the facets of patient involvement in research that is becoming clearer is the manner in which the invitation to participate is extended. Consent is a necessary but only a partial justification for enrolling subjects. Much of the subject matter of this conference was directed toward the nature and quality of disclosure and consent. The "process" question arises in several different forms. Should, for instance, local IRBs be authorized to permit research to be conducted whenever the disclosure/consent issue is lacking or ambiguous?

False and tacit assumptions are present in many cases. The patient may perceive the investigator as purely a physician, while the investigators are significantly constrained by external research funding/deadlines and expected productivity by their employers. The author points out that since the 1940s medical research has become a medical-industrial enterprise. Physician-investigators and their careers are currently at the mercy of their universities and the National Institute of Health. Maintaining a moral stance in the face of such pressures is a serious challenge, especially if taken in conjunction with the frustration such physicians endure because of the less-than-adequate results of medications presently available.

Dr. Katz believes that the divalent role of physician-researchers needs to be explained very clearly in the consent dialogue and that the patients welfare in research be compared to standard therapy. In simple terms, how participation in research differs from the usual and ordinary provision of health care, needs to be explained very clearly and lucidly. It is not necessary for the subjects to understand the arcana of medical science. Unfortunately, many documents currently in use are incomprehensible and contain a wealth of distracting and useless technical information. IRBs do not have the authority to modify the consent forms appropriately. Such forms often are merely a record that some informed consent event or process actually occurred. Dr. Katz believes consent forms should represent the final event at the end of the dialogue with patients.

Dr. Katz chaired the sub-committee appointed to review the Tuskegee Syphilis Study. That body recommended a permanent structure, the National Human Investigation Board, which was viewed as separate and independent of HEW and NIH, the funding agencies. Given the patchwork of regulatory rules that have transpired, it is unfortunate that this suggestion was never implemented. Society should be forced to make these difficult moral choices in bright sunlight and a thoroughly ventilated atmosphere such as pervaded this conference.

Adil E. Shamoo, PhD
School of Medicine
University of Maryland
Baltimore

Marie M. Cassidy, PhD
George Washington University
Medical Center
Washington, DC

Accountability in Research Using Persons With Mental Illness

Adil E. Shamoo† and Dianne N. Irving‡

†*Center for Biomedical Ethics, and the Department of Biochemistry, School of Medicine, University of Maryland at Baltimore, 108 N. Greene St., Baltimore, Maryland 21201*
‡*Department of Philosophy, De Sales School of Theology, Washington, D.C. 20017*

Although medical research involving the use of persons with mental illness is critically important, in order for the research to be ethical and legal there are certain considerations and restrictions which should be immediately readdressed in order to insure that the welfare of these vulnerable research subjects is protected, and their best interests are assured. A brief historical examination of medical research codes, guidelines, recommendations and Federal Regulations reveals the various considerations and restrictions on informed consent and accountability applicable to the use of persons with mental illness in medical research. Several concerns are raised about how these considerations and restrictions have been interpreted, and specific recommendations are offered to improve them immediately by means of representation from consumers and/or their families, and organizations, e.g., NAMI members.

Keywords: mental illness, Federal Research Regulations, informed and proxy consent, accountability, NAMI

INTRODUCTION

The use of human subjects in biomedical research has been discussed, debated and analyzed for decades now, especially since the findings of the medical research atrocities committed during World War II (e.g., Appelbaum, 1987; Barber, 1973; Beecher, 1966; Frankel, 1975; Ivy, 1948; Jonas, 1984; Katz, 1972; Levine, 1986; Lifton, 1986; Mitscherlich, 1962; Muller-Hill, 1988; Proctor, 1988)—culminating in the Nuremberg Code (1947). Since then further refinements and protections have been added through the Declaration of Helsinki (1965, amended 1989), and here in the United States, the National Commission (1978), the President's Commission (1979–1983), and a series of regulations at the Federal level.

The Federal Regulations 45CFR46 (1981,1983,1989,1991), with subsequent amendments and additions, clearly required the establishment of research protocol review by institutional review boards (IRB's) at each institution conducting research

This article had previously been published in the journal: *Accountability in Research*, Volume 3, pp. 1–17 (1993).

supported by government funds, or submitted in support of a government-required approval such as that for drug testing. Based on detailed studies of the National Commission and the President's Commission, and promulgated through the series of Federal Regulations, there were specific protections for special groups who were considered "vulnerable," e.g., persons with mental illness, pregnant women and fetuses, prisoners and children.

In the 1983, 1989 and 1991 Federal Regulations, special subparts of these regulations (Subparts B,C and D) were detailed for all of these vulnerable subjects *except* for those persons with mental illness. This raises the question as to why persons with mental illness—who are unquestionably a "vulnerable" group—were not considered as human research subjects needing special protections in a specific subpart of the Federal Regulations?

The official response seems to have been that the Federal Regulations do provide for such protections—by authorizing local IRB's to determine if any such protections are necessary. But the question remains as to why such extensive, well-debated and publicly-scrutinized recommendations such as those reported by the National Commission and the President's Commission were virtually dropped for *only* that vulnerable group of human research subjects who were persons with mental illness—and not, then, also dropped for the other vulnerable groups? Additionally, what was the justification for delegating to local IRB's the essential responsibilities for affording protections for persons with mental illness—protections which were otherwise provided through the recommendations of the National Commission? What was the justification for allowing the IRB's to in turn redelegate this authority in certain cases to others?

The best description of what happened to the issue of the use of persons with mental illness in research is summarized in a section on "special populations" in a book by Paul Applebaum, M.D., Charles W. Lidz, Ph.D., and Alan Meisel, J.D.:

Special protections for mentally ill and mentally handicapped persons were the subject of extensive recommendations by the National Commission for the Protection of Human Subjects of Biomedical and Behavioral Research (53). None of those recommendations were implemented. This outcome was the result in large part of opposition from researchers on mental disorders, who claimed that the populations in question were no more vulnerable than most persons with severe medical disorders and that the suggested limitations would seriously restrict research on mental disorders (54). Mentally impaired subjects continued to be covered under the general regulations, which, in the case of incompetency, allow consent to be obtained from a subject's legally authorized representative, presumably a guardian or equivalent person with the power to consent on the subject's behalf under the laws of the jurisdiction. (Appelbaum, Lidz and Meisel, 1987, p. 228).

Another author noted the express protections afforded persons with mental illness under the earlier recommendations of the National Commission, which he assumed would be carried over into the Federal Regulations. Dennis M. Maloney stated:

Due at least in part to the freeze on government regulations, these proposed rules remain the ones for us to heed at present. It is not currently known whether these rules will be revised or made final by HHS in view of federal policy to limit regulatory activities and reduce government paperwork. In the author's opinion, the proposed rules that do exist, as they appear appended to this chapter, should still serve as a useful guide for researchers in this area. Unless unusually substantial changes are made in the proposed rules, final rules (when published) should draw heavily on the proposed rules as appended to this chapter. Once again, the cautious researcher should consult federal agencies (e.g., the NIH Office for the Protection of Human Subjects). (Maloney, 1984, p. 377).

It would appear that the issue of using persons with mental illness as human research subjects has been lost in the shuffle, due in part to the lobbying efforts of some researchers on mental disorders, and in part to the relatively small degree of constituency and advocacy for persons with mental illness when the above developments occured during the mid to late seventies. Especially given their vulnerability and the reality of their use in on-going research, the situation of these persons with mental illness remains a concern. As one historian has recently and succinctly put it: "...it is the socially powerless and disadvantaged who are most likely to be subjected to unethical research" (Gillespie, 1989, p. 13).

The aim of this article is thus threefold: first, to indicate briefly some of the history relevant to the use of human subjects in medical research in general, and of persons with mental illness in particular; second, to draw attention to two of the most significant inadequacies in the Federal Regulations which deserve further examination and discussion; and third, to propose recommendations in which members of the National Association for the Mentally Ill (NAMI) could be useful in carrying out these regulations on many levels of participation—including membership on IRB boards, and in the role of the "consent auditor" as proposed under the recommendations of the National Commission Report (National Commission, 1979, p. 6).

INFORMED CONSENT AND ACCOUNTABILITY IN TWO MAJOR HISTORICAL DOCUMENTS

In reviewing briefly two of the most well respected historical documents concerning the use of human subjects in medical research, it should be noted that some restrictions present in early documents have disappeared from the more recent ones. On the other hand, several exceptions included in recent documents were never contained in the earlier ones. This does not, we would argue, presume that what is most recent is thereby what is most acceptable or sophisticated. With respect to readdressing the current Federal Regulations, it would be prudent to reconsider and reevaluate again both old and new restrictions and exceptions.

A. The Nuremberg Code

Although no reference is made to the Nuremberg Code (1947) in the current Federal Regulations, these codes were the basis for later codes and regulations. Given that this multi-national tribunal in particular was so close to the massive *empirical* evidence presented to it on the scientific fraud and human abuses of recent medical research which used human subjects, we should note the issues which they so clearly judged to be critical, and which they so strongly considered should become incorporated in any country's policies when human beings are allowed to participate in the medical research enterprise.

It is often not realized that the ten principles of the Nuremberg Code were actually part of the "judgment" handed down during the Nuremberg Trial's decision in *United States v. Karl Brandt*:

...The great weight of the evidence before us is to the effect that certain types of medical experiments on human beings, when kept within reasonably well-defined bounds, conform to the ethics of the *medical profession* generally. The protagonists of the practice of human experimentation justify their views on the basis that such experiments yield results for the good of society that are unprocurable by other methods or means of study. *All agree, however, that certain basic principles must be observed in order to satisfy moral, ethical and legal concepts.* (in Katz, 1972, p. 305) (emphasis added)

The court proceeded then to list ten basic principles, the core of which includes the following points.

The very first principle relates specifically to the *human subjects* who might participate in biomedical research, and is by far the lengthiest statement. Two issues are critical. First, only *competent* subjects may participate; and second, they must have *legal* capacity to consent. The former calls into question whether or not persons with mental illness may be used as research subjects at all. The latter is emphasized as the means by which the court could determine "criminal culpability and punishment" (in Katz, 1972, p. 305; Katz, 1992, pp. 227–239).

In addition, this first principle requires that the subject: be able to exercise free power of choice without restraints; have sufficient knowledge and comprehension of the information; be advised of the nature, duration and purpose of the experiments, the method and means used, what risks to expect, and the anticipated effects on his or her "health or person." Interestingly, the duty and responsibility (legal) for the consent process fell to the investigator—a "personal duty and responsibility which may not be delegated to another with impunity" (in Katz, 1972, p. 305). Presumably, the purpose here was for any court (such as the Nuremberg Court) to be able to determine the *accountability* of any harm caused to a competent legally consenting human subject.

These two points—competency and accountability—continue today to be the crux of the issues concerning the use of persons with mental illness in medical research. The debates need to be redirected to the issues of whether or not the mentally infirm may be used as human research subjects; and if so, what are the limits and restrictions. Who is at least *legally* accountable for insuring proper consent from and for causing harm to these vulnerable human subjects during and after their participation in medical research?

There seems to be no concensus even now on what constitutes "competency," even within the field of psychiatry which is, at least professionally, the ultimate judge on these issues, although the literature abounds with conflicting theories and suggestions. What is often not discussed is the issue of accountability, especially in these days of "consensus" building and committee or team decisions which in effect dilute the responsibility and therefore the accountability for those decisions. These two major issues of competency and accountability need to be settled in order for any regulations or guidelines to carry any credibility—either ethically, professionally, legally or socially. If no "concensus" on the theoretical level can be attained, then perhaps the issues should be referred to the electorate, if credible research using persons with mental illness is to continue. What seems to have happened is that research has been allowed to progress *before* these and other such critical issues have been resolved.

To return to the Nuremberg Code, with reference to the *experiments* themselves, it requires that they: be designed and based on the results of animal experiments as well as on a knowledge of the natural history of the disease or problem under study; not be random or unnecessary in nature; be performed to obtain fruitful results for the good of society which are not procurable by other means (*if* they *first* satisfy moral, ethical and legal demands); should avoid all unnecessary physical and mental suffering and injury; should not be conducted if there is a reason to believe death or disabling injury will occur; should not contain degrees of risk greater than those determined by the humanitarian importance of the problem; should be performed with proper preparations and in adequate facilities to protect the subject against even the "remote possibility of injury, disability or death" (in Katz, 1972, p. 305).

Clearly, the Code requires that no experiment should be performed on human subjects which are scientifically ill-designed—or ill-proposed. The most basic of ethical considerations requires that the experiment be designed and executed on scientifically sound bases (Shamoo, 1991). There is also a limit to the amount of physical or mental injury to the human subject which is acceptable, as well as concern expressed for the quality of the facilities in which these experiments would take place. These points are not to be lost in current debates about research fraud and misinformation (Shamoo and Annau, 1990). No human patient should be subjected to even minimal invasion, injury or risk simply because the scientific investigator did not really know what he was doing, or if he was mishandling his data.

The final issue the Codes were addressing concerns the *medical researcher* him or herself. The experiment should be conducted "only by scientifically qualified persons"...with the "highest degree of skill and care" (see Irving, 1993; Shamoo, 1991). And the researcher should be prepared to terminate the experiment at any stage if it is believed injury to, disability of, or death to the human research subject may result (in Katz, 1972, p. 305–306).

It is precisely here that the connection between informed consent and account-ability can be justified. It is the *medical researcher* who is responsible for the design and the execution of the experiment. This is his or her area of professional expertise—an expertise not possessed by the human research subject or others who may be tangentially involved. It is precisely *because* he has this expertise that he is allowed to design and to perform these experiments on human subjects. It is because of this expertise that he is the one charged with informing the human research subject about the facts and details of his scientific experiments. And it is on the basis of the medical researcher properly communicating this scientific information to his patient that the patient can become truly *informed* and thereby *freely choose* to participate in the researcher's project. If the responsibility for the design and for the execution of the experiment rests on the scientific expertise of the researcher, as well as the responsibility for the quality of the scientific informa-tion which is required to be transmitted to the human research subject before that person can give truly informed consent, then it is the *researcher* who is *accountable* for the safety of his or her human research participant as well as for the quality of the informed consent process.

Therefore, the medical researcher as a *research expert* is accountable for his own work. But what precisely is a "medical research expert"? Generally, scientists have little if any medical course work; physicians receive little if any course work in basic or clinical research skills or techniques during their medical school education, and are their separate goals mutually inclusive? Generally, scientists are seeking basic facts and data about nature; physicians are seeking how to cure or care for their patients. Could these inherent differences be a serious source of conflict of interests (Irving, 1993)? Finally, how to assure the integrity and accountability of the expertise of the medical researcher, the quality of his research, the reliability of his data—and the physical and mental integrity of the consenting human subjects on whom he performs his experiments—are additional issues that need to be readdressed (Shamoo 1991, 1992, 1993).

In sum, the Nuremberg Code of 1947 clearly identifies at least two critical issues of informed consent and accountability before *competent* human subjects are permitted to participate in medical research.

B. The Declaration of Helsinki

The Declaration of Helsinki (1964) begins a series of refinements on the Nuremberg Codes. There is more of a recognition that the researcher is not just a "scientific" researcher in medicine, but that he or she is also a *physician*, with a special physician-patient relationship and professional responsibility, which must be taken into account when doing medical research on any human subjects. Hence, "clinical" research is the focus, and there is a considered effort to distinguish between therapeutic and experimental research. It is here that the potential *conflict* between "researcher" and "physician" begins to take shape, as well as the issue of the use of *incompetent* human research subjects.

The Declaration begins by "binding" the doctor with the words, "The health of my patient will be my first consideration." First, it binds the doctor to the International Code of Medical Ethics which states that *"any act or advice which could weaken physical or mental resistance of a human being may be used only in his interest"* (in Katz, 1972, p. 312). The implication is clear: clinical research may only be performed when there is a *direct therapeutic benefit* to the patient. The Declaration proceeds to make the distinction between therapeutic and experimental (or purely scientific) research, a distinction which is based on the purpose or the aim of the research. The aim of clinical research can be directed to the therapeutic value it may have for that patient; or it can be directed to the acquisition of purely scientific knowledge, with no therapeutic value to that particular patient (in Katz, 1972, p. 312). Although the distinction is an important one, and one which needs to be reconsidered in the current Federal Regulations, the Declaration actually seems to contradict its own statement (above) on this critically important distinction.

For example, the Declaration proceeds to address two broad categories of clinical research: clinical research combined with professional care; and clinical research which is *non-therapeutic*! In the former category it states that "the doctor can combine clinical research with professional care, the objective being the acquisition of new medical knowledge *only* to the extent that clinical research is justified by its therapeutic value for the *patient*." Yet, under the second category it states that "in the *purely scientific* application of clinical research carried out on a human being, it is the duty of the doctor to remain the protector of the life and health of that person on whom clinical research is being carried out" (in Katz, 1972, p. 312).

This distinction made between therapeutic and experimental research is an important one, but confusing and misleading as a guide to clinical researchers. Is the Declaration restricting clinical research to: therapeutic research only for the direct benefit of the patients; or allowing therapeutic research which would only benefit similar "classes" of patients; or for purely scientific knowledge? These are three different aims. Nevertheless, it does afford us an important issue to be grappled with when reconsidering recommendations for the Federal Regulations. This issue has been vigorously debated in the recent past, yet has resulted in no real consensus. The issue needs to be redebated—particularly considering the fact that all three types of research are *on-going*.

The Declaration also makes another important distinction: between consent for research by competent or by incompetent patients. Under the first category (clinical research combined with professional care), the requirements for competent patients are similar to those of the Nuremberg Code. Again, under the second category (clinical research which is non-therapeutic), the "nature, purpose and the risks of clinical research must be explained to the subject *by the doctor*." Clinical research "cannot be undertaken without his [the patient's] free consent, after he has been fully informed"; and the subject of clinical research "should be in such mental, physical, and legal state as to be able to exercise fully his power of choice." Finally, the consent should as a rule be obtained in writing (in Katz, 1972, p. 313).

But what is *new* is the express possibility of the inclusion of *incompetent* patients in the research enterprise (Katz, 1992, pp. 227–239). That is, the Declaration continues that if there is legal incapacity, "consent should also be procured from the legal guardian." Thus with proxy consent by a legal guardian, an incompetant human subject may participate in clinical research. Once again, the issue of consent in the Declaration is tied to the issue of accountability. The issue is not whether or not a patient is "medically" competent, as might be determined by a psychiatrist or by a behavioral psychologist; but whether he is *legally* competent. And legally, we know that a third party may give consent only for activities which are for the *direct benefit* of the patient. This would presumably preclude the participation of persons with mental illness in any medical research other than that which would be for his or her direct benefit.

Given the theoretical turmoil within the psychiatric and psychological professions (whether understandable or not), at least *legal* accountability can be assigned. Here the patient must be legally determined to be incompetent so that proper *legal* guardianship may be provided. Presumably this guardian is also accountable to the courts for any decisions he or she makes on the patient's behalf.

Again, in the Declaration accountability will ultimately rest with the researcher himself. The researcher is to consider the same requirements for scientifically sound, well designed and properly executed experiments as in the Nuremberg Code in order to limit the risk of harm to his patients. Because of his or her expertise, it is the clinical researcher who is vested with the responsibility to inform the patient of the relevant scientific facts in the consent process, and it is the clinical researcher who has ultimate responsibility for the project: "...the responsibility for clinical research always remains with the research worker; it never falls on the subject, even after consent is obtained." Indeed, here the doctor is even cautioned about the use of experimental drugs: "Special caution should be exercised by the doctor in performing clinical research in which the personality of the subject is liable to be altered by drugs or experimental procedures" (in Katz, 1972, p. 312). This "special caution" should not be lost, we would argue, on researchers who use persons with mental illness as research subjects.

Thus, in addition to the initial general issues of informed consent and accountability in the Nuremberg Code, the Declaration begins to open up the controversy by making two new important distinctions: first, between the permissibility of therapeutic and/or experimental research; and second, between the consent requirements for competent as opposed to incompetent research subjects. It is in the recommendations of the National Commission where much greater scrutiny of these and related issues are found.

RECOMMENDATIONS OF THE NATIONAL COMMISSION

The Belmont Report (1979) of the National Commission identified the need for special protections for persons with mental illness when they participate as research subjects, especially with respect to their compromised ability to give informed consent. This report also indicates the need for "other parties in order to protect the subject from harm":

Special provision may need to be made when comprehension is severely limited—for example, by conditions of immaturity *or mental disability*. Each class of subjects that one might consider as incompetent (e.g., infants and young children, mentally disabled patients, the terminally ill and the comatose) should be considered on its own terms... Respect for persons also requires seeking the permission of other parties in order to protect the subjects from harm. Such persons are thus respected both by acknowledging their own wishes and by the use of third parties *to protect them from harm*.

The third parties chosen should be those who are most likely to understand the incompetent subject's situation and *to act in that person's best interest*. The person authorized to act on behalf of the subject should be given an opportunity *to observe the research* as it proceeds in order to be able to withdraw the subject from the research, if such action appears in the subject's best interest. (National Commission, 1979, p. 6) (emphasis added)

These recommendations of the National Commission are quite explicit and clear about the need for vulnerable human subjects, specifically persons with mental illness, to be protected from harm, by third parties, who may even observe the research and act in these subjects' "best interests." In current Federal Regulations this language will *shift* to persons "knowledgeable about" the patients' illness, and to surrogates who make "substituted judgments" about what the patient would decide if he/she were competent.

FEDERAL REGULATIONS (1981–1989)

From 1981–1989, the Federal Regulations were subdivided into four major subparts: A, B, C and D. Generally speaking, Subpart A covered protection for research subjects who were competent and who could therefore give informed consent. The additional subparts provided special protections for several groups of research subjects who were considered "vulnerable," particularly insofar as their ability to give true informed consent was in some way compromised. Thus Subpart B provided extra layers of protections for fetuses, pregnant women and *in vitro* fetuses fertilized *ex utero*. Subpart C included prisoners. And Subpart D referred to children.

However, no special subpart was provided for that category of research subjects who are persons with mental illness. The only reference to these human subjects comes in Subpart A, in a discussion concerning IRB membership:

If an IRB regularly reviews research that involves a vulnerable category of subjects, including but not limited to subjects covered by other subparts of this part, the IRB shall include one or more individuals who are *primarily concerned with the welfare* of these subjects. (Code of Federal Regulations 45 CFR 46, 1981, 1983, 1989, p. 7) (emphasis added)

To our knowledge, there is no strong evidence to indicate that NIH has required IRB's to include either persons with mental illness, members of their family, NAMI or any of the state AMI's to serve on these IRB's on a permanent basis. Our interpretation is that they have construed "individuals who are primarily concerned with the welfare of these subjects" to mean psychiatrists, nurses, ethicists, and lawyers. If this is their interpretation, then we obviously disagree strongly with such an interpretation. Regardless of the interpretation, the fact remains that to our knowledge, no NAMI members are involved at all. Therefore, we suggest that NIH has not appropriately enforced the Federal Regulations as stated above with respect to vulnerable patients. This is probably due in large part to the fact that very few have objected to their policy—and thus the policy has remained in effect to the present time.

Three things should be pointed out at this point. First, despite the clear concern for the use of persons with mental illness in the National Commission's Report, these series of Federal Regulations do not identify them as requiring any specific protections in a separate subpart as is afforded the other vulnerable groups. Second, supposedly they will be provided sufficient protections through an IRB—a point to be seriously debated. Third, even when simply under an IRB protection, the IRB is to include one or more individuals who are "primarily concerned with the *welfare* of these subjects." The language descriptive of these additional IRB members will change—we think significantly—in the present 1991 Federal Regulations.

THE PRESIDENT'S COMMISSION REPORT

In 1983, the President's Commission issued its final recommendations in their volume, *Summing Up*. The Commission expressed obvious concern for the lack of attention to protections for certain vulnerable human subjects in its section on "Protecting Human Subjects":

(5) The National Commission's recommendations on research involving children *and the mentally disabled* should be acted upon promptly. Ethical concerns about these individuals revolve around the issue of informed consent. In order for research on the causes, treatment, and prevention of pediatric

diseases and of emotional and cognitive disorders to proceed in an ethically acceptable manner, the National Commission had urged the adoption of special protections for children and the mentally disabled. Although the Secretary, HEW (now HHS), was to respond promptly to that Commission's recommendations, it has been four years since those recommendations have been submitted …

The Department has partially responded to Recommendation (5) regarding rules to protect children and the mentally disabled who are asked to participate in research. After further encouragement from the Commission during 1982, HHS on March 8, 1983, published in the *Federal Register* regulations for research involving children. Although the Commission has continued to urge HHS to act expeditiously to remove regulatory ambiguities and impediments that may exist to research with mentally disabled subjects, Secretary Schweiker has informed the Commission that *no regulations will be issued by HHS research involving persons institutionalized as mentally disabled.* (Presidents' Commission 1983, p. 54, 56) (emphasis added)

Thus, despite these strong recommendations of both the National Commission and the President's Commission, under present Federal Regulations persons with mental illness may participate in both therapeutic and experimental research as human subjects with *only* those protections afforded them by an IRB. Furthermore, even those protections apply only to those who are *institutionalized*, and neglects those persons with mental illness who are not institutionalized but participate in research as human subjects in out-patient care, physicians offices, nursing homes, etc.

It is clear to any objective observer that those inflicted with serious diseases of the brain (which affect one's ability to think—and therefore affect their ability to give true informed consent) are more vulnerable than those with serious diseases of other organs of the body. It is also clear that all other special groups had strong advocacy groups at the time. The mentally ill cannot advocate for themselves, and thus this responsibility was then, in part, assumed by the researchers in psychiatry—those who would have benefited themselves by using the mentally ill in their research protocols. Is this not a conflict of interests? It is also important to note that the National Alliance for the Mentally Ill (NAMI) was only started in 1980, and thus not available for discussions on these issues. Thomas Jefferson's comments on the lack of efficiency inherent in a democratic form of government could be applicable here. The lack of an advocacy group for persons with mental illness at the time of the hearings for the proposed Federal Regulations contributed to the present situation, where persons with mental illness are not even recognized as a vulnerable group at all. Additionally, all of these vulnerable groups may be used in research with *"certain exemptions"* from the standard protections! This unfortunate situation is captured by the remarks of Professor Jean-Louis Baudouin, QC, Professor of Law at the University of Montreal, Canada. She summarized the American position on the use of persons with mental illness in an article on medical ethics in the journal *Medicine and Law* (1990): "One is surprised, however, to find no specific provisions relating to mentally handicapped persons."

PRESENT NIH CLINICAL CENTER POLICY

In 1986 the Clinical Center at NIH approved a new policy for the consent process in clinical research with patients who are or will become "cognitively impaired." It is certainly laudable that the Clinical Center, on its own initiative, took extra steps to afford more protection for cognitively impaired research subjects, yet several concerns remain.

Aside from obviating the current debate as to what constitutes "cognitively impaired" as opposed to "organically impaired," the policy, which was *adopted in 1987*, included a new effort to assure that the consent process is appropriate. This new protection was afforded by a Durable Power of Attorney (DPA). They state that their use of the DPA was modeled after a Maryland State law which, they claimed, extended the use of the DPA (originally used for purposes of property management) to allow *competent* Maryland residents to select a surrogate in advance who could later consent to procedures in *health care* if that resident should become incompetent to make these decisions in the future (Clinical Center, 1987; Fletcher, 1985, p. 1).

Under this new Clinical Center policy (which applies only to intramural research in the Clinical Center), the Institute Clinical Research Subpanel (ICRS)—or the CC's IRB—may give approval for a subject who is "not seriously impaired" to give *informed consent* to select a surrogate decisionmaker. The DPA is the record of the impaired subject's choice of a surrogate. That is, human subjects *who are or will become cognitively impaired* may appoint a surrogate to make decisions for the subject about his or her participation in medical *research* at the NIH. Thus, the surrogate would provide the *"best substituted judgment* that the subject would consent to the research if he or she were not impaired" (Clinical Center, 1987, p. 2; Fletcher, 1985, p. 1–3).

We have several concerns with this present Clinical Center policy. First, in 1987, when this policy was written, the Maryland State DPA statute was generally applied to the management of financial affairs only. (a) It was unclear if it could apply to *health care* decisions. In 1987 one Maryland attorney general stated in an opinion that the generic DPA could possibly apply to health care decisions if the document specified that intent. (b) It's application to *medical research* decisions had not really been addressed, although it could possibly have applied if executed by a *competent* person and so specified in the document. (c) It could only be executed by *competent* persons, not by *never or presently incompetent* persons—i.e., the category of research subjects participating in medical research at the Clinical Center. In October of 1993, a new Health Care Decisions Act will go into effect which specifically authorizes the use of the DPA for health care decision makers. But to use it for never competent persons, or persons who lack decision making capacity at the time the document is executed, would not be consistent with either the current Maryland law or the new (1993) Health Care Decisions Act (D. Hoffmann 1993).

Second, there is already a growing concern about the acceptance of the use of the concept of "substituted judgment" in many health care contexts (e.g., Case Studies, 1992; Nelson, 1992; King, 1992; Truog, 1992; Capron, 1992; Santurri, 1982; Dresser, 1984). Third, this policy applies only to research with cognitively impaired human subjects, and does not include persons with any of the other mental illnesses that are being studied.

Fourth, one has to wonder how a subject who always has been and who is still cognitively impaired can competently give *informed consent* to choose a surrogate who will make such important decisions and "substituted" judgments for him/her. There are some who would argue that:

... It may seem anomalous to say that a subject is, at the time of appointment, incapable of consenting for herself. However, it is not unusual to see situations in which a subject may be capable of understanding that someone may act for her in making a decision and of naming a trusted person to perform that function, even though the subject is not capable of understanding the risks and benefits associated with a complicated research protocol. (Fletcher, 1985, p. 2)

This analogy between a subject's ability to choose who to trust, and the ability to know that that person would—or could—know if the subject would choose to participate in research if the subject were competent, would still remain an "anomaly" to many.

This rationale is similar to that used in determining the capacity of the research subject. For example, in such determinations, the assessment of a patient's *actual functioning* in decision-making situations is given precedence over either the outcome of a patient's decision or the patient's status (Fletcher, 1985, p. 4). The relevant issue, they claim, is whether the individual is capable *of making* a particular decision. However convincing this may seem to some, it is clearly very possible that there is a distinct difference between the mere ability *to make* a decision, and the *kind* of decision that one is making. Arguments about imposing other's preferences on these subjects aside, there is at least a broad area of decision making which is considered in most other contexts of life as "competent" and "rational," and past that point the *kind* of decision that is made is definitely relevant to the determination of competency (unless one is willing to entertain the notion that a person is "competent to be irrational"). In the genuine concern to restore and protect the individual rights and dignity of incompetent persons, perhaps care is needed not to extend their "autonomy" so far that these persons are instead actually injured or neglected (see Shamoo and Irving, 1993; Ganzini, Lee and Bloom, 1993).

Another determination that is used is that, except as noted, individuals should be assumed to possess decisional capacity unless otherwise demonstrated; incapacity should be found to exist only when the individual lacks the ability to make decisions that promote his/her well-being in conformity with his/her own previously expressed values and preferences. However, it is not unusual for an incompetent person to appear to be competent when they are not; many learn how to play that game. In addition, it does not follow that if a person's expressed values are to slam his head against the floor every five minutes, and that therefore if he expresses that he now desires not to do so, that that person is incompetent. These directives as to the determination of competency or incompetency need to be seriously addressed by as large a body of responsible experts as possible.

Fifth, of equal concern is the provision in the Clinical Center policy that when the subject is so *seriously impaired as to be incapable of understanding the intent or meaning of the DPA process*, then a *next-of-kin surrogate* may be *chosen by the physician*, with an additional consultation with the *bioethicist* from the CC Bioethics Program—who may delegate this responsibility to *other* consultants approved by the Director of the Clinical Center. This includes participation in therapeutic research. This NIH practice would probably not be consonant with most past or present state laws in this area which provide a definite ranking of family members for the use of substituted consent. Under the current substituted consent law in Maryland (Health Gen. Sec. 20–107) *family* members in a certain order could make *health care* treatment decisions for an incapacitated patient, but it is "highly unlikely that *medical research* would be considered such treatment under the statute." The law will be repealed in October 1993; and the new Health Care Decisions Act would not likely be interpreted to allow an agent, including family members, to consent to the use of *incapacitated* patients in *medical research* unless the patients were *competent* at the time the document was executed and unless specified explicitly in the document. For *never or presently incompetent* patients, even *family* members probably could not consent for such purposes. In any event, under most state laws, the choice of who the surrogate would be would not depend on the physician, or the CC bioethicist, or the Director of the Clinical Center, etc.—but on the family ranking as already specified in the law (D. Hoffmann, 1993). Certainly some clear rationale openly discussed and accessible to the public should preceed a policy which would use institutionalized seriously cognitively impaired human subjects in any kind of medical research.

It would appear that what has transpired up to the present, then, is a constant *shifting* of the *accountability* involved in the use of persons with mental illness in research. No Federal Regulations have incorporated the earlier calls for special protections for this vulnerable group. And even the purported protections afforded by an IRB have now been transfered in part to surrogate decision-makers under the "legal protection" of a dubiously valid DPA—a surrogate who may be appointed by a cognitively impaired person, or his/her physician to participate in medical research, with the approval of the CC bioethicist, who may delegate this authority to "other consultants." Who, then, is clearly legally responsible and accountable for any harm which the research subject sustains as a result of the experiment? And who is responsible, now, for the "welfare" and the "protection" of the research subject?

THE PRESIDENT FEDERAL REGULATIONS (1991)

In 1991 the Federal Regulations were amended to incorporate the relevant provisions required in the Common Rule (1991). These present regulations make no changes in reference to any special protections for persons with mental illness, other than, as before, an IRB. However, there is a subtle change in the language concerning IRB membership when the use of vulnerable subjects is to be considered:

…If an IRB regularly reviews research that involves a vulnerable category of subjects, such as children, prisoners, pregnant women, or handicapped or mentally disabled persons, consideration shall be given to the inclusion of one or more individuals who are *knowledgeable about and experienced in working with* these subjects. (Code of Federal Regulations 45 CFR 46, 1991, p. 7) (emphasis added)

Our concern is the shift from the language "shall include one or more individuals who are *primarily concerned with the welfare* of these subjects" as used in the 1981–1989 Federal Regulations, to "the inclusion of one or more individuals who are *knowledgeable about and experienced in working with* these subjects."

The primary goal, as stipulated in a long series of recommendations and regulations, is the respect, protection, and welfare of these vulnerable human subjects. Such respect and protection may not necessarily be afforded if the IRB member is now simply someone who is "knowledgeable about and experienced with theses subjects." Clearly, if this role is now played by a physician researcher, for example, who has personal interests in the outcome of the experiment, there is a serious concern for a conflict of interests.

So the question is not only why were persons with mental illness *dropped* as a special category of vulnerable subjects in the Federal Regulations, but also who is *accountable* for these vulnerable patients, and who is protecting them and truly looking out for their *welfare*? At present, no regulations cover research subjects with mental illness who are not institutionalized; those who are institutionalized are only covered by an IRB, with a board member who only has to be "knowledge-able about and experienced with" these kinds of human subjects; only an IRB policy at the NIH Clinical Center covers in more detail such protections, but those protections apply only to research subjects who are cognitively impaired, involves the use of a dubious DPA which has been authorized by a cognitively impaired subject, and if they are severely impaired may be authorized by the physician researcher him/herself.

DISCUSSION

It is time, we think, for a full reexamination of several of the concepts and rationales which have guided the use of persons with mental illness in both therapeutic and experimental research for many years. Such a reexamination should include at least such questions as:

1. Are vulnerable groups of human subjects, e.g. persons with mental illness, really being *exploited* because of: their institutionalization; the impaired sta-tus of their ability to give true informed consent; sometimes the absence of any family members to protect them, e.g., the homeless, elderly; undue pressure from other members of families with vulnerable subjects to "eradi-cate" the illness for future generations; bench or physician researchers who benefit professionally from the research, etc.?
2. What should be the *criteria* by which to determine if a patient is "competent" or "incompetent"; and *who* should determine these criteria? Who should be allowed to make the determination that a *particular* patient is competent or incompetent? How *accountable* should they be for their diagnosis?
3. How should diseases which afflict persons with mental illness be *classified*, and *who* should decide? How *accountable* should they be for their classifi-cations?
4. It is true that persons with mental illness *must* be used in therapeutic research for their *class* of illnesses because such information is "unprocurable by other means"? Is "unprocurable by other means" an absolute societal value; or are there possibly some sorts of experiments that shouldn't be done or informa-tion which must remain "unprocured" simply because to obtain it would involve the unethical or illegal use of human subjects?
5. Should persons with mental illness be allowed to participate in medical research which is: only for their *direct* benefit, and/or for the benefit for

classes of individuals with that same disease, and/of for the benefit of purely scientific knowledge and the greater good of society? Who decides these issues? Who is *accountable* for these *conceptual* decisions?

6. Is it medically, ethically or legally accurate or realistic to claim that persons with mental illness are competent to *choose* a surrogate to make a "substituted judgment" as to whether or not they [the patients] would freely participate in therapeutic or experimental research if they [the patients] were competent?

7. Should the appropriate "third party" role be to give a *substituted judgment* about what the surrogate thinks the patient would want to do, or to give an informed judgment about the *best interests* of theses subjects?

8. Can a Maryland DPA law, based on the protection of the property of a citizen, and now even questionably applicable to only presently *competent* persons to choose a surrogate to make *health care* decisions after they have become incompetent, be appealed to and be the justification for the on-going practice at the Clinical Center of allowing *cognitively impaired* patients to choose someone to make a substituted judgment for them to participate in *medical research*?

9. Precisely who is legally responsible for *injuries* sustained not only during the course of experiments, but for symptoms which may appear some time afterwards (the entire IRB, the CC Director, the researcher, the attending physician, the surrogate holding the DPA, the next-of-kin, the "other consultants" appointed by the CC bioethicist)?

10. Should the role of some one who is *knowledgeable about* these vulnerable patients be separated from the role of some one who is concerned about the *welfare* of these patients? Perhaps there should be both kinds of members represented on the IRB boards.

RECOMMENDATIONS

1. Public and academic comment should begin again on many of the issues surrounding the use of persons with mental illness in therapeutic and experimental research.

2. We strongly urge the NIH OPRR to *immediately* enforce and monitor compliance by placing individuals concerned with the *welfare* of persons with mental illness on all IRB committees dealing with research using these vulnerable subjects. Moreover, all such individual IRB guidelines should *immediately* reflect this change. *NAMI* members should be tapped for this function. This could include the use of NAMI members as "consent auditors," as monitors of the experiments themselves, and as possibly legally approved surrogates to advocate for the welfare and best interests of persons with mental illness who have no reliable family members, next of kin or friends.

3. NIH should immediately incorporate the relevant recommendations of the National Commission with its original language into all of its intramural *and* extramural research programs.

4. The Common Rule of the Federal Regulations (45 CFR 46) should be *amended*

to include the recommendations of the National Commission. We further recommend that the rules be applied to *all* persons with mental illness— without separating them into institutionalized and non-institutionalized groups. Both groups are vulnerable and need protection—whether in an institution, a regular health care facility, an out-patient clinic, a nursing home or a physician's office.

The authors strongly support most research efforts, and especially research that improves the treatment of persons with mental illness. Our call for recognizing them as a vulnerable group requiring special protection from potential abuse in no way diminishes that support, nor is it incompatible with our strong support for research in general. But we all have a moral obligation to ourselves and to our society to ensure proper, ethical and humane treatment for persons with mental illness when they are used as subjects in *any* kind of research.

ACKNOWLEDGMENTS

This paper is an outgrowth of Dr. Shamoo's report written on the subject on request of the National Alliance of the Mentally Ill [NAMI], Science Review Committee. Dr. Shamoo is a member of the Committee. The authors gratefully acknowledge comments on this paper by Benedict Ashley, James Burtchaell, Arthur Caplan, John Fletcher, Diane Hoffmann, Jay Katz, Eric Meslin, Alan Stone, David Thomasma, and Robert Veatch.

NOTES

1. Appelbaum, P., Lidz, C. and Meisel, A. (1987) *Informed Consent—Legal Theory and Clinical Practise*. Oxford: Oxford University Press.
2. Barber, Bernard. (1973) *Research on Human Subjects*. New York: Russell Sage Foundation.
3. Baudouin, J. (1990) *Medicine and Law* 9:1052–1061.
4. Beecher, H.E. (1966) Ethics and clinical research. *New England Journal of Medicine* 274:1354–60.
5. Capron, A.M. (1992) Where is the sure interpreter? *Hasting Center Report* 22:26–27.
6. Case Studies. (1992) Substituting our judgment. *Hastings Center Report* 22.
7. Dresser, R. (1984) Bound to treatment: the Ulysses contract. *Hastings Center Report* 14, 3:13–16.
8. Fletcher, J., Dommel, F. and Cowell, D. (1985) A trial policy for the intramural programs of the National Institutes of Health: consent to research with impaired human subjects. *IRB* 7, 6:1–6.
9. Frankel. (1975) The development of policy guidelines covering human experimentation in the United States. *Ethics in Science and Medicine* 2:43–59.
10. Ganzini, L., Lee, M.A., Heintz, R.T. and Bloom, J.D. (1993) Is the Patient Self-Determination Act appropriate for elderly persons hospitalized for depression? *Journal of Clinical Ethics* 4, 1:46–50.
11. Gillespie, R. (1989) Research on human beings: an historical overview. *Bioethics News* 8, 2:4–15.
12. Hoffmann, D. (1993), Associate Professor of Law, University of Maryland, Baltimore, MD (personal communication, May 13, 1993).
13. Irving, D. (1993) The impact of scientific misinformation on other fields: philosophy, theology, biomedical ethics, public policy. *Accountability in Research* 2, 4:243–272.
14. Irving, D. (1993) Philosophical and scientific expertise: an evaluation of the arguments on 'person-hood.' *Linacre Quarterly* 60, 1:18–46.
15. Ivy, A.C. (1948) The history and ethics of the use of human subjects in medical experiments. *Science* 108:1–5.
16. Jonas, H. (1984) *The Imperative of Responsibility*. Chicago: University of Chicago Press.
17. Katz, J. (1992) The consent principle of the Nuremberg Code: its significance then and now; in

George J. Annas and Michael A. Grodin (eds.) *The Nazi Doctors and the Nuremberg Code: Human Rights in Human Experimentation*. New York: Oxford University Press.

18. Katz, J. (1972) *Experimentation with Human Beings*. New York: Russell Sage Foundation, pp. 305–306 and 312–313.
19. King, N.P. (1992) Transparency in neonatal intensive care. *Hastings Center Report* 22:18–25.
20. Levine, R.J. (1986) *Ethics and Regulation of Clinical Research*. New Haven: Yale University Press.
21. Lifton, R.J. (1986) *The Nazi Doctors*. New York: Basic Books, Inc.
22. Maloney, D. (1984) *Protection of Human Research Subjects*. New York: Plenum Press.
23. Mitscherlich, A. and Mielke, F. (1962) *The Death Doctors*. London: Elek Books.
24. Muller-Hill, B. (1988) *Murderous Science*. Oxford: Oxford University Press.
25. National Commission for the Protection of Human Subjects of Biomedical and Behavioral Research, *The Belmont Report* (1978). U.S. Department of Health, Education, and Welfare, Washington, D.C.
26. Nelson, J.L. (1992) Taking families seriously. *Hastings Center Report* 22:6–12.
27. President's Commission for the Study of Ethical Problems in Medicine and Biomedical and Behavioral Research (1983) *Summing Up*. Washington, D.C.: U.S. Government Printing Office.
28. Proctor, R.N. (1988) *Racial Hygiene*. Massachusettes: Harvard University Press.
29. Santurri, E.N. (1982) Substituted judgment and the terminally-ill incompetent. *Thought* 57, 227: 484–501.
30. Shamoo, A.E. (1993) Role of conflict of interests in public advisory councils. Chapt. 17, pp. 159–174; in "Ethical Issues in Research," edited by D. Cheney, University Publishing Group, Inc. Frederick, Maryland, 21701.
31. Shamoo, A.E. (1992) Role of conflict of interest in scientific objectivity: a case of a Nobel Prize work. *Accountability in Research* 2:55–75.
32. Shamoo, A.E. (1991) Policies and quality assurances in the pharmeceutical industry. *Accountability in Research* 1:273–284.
33. Shamoo, A.E. and Annau, Z. (1989) Data audit: historical perspectives; in *Principles of Research Data Audit*. ed. A.E. Shamoo. New York: Gordon and Breach, Science Publishers, Inc., Chapter 1, pp. 1–12.
34. Shamoo, A.E. and Irving, D.N. (1993) The PSDA and the depressed elderly: intermittent competency revisited. *Journal of Clinical Ethics* 4, 1:74–80.
35. Truog, R.D. (1992) Triage in the ICU. *Hastings Center Report* 22:13–17.
36. *United States Code of Federal Regulations: Protection of Human Subjects* 45 CFR 46 (revised Jan. 12, 1981; revised Mar. 8, 1983; reprinted July, 1989), (a), p. 7.
37. Veatch, R. (1987) *The Patient as Partner: A Theory of Human Experimentation Ethics*. Bloomington, IN: Indiana University Press.

Proposed Regulations for Research Involving Those Institutionalized as Mentally Infirm: A Consideration of Their Relevance in 1995

Robert J. Levine

Professor of Medicine and Lecturer in Pharmacology, Yale University School of Medicine, New Haven, CT 06520

In 1978 the National Commission for the Protection of Human Subjects of Biomedical and Behavioral Research published its Report, *Research Involving Those Institutionalized as Mentally Infirm*, which included its recommendations for federal regulations to provide special protections for this class of human subjects (National Commission, 1978). The Department of Health, Education and Welfare, forerunner of the Department of Health and Human Services (DHHS), responded by publishing proposed regulations (DHEW, 1978). These proposals were never converted to final regulations; to this day there are no federal regulations addressed to the special needs of this population.

In 1981 and again in 1983, the President's Commission for the Study of Ethical Problems in Medicine and Biomedical and Behavioral Research insisted that it was necessary to promulgate final regulations in this field (President's Commission, 1981, pp. 74–76; 1983, pp. 23–29). The DHHS Secretary "responded that, while continuing to consider specific issues regarding protections for institutionalized mental patients, the Department is not intending to issue additional regulations in the near future." The Secretary provided two justifications for this inaction (President's Commission, 1983, p. 26): "first, that the rules proposed by the Department in November 1978 had produced a 'lack of consensus' and, second, that the basic regulations on human subjects research adequately respond to the recommendations made by the National Commission to protect persons institutionalized as mentally disabled."

I share the opinion of the President's Commission that this response is unacceptable. The basic federal regulations on human subjects research do not provide adequate guidance for the conduct of research involving those institutionalized as mentally infirm; they are not an adequate response to the National Commission's recommendations.

But what about the Secretary's second justification for the lack of action, that publication of the proposed regulations was greeted by a "lack of consensus."

This article will also appear in the journal: *Accountability in Research*, Volume 4, pp. 177–186 (1996).

Unfortunately, this claim is correct. The reason for this lack of consensus is that the DHEW regulation writers departed substantially from the recommendations of the National Commission in ways I shall soon explain. First, I want to make it clear that, had they adhered to the National Commission's recommendations, consensus could have been achieved.

About one year earlier the National Commission had published its recommendations for regulations for the special protection of children as research subjects (National Commission, 1977). These recommendations were very similar to those it made subsequently for those institutionalized as mentally infirm. DHEW regulation writers attempted to introduce in their proposed regulations for research involving children several provisions that were not recommended by the National Commission. In response to the National Commission's protests these changes were removed. Consequently, the final regulations for research involving children correspond very closely to the National Commission's recommendations (Levine, 1986, Chapter 10). Most commentators on the final regulations expressed their support for them as accurately reflecting a widely held consensus in the field.

The proposed regulations for those institutionalized as mentally disabled were published after the National Commission had been dissolved. I believe that if the National Commission had still been available to protest DHEW's unwarranted changes, the outcome would have been much like it was with the children's regulations. Final regulations would have been promulgated that corresponded closely to the National Commission's recommendations and most informed and interested people would have been satisfied with them.

THE PROPOSED REGULATIONS

Let us consider some of the changes introduced by the DHEW regulation writers. DHEW changed not only the name of the class of persons to "those institutionalized as mentally disabled," but also its definition (DHEW, 1978). DHEW accepted the National Commission's recommendation that consent auditors should be appointed by the IRB. The National Commission had recommended that the need for such agents should be determined as a discretionary judgment of the IRB, except when procedures presenting more than minimal risk were to be performed for research purposes; only in such circumstances would the appointment of consent auditors have been mandatory. DHEW, by contrast, stated that it "is giving consideration to requiring that a consent auditor monitor all research covered by these Regulations, including research involving no more than minimal risk" (DHEW, 1978, p. 53952). In addition, DHEW proposed to create the position of advocate, an individual appointed by the IRB but having no other tie to any institution or agency with an interest in the research. The advocate would have been "construed to carry the fiduciary responsibilities of a guardian ad litem" (Section 46.503k). "Consideration is being given to mandating that … whenever the consent auditor determines that a subject is incapable of consenting, the subject may not participate without the authorization of the advocate" (Section 46.505b). The Commission had not recommended advocates.

Thus, in order to involve some of those institutionalized as mentally infirm (or disabled) in research, it would have been necessary for an investigator to first seek the approval of 1) DHEW, 2) the IRB, 3) the subject, 4) the subject's "legally authorized representative" (who, in many cases will be a guardian ad litem), 5) the consent auditor, 6) the advocate (who carries the fiduciary responsibility of a guardian ad litem) and 7) a court of competent jurisdiction. It is unsurprising that the proposed regulations evoked from "the public" a massive and generally disapproving response (Reatig, 1981), what the Secretary subsequently called a "lack of consensus."

Natalie Reatig suggests that the tendency to excessive bureaucracy expressed in the proposed regulations reflects their authors' preoccupation with the most seriously impaired "back ward" population, the relatively few for whom such procedural protections might be appropriate (Reatig, 1981). She argues that there are many persons for whom modest levels of consent auditing could serve some useful purpose; these are generally those described as vulnerable (Levine, 1986, Chapter 4). Thus, in order to avoid stigmatization of the mentally infirm, she calls upon DHHS to issue guidelines (not regulations) suggesting to the IRB when and how it should consider the application of additional procedural protections.

Reatig further calls for a second set of guidelines that would require special procedural protections for institutionalized persons, but only when there is reason to believe that specific individuals or groups of persons have impaired rationality or comprehension. Although this second set of guidelines would be more directive than the first, they should still allow the IRB sufficient flexibility to meet the needs of particular individuals or groups.

THE NATIONAL COMMISSION'S RECOMMENDATIONS

Recommendations for All Persons Having Limited Capacity to Consent

In general, the National Commission concluded that persons having limited capacity to consent are vulnerable or disadvantaged in ways that are morally relevant to their involvement as research subjects (Levine, 1986, Chapter 10). Therefore, the ethical principle of justice is interpreted as requiring that we facilitate activities that are expected to yield direct benefit to the subjects and that we encourage research designed to develop knowledge that will be of benefit to the class of persons of which the subjects are representative. However, we should generally refrain from involving the special populations in research that is irrelevant to their conditions as individuals or at least as a class of persons.

The principle of respect for persons is interpreted as requiring that we show respect for a potential subject's capacity for self-determination to the extent that it exists. Some who cannot consent can register knowledgeable agreements (assents) or deliberate objections. To the extent that the capacity for self-determination is limited, respect is shown by protection from harm. Thus, the National Commission recommended that the authority accorded to members of the special populations or their legally authorized representatives to accept risk be strictly limited; any proposal to exceed the threshold of minimal risk requires special justification.

Many individuals who have limited capacity to consent have not been identified by Congress or DHHS as members of the "special populations." The norms recommended by the National Commission for identified special populations were intended by the Commission to be applied, as appropriate, to others. For example, some rules recommended for those institutionalized as mentally infirm may be appropriate for some other prospective subjects with limited capacities for comprehension who are not institutionalized; for further discussion of vulnerable subjects not identified as members of the special populations, see Levine (1986, Chapter 4).

Recommendations Specifically Addressed to Those Institutionalized as Mentally Infirm

Recommendation 1. For all research involving those institutionalized as mentally infirm, the IRB must determine that:

"1B: The competence of the investigator(s) and the quality of the research facility are sufficient for the conduct of the research...." Although the international ethical codes require that researchers should be competent, the Commission assigns to the IRB responsibility for determining that they are only for this class of subjects (Levine, 1986, Chapter 2).

"1D: There are good reasons to involve institutionalized persons in the conduct of research...." The investigator must satisfy the IRB that it is appropriate to involve those institutionalized as mentally infirm in the research project. The IRB should consider whether the research is relevant to the subjects' emotional or cognitive disability, whether individuals with the same disability are reasonably accessible to the investigator outside the institutional setting, or whether the research is designed to study the nature of the institutional process or the effect of some aspect of institutionalization on persons with a particular disability. For an example of a well-reasoned justification involving institutionalized persons exclusively as subjects of research not designed to evaluate therapeutic interventions, see Marini (1980). Following the National Commission's guidelines, he argues that some types of research on some types of behavior (in this case, aggression) should be done only on institutionalized persons.

The IRB is instructed to review with special care any proposal to institutionalize subjects or to extend their stay in an institution solely for research purposes; this is to be permitted only if subjects "knowledgeably agree." In addition, careful consideration should be given to the possibility of using alternate means to solve the research problem.

1H: Adequate provisions are made to assure that no prospective subject will be approached to participate in the research unless a person who is responsible for the health care of the subject has determined that the invitation to participate in the research and such participation itself will not interfere with the health care of the subject....

In general, one should not invite patients to become research subjects without authorization of the physicians responsible for their care (Levine, 1986, Chapters 4 and 5). However, this is the only recommendation for a regulation that would require such consultation.

In the commentary, the National Commission further elaborates that when the potential subject's physician or other therapist is involved in the proposed research, independent clinical judgment should be obtained regarding the appropriateness of including that patient in the research. This is intended to reduce conflicts of interest between the objectives of health care and those of research, while still permitting clinicians, who may be especially knowledgeable regarding promising avenues of research, to apply their expertise in both enterprises. This recommendation is addressed to the same problem as Principle I.10 of the Declaration of Helsinki; however, the National Commission would not require that a third party obtain informed consent as does Helsinki. Further discussion of separating the roles of clinician and investigator can be found in (Levine, 1986, Chapter 5).

Recommendation 2. For research that presents no more than minimal risk, the IRB must determine that:

2B: Adequate provisions are made to assure that no subject will participate in the research unless: (I) the subject consents to participation; (II) if the subject is incapable of consenting, the research is relevant to the subject's condition and the subject assents or does not object to participation....

Persons who object to participation may not participate in any research except as specified in Recommendation 3. For research activities covered by both Recommendations 2 and 3, consent auditors should be appointed at the discretion of the IRB.

Recommendation 3. For research in which more than minimal risk is presented by an intervention that holds out the prospect of direct benefit for the individual subjects or by a monitoring procedure required for the well-being of the subjects, the IRB must determine that:

3D: ... [N]o adult subject will participate ... unless: (I) the subject consents to participation; (II) if the subject is incapable of consenting, the subject assents to participation (if there has been an adjudication of incompetency, the permission of a guardian may also be required by state law); (III) if the subject is incapable of assenting a guardian of the person gives permission (if a guardian of the person has not been appointed, such appointment should be requested at a court of competent jurisdiction) or the subject's participation is specifically authorized by a court of competent jurisdiction; or (IV) if the subject objects to participation, the intervention holding out the prospect of direct benefit for the subject is available only in the context of research and the subject's participation is specifically authorized by a court of competent jurisdiction....

These provisions are similar to those authorizing similar activities for research involving children. In addition, they reflect a concern that the guardian from whom permission is sought is truly a "legally authorized representative." Moreover, they recognize that there will at times be differences of opinion as to what constitutes a benefit for a particular patient. In jurisdictions that grant institutionalized individuals an unqualified right to refuse therapy, their objection to participation in research will be binding.

Recommendation 4. For research in which minor increments above minimal risk are presented by interventions that do not hold out the prospect of direct benefit for individual subjects, the recommendations are very similar to those for research involving children. The only substantive difference is that there is no requirement that the procedures be reasonably commensurate with those inherent in the subject's life situation. As is also required for research involving children, the IRB must determine that:

4C: the anticipated knowledge (I) is of vital importance for the understanding or amelioration of the type of disorder or condition of the subjects, or (II) may reasonably be expected to benefit the subjects in the future....

The requirements for consent and assent are the same as those specified in Recommendation 3. For this class of research, the appointment of a consent auditor by the IRB is mandatory.

Recommendation 5. For research in which more than minor increments above minimal risk are presented by interventions that do not hold out the prospect of direct benefit to the subjects, review by a National Ethical Advisory Board is required; the criteria for its approval are substantially identical to those prescribed for children.

DISCUSSION

In this discussion I shall consider a class of persons defined as *those who by reason of mental or behavioral disorder are vulnerable in that they are not capable of giving adequately informed consent.* I shall not consider further the classes defined as "those institutionalized as mentally infirm (or disabled)."

There are some features of the National Commission's recommendations that should be incorporated in any guidelines or regulations for the ethical conduct of research on this class of persons. However, it must be recognized that these recommendations have two general limitations. Firstly, they are addressed only to a subset of the population with mental or behavioral disorders—those who are institutionalized. Secondly, they reflect the attitude of protectionism that dominated all consideration of the ethics of research involving human subjects from the 1940s through the early 1980s.

Let us consider briefly each of these limitations.

Institutionalization

By the time the National Commission was established in 1974, many commentators had expressed concern that institutionalized persons in general were used disproportionately and unfairly as research subjects because they were administratively convenient to the researchers (Levine, 1986, Chapter 4). Consequently, there were those who argued that an individual who is institutionalized as mentally infirm should not participate in research for which noninstitutionalized persons would be suitable subjects (National Commission, 1978, pp. 115–117).

This argument was grounded in the requirement of the principle of justice that we protect vulnerable persons from harm and that we minimize the burdens imposed upon those who already bear more than their fair share. It was argued that those institutionalized as mentally infirm are particularly vulnerable to exploitation; this vulnerability arises from their relative incapacity to protect themselves through the usual negotiations for informed consent. In addition, some of them tend to be totally dependent upon the institution in which they reside. Moreover, persons who are institutionalized as mentally infirm already carry burdens—often heavy burdens—imposed by their disabilities; therefore, it is unjust to ask them to assume additional burdens. Those outside the institution may be burdened by similar incapacities (diseases and social maladjustments); however, they are relatively less vulnerable because they are more likely to have caring persons to assist and protect them.

Opponents to these arguments claimed that it was incorrect to assume that participation in research is always a burden or that being in an institution is always a damaging experience. At times, participation in research may have beneficial effects; e.g., research subjects may receive additional attention and may be afforded opportunities for interaction with people from outside the institution. The tasks associated with research may be interesting and a welcome change from the boredom of institutional life. In addition, deinstitutionalization of mental patients is not always associated with satisfactory integration into a community of caring persons; in some cases it may be perceived as abandonment to ghettos where they have no one to look after their personal, health, and social needs.

The Commission addressed these issues by recommending two requirements. Firstly, investigators proposing to recruit subjects from an institution should be required to justify their involvement. Secondly, institutionalized individuals should be permitted to participate in research that is not relevant to their conditions only if they are capable of consenting and the research presents no more than minimal risk. For more extensive discussion of the problems presented by the fact of institutionalization, see (Levine, 1986, Chapter 12).

Protectionism

All international codes of research ethics and virtually all national legislation and regulation in the field of research involving human subjects project an attitude of protectionism. Their dominant concerns are the protection of individuals from injury and from exploitation. Indeed, when the United States first promulgated federal regulations in this field, it called them "Policy for the *Protection* of Human Research Subjects" and assigned responsibility for assuring compliance with them to the "Office for *Protection* from Research Risks."

There are important historical reasons for this attitude. The first international code of research ethics, the Nuremberg Code, was drafted as a reaction to the atrocities committed by Nazi research physicians. Many of their experiments entailed the deliberate killing or injuring of prisoners whose right to consent, or to refuse, was ignored. Thus, the field of research ethics began as a search for secure defenses against a repetition of the most egregious assaults on the rights and welfare of human beings ever committed in the name of science. The perception of research as perilous was reinforced by news of the thalidomide disaster and, in the United States, through public exposure of activities in which there had been violations of the standards of Nuremberg, e.g., the Tuskegee syphilis studies.

In the past decade, however, we have seen a dramatic change in the popular vision of research. Although this change may seem abrupt, even in the 1950s and 1960s, there were some writers who recognized that some people regarded their participation in research as beneficial (Levine & Lebacqz, 1979). This message began to gain clarity early in the 1980s. When Barney Clark received the first artificial heart in 1982, it became apparent that many people with end stage heart disease saw this device as a very valuable and very scarce benefit (Levine, 1984). The only alternative was certain death. Later in the 1980s, during the placebo-controlled trials of AZT in the treatment of patients with AIDS, the message became even more clear as a new voice emerged—the very articulate and increasingly powerful voice of the AIDS activists. They correctly pointed out that enrollment in this clinical trial was the only way for HIV-infected persons to get even a 50 percent chance of receiving the only therapy that offered any hope of delaying death or the onset of opportunistic infections. The widely held perception that participation in this clinical trial was beneficial was so powerful that some patients even faked the eligibility criteria for inclusion (Levine, C. et al., 1991). In fact, the perception of benefit was so strong, even among physicians, that some primary care physicians cooperated with patients in falsifying inclusion criteria.

Thus, in a remarkably short time, the popular image of research has changed dramatically. Investigational drugs, once regarded as dangerous, potential "thalidomides," are now described as "promising new therapies" to which persons suffering with serious illnesses should have access. Indeed, policies and programs have been set up to assure such access even during the conduct of clinical trials designed to find out whether they are safe and effective; we call these programs "expanded access" and "parallel track." (Levine, 1987; DHHS, 1992).

People are now demanding access to randomized clinical trials, which previously were envisioned as experiments in which persons are stripped of the "good of personal care" and treated as "impersonal statistical objects"(Fried, 1974). The pendulum has come full swing. I believe that the current tendency to see research as largely beneficial and benign is just as wrong-headed as the earlier tendency to view it as primarily dangerous and exploitative. I believe it is necessary to maintain a balanced perspective. Our policies should reflect the need to encourage the conduct of ethical research while maintaining the necessary vigilance to safeguard the rights and welfare of the subjects.

As a society, we have begun to act on our recognition that the policies designed to protect vulnerable individuals may inadvertently create barriers to the development of therapies essential to combat their diseases. Policies designed to protect children, for example, resulted in making them, in the words of the pediatrician, Harry Shirkey, "the therapeutic orphans of our expanding pharmacopoeia." (Shirkey, 1968). Such deprivation of vulnerable populations of the benefits of new therapeutic products is now recognized as a class injustice (Levine, 1986, Chapter 10). In the light of this recognition, we are revising our policies to eliminate unjustified obstacles to the involvement of children in research. For example, in view of the special problems presented by pediatric AIDS, we no longer insist on completing studies in adults before beginning studies on children.

Similar revisions in policy are in various stages of development for other populations who previously were excluded from research "for their own good" (Levine, 1994). These include women, adolescents, racial and ethnic minorities, prisoners and others. I believe that we must carefully review any policy that might be proposed for the ethical conduct of research involving persons who by reason of mental or behavioral disorders are vulnerable in that they are not capable of giving adequately informed consent. There must be an appropriate balance between the need to encourage the conduct of ethical research and the necessity to safeguard the rights and welfare of the subjects.

RECOMMENDATIONS

1. There should be a federal policy for the ethical conduct of research on persons who by reason of mental or behavioral disorders are vulnerable in that they are not capable of giving adequately informed consent.

2. This policy should be formed by a national advisory body patterned after the National Commission for the Protection of Human Subjects of Biomedical and Behavioral Research. It should have similar composition, independence, administration and financial support.

3. A good starting point for the development of this policy would be to review the recommendations of the National Commission. There are many features of these recommendations that are sound and remain appropriate. However, it will be necessary to build upon these recommendations so that they will be:

    A. applicable to human subjects who are not institutionalized, and

    B. suitably responsive to changes since 1978 in the public perception of research. A balance should be struck between the excessive protectionism of the 1970s and the excessive tendency of the present time to envision research as benign and beneficial.

REFERENCES

DHEW: Protection of Human Subjects: Proposed Regulations on Research Involving Those Institutionalized as Mentally Infirm. Federal Register 43 (No. 223) (November 17, 1978) 53950–53956.

Department of Health and Human Services: Expanded availability of investigational new drugs through a parallel track mechanism for people with AIDS and other HIV-related disease. Federal Register 57(No. 73): 13244–13259, April 15, 1992.

Fried, C.: *Medical Experimentation: Personal Integrity and Social Policy.* American Elsevier Company, New York, 1974.

Levine, C., Dubler, N.N., and Levine, R.J.: Building a new consensus: Ethical principles and policies for clinical research on HIV/AIDS. *IRB: A Review of Human Subjects Research* 13 (No. 1&2):1–17, January–April, 1991.

Levine, R.J.: Total artificial heart implantation—eligibility criteria. *JAMA* 252:1458–1459, 1984.

Levine, R.J.: *Ethics and Regulation of Clinical Research.* Urban & Schwarzenberg, Baltimore, MD, 1981, Second edition, 1986.

Levine, R.J.: Treatment use and sale of investigational new drugs. *IRB: A Review of Human Subjects Research* 9(4):1–4, July/August, 1987.

Levine, R.J.: The impact of HIV infection on society's perception of clinical trials. *Kennedy Institute of Ethics Journal.* 4 (No. 2):93–98, 1994.

Levine, R.J. & Lebacqz, K.: Some ethical considerations in clinical trials. *Clinical Pharmacology & Therapeutics* 25:728–741, 1979.

Marini, J.L.: Methodology and ethics: Research on human aggression. *IRB: A Review of Human Subjects Research* 2 (No. 5) (May 1980) 1–4.

The National Commission for the Protection of Human Subjects of Biomedical and Behavioral Research: *Research Involving Children: Report and Recommendations.* DHEW Publication No. (OS) 77-0004, Washington 1977.

The National Commission for the Protection of Human Subjects of Biomedical and Behavioral Research: *Research Involving Those Institutionalized as Mentally Infirm: Report and Recommendations.* DHEW Publication No. (OS)78-0006, Washington, 1978.

The President's Commission for the Study of Ethical Problems in Medicine and Biomedical and Behavioral Research: *Protecting Human Subjects: The Adequacy and Uniformity of Federal Rules and Their Implementation.* U.S. Government Printing Office, Stock No. 040-000-00452-1, Washington, 1981.

The President's Commission for the Study of Ethical Problems in Medicine and Biomedical and Behavioral Research: *Implementing Human Research Regulations: The Adequacy and Uniformity of Federal Rules and of Their Implementation.* U.S. Printing Office, Stock No. 040-000-004781-8, Washington, 1983.

Reatig, N.: Government Regulations Affecting Psychopharmacology Research in the United States: Implications for the Future. In: *Human Psychopharmacology: Research and Clinical Practice,* Vol. II, ed. by G.D. Burrows and J.S. Werry, JAI Press, Inc. Greenwich, CT, 1981.

Shirkey, H.C.: Therapeutic orphans. *Journal of Pediatrics* 72 (1968) 119–120.

Informed Consent to Research on Institutionalized Mentally Disabled Persons: The Dual Problems of Incapacity and Voluntariness[1]

George J. Annas[†] *and Leonard H. Glantz*[‡]

[†]*Utley Professor and Chair, Health Law Department, Boston University Schools of Medicine and Public Health, 80 East Concord St., Boston, Massachusetts*
[‡]*Professor, Health Law Department, Boston University Schools of Medicine and Public Health, Boston, Massachusetts*

Institutionalized mental patients are perhaps the most isolated and underprivileged members of our society. They have often been victims of social injustices, including horrible facilities, poor or nonexistent treatment and education, indiscriminate sterilization, and deprivation of basic legal protections, including the performance of unethical and/or illegal human experimentation.

The problem of institutionalized mental patients providing informed consent to experimentation combines the issues faced in regard to children and prisoners. The problem is two-pronged, concerning both the legal capacity of the individual to consent and the inherently coercive nature of institutionalization. The major questions may be highlighted by reference to the Nuremberg Code. (Appendix A) Do mental patients in general have the legal capacity to consent? Is a particular mental patient sufficiently competent to enable an "understanding and enlightened decision?" Is proxy consent ever valid, and if so, under what circumstances? Does the fact of institutionalization create a coercive situation which effectively removes the individual's ability "to exercise free power of choice?" In this paper we examine the legal issues regarding informed consent to research on institutionalized mentally disabled persons, and identify issues that must be resolved before adopting regulations to protect the rights and welfare of these individuals.

Although "desinstitutionalization" is now the rule, large institutions historically carried the responsibility for caring for the mentally deficient individual who either cannot function in the community or whose family has decided not to have him remain at home.[2] Dehumanization has been amply demonstrated in such residential facilities.[3] Institutionalized mental patients have traditionally been subjects of experiments, and not necessarily because the research has special applicability to this group.[4] Research frequently requires that a convenient, stable subject population be followed over a period of time. Thus, the institutionalized have been particularly attractive to investigators because they constitute a "controlled" or "captive" community, with a relatively uniform diet, schedule of sleeping hours, and daily routine, and since they were often wards of the state, they formed an inexpensive pool of experimental subjects.[5] In addition, people in institutions are often easily manipulated, either due to their own mental deficiencies or the lack of interest in their welfare demonstrated by their legal guardians or facility administration and staff.

CAPACITY TO CONSENT

In general, "every human being of adult years and sound mind has a right to determine what shall be done with his own body."[6] Thus, competent adults have the right to choose their own course of care and to be apprised of the facts necessary to make an informed choice.[7] The doctrine of informed consent, based on individual autonomy and promoting rational decision-making in the context of a fiduciary doctor-patient relationship, has been developed to insure that patients are provided with specific information about the proposed treatment, alternatives, risks and benefits, and likely outcomes, before being asked to agree to undergo it.[8] In the experimental setting, legal and ethical principles developed since World War II, and articulated in the Nuremberg Code, require that consent to experimental interventions be voluntary, competent, informed and understanding.[9] As will be discussed, there may be some circumstances in which others can consent for either therapy or research on an individual, but proxy consent can only be used if the research subject is incompetent, and many, if not most, mentally disabled persons are competent.

Institutionalization in a facility for the mentally deficient and legal incompetence, are not synonymous.[10] Thus, the institutionalized individual is often deemed to have the same legal ability to exercise his rights as a free-living person.[11] Competent adult mental patients have the same constitutional right to refuse medication, including psychotropic drugs, as other adults have. This is a relatively new notion in mental health law, and apparently in mental health therapy as well.[12] On the other hand, it is perfectly consistent with the treatment refusal cases in the medical care area, and there is no *a priori* reason why the constitutional rights of medical and mental patients should differ.

As Judge Brotman of the United States District Court of New Jersey put it, "Individual autonomy demands that the person subjected to the harsh side effects of psychotropic drugs have control over their administration."[13] Judge Brotman also defined four factors that could be weighed by a court in forcing treatment in a non-emergency: (1) the patient's physical threat to patients and staff at the institutions; (2) the patient's capacity to decide on his particular treatment; (3) whether any less restrictive treatments exist; and (4) the risk of permanent side effects from the proposed treatment. In a further proceeding, Judge Brotman ruled that for involuntary patients, the weighing of these four criteria need not to be done by a judge, but could be done by an "independent psychiatrist" at an informal hearing at which the patient could be represented by counsel.

A later ruling by Judge Joseph Tauro of the U.S. District Court of Massachusetts made no distinction between voluntary and involuntary patients, defined emergency treatment very narrowly, and required consent of the court or a court-appointed guardian before incompetent patients can be medicated against their will in nonemergencies.[14] The defendants argued that mental patients are *de facto* incompetent to make treatment decisions and that the state must therefore act as *parens patriae* and make such decisions. The court disagreed, and found that most mental patients "are able to appreciate the benefits, risks, and discomfort that may reasonably be expected from receiving psychotropic medication." The court found this "particularly true" in patients who had experienced the medication, and noted that the "therapeutic alliance" between psychiatrist and patient demanded an understanding and acceptance of the treatment program by the patient. If a physician wanted to treat a patient against his will, a probate court had to be petitioned for the appointment of a guardian. If the probate court found the patient unable to give informed consent to treatment, it would appoint a guardian who could give consent.

As in the New Jersey case, the court found that forced medication violated the patient's constitutional right to privacy in the sense of self-determination. It accordingly required physicians to obtain informed consent prior to administering psychotropic drugs "that may or may not make the patient better, and that may or may not cause unpleasant and unwanted side effects."[15]

In general, therefore, as concerns a therapeutic (beneficial) biomedical or behavioral procedure, informed consent of the subject is needed prior to its performance unless the mental patient is a minor or has been judicially declared incompetent. If the person is incompetent, consent can be obtained from a parent or legal guardian. This substitute consent is valid since, by definition, a therapeutic procedure is for the benefit of the individual. For example, if an incompetent mental patient needed an appendectomy, the proxy consent of his guardian would be sufficient. For nontherapeutic procedures that have risks, informed consent may only be secured from a competent patient himself, because proxy consent is not sufficient for risky nonbeneficial procedures.

BARRIERS TO COMPETENCE

Effects of "Institutionalization"

Determining whether an institutionalized individual is competent to consent is complicated by various factors. In the first place, the very fact that the individual is institutionalized may have a practical effect on the issue of competence. This is due to the results of a process termed "institutionalization." People who are cordoned off from the outside world are often effectively stripped of their concept of "self", a perception which is vital in order to satisfy the demands of informed consent. Erving Goffman, in *Asylums*, discusses the effects of "total institutions."

In total institutions there is a basic split between a large managed group, conveniently called inmates, and a small supervisory staff. Inmates typically live in the institution and have restricted contact with the world outside the walls; staff often operate on an eight-hour day and are socially integrated into the outside world. Each grouping tends to conceive of the other in terms of narrow hostile stereotypes, staff

often seeing inmates as bitter, secretive, and untrustworthy, while inmates often see staff as conde-
scending, high-handed, and mean. Staff tends to feel superior and righteous; inmates tend, in some
ways at least, to feel inferior, weak, blameworthy, and guilty.[16]

Further complicating this situation is the inherent duress within the institution
whenever an attempt is made to obtain consent. Physicians are often able to
"engineer" consent from their patients/subjects by manipulation of their fiduciary
relationship. In addition, a patient will often be swayed by hopes of influence with
institutional authorities or release from an indeterminate commitment—even if
these things were never promised nor even mentioned by the physician in his
discussions with the individual. The supreme inducement to consent is the hope of
obtaining freedom. The institutional setting makes it difficult for one not to feel
some coercion or encouragement to consent merely in being approached for the
particular procedure. This is particularly true for those individuals who see little
or no hope of their eventual release, but who are assured that this particular
treatment may make this possible.[17]

Ability to "Comprehend"

Another troubling factor influencing the issue of competence is the fact that there
are numerous levels of mental retardation and mental illness, ranging from rather
mild to severe, found within each facility. Mentally retarded residents of institutions who are either mildly or borderline retarded are capable of a relatively
independent life, as opposed to the severely and profoundly retarded, who range
from those who may function under sheltered conditions to those who are completely helpless. The same holds true for the difference in the level of functions
of the various groups of mentally ill.

A mentally ill patient may be competent to consent one day and yet become
incompetent the next. An acute onslaught of particular forms of mental illness are
often possible, so that a patient's condition can change dramatically in a very short
period of time. Finally, it is not always easy to distinguish competence from
incompetence. Although a particular patient may not have been judicially determined to be incompetent, from a practical viewpoint it may be impossible to gain
adequate consent from him. Certainly if a patient is psychotic or hallucinating and
cannot assimilate information about a proposed procedure, he does not have the
capacity to reach a decision bout the matter in question. Some mental patients are
incapable of evaluating information in what most people would call a rational
manner. A treatment decision might ordinarily be based on considerations of
perceived personal objectives, or long-term versus short-term risks and benefits.
But there are patients whose acceptance or rejection of a treatment is not made in
relation to any "factual" information. To add to this dilemma, although a mental
patient may refuse to give his consent to a procedure, his refusal may only be a
manifestation of his illness, having little resemblance to his actual desires.[18]

A possible solution to this problem has been offered by one commentator, who
suggested that persons incapable of giving consent should be treated with the least
intrusive therapies until they learn "to appreciate the value of treatment and those
who offer it."[19] Another possibility would be to bring all people falling within this

category to court for a competency hearing. If found incompetent, a guardian could then be appointed. Problems with this approach include the fact that it would be burdensome and time-consuming; it may also be "overkill." Should we subject the patient to the stigma of the incompetency label, together with the removal of many of his rights, under these circumstances?

Another way of approaching the predicament would be to classify, on a procedure-by-procedure basis, legally competent patients who are potential subjects into two groups: those having the capacity to give consent and those not having that capacity.[20] Those in the second group would be subjects in the experiment only if a neutral decision-maker decided it was in their best interest. Competence should be defined as the capacity to understand the nature of the procedure, to weigh the risks and benefits, and to reach and communicate a decision.[21] A reviewer should be obliged to honor the patient's decision as long as the person understands the nature of the procedure, its risks and benefits, and the possible alternatives.

PROXY CONSENT

Individuals who are incompetent are precluded from making legally binding determinations concerning medical care. The fact that the person is a minor usually means that the person does not have the intelligence and capability to comprehend fully the nature and purpose of a procedure or to engage in the weighing of risks and benefits which is involved in the decision-making process. The same is true for someone who, as the result of a judicial hearing, has been declared incompetent to manage his own affairs, and has therefore had a guardian appointed for him. Others, the parents for the child and the court-appointed guardian for the adjudicated mentally incompetent adult, assume this function for him. The purpose is to protect the incompetent individual from harm that might result from either his own lack of knowledge or from coercive methods used to obtain his consent. However, under the common law, guardian consent on behalf of an incompetent may only be granted or withheld on the sole basis of the incompetent person's welfare. Even the judgment of the guardian regarding the incompetent's best interests is not always conclusive, and courts will intervene, under the doctrine of parens patriae, to protect the welfare of the incompetent person.[22]

Proxy consent presents serious problems for institutionalized mentally ill and mentally retarded persons. For example, does the parent/guardian have both the motivation and capability to represent the best interests of the institutionalized person? Implicit in guardianship is the assumtion that the guardian has the ability to care for and deal with the incompetent person and represent him in his dealings with society in general and the institution in particular.[23] There may be a conflict of interest between the guardian and ward so as to preclude adequate representation of the institutionalized person's interests. For example, the parent/guardian may have been the individual who originally "voluntarily" placed a minor/incompetent in the facility. There are also many societal pressures that operate to induce this, including mental and physical frustration, economic stress, hostility toward the individual stemming from added pressures, and perceived stigma of mental deficiency.[24] Often the individual is institutionalized less for his own benefit than for the comfort of others. The particular guardian may also be unable to deal effectively with the public and private institutional providers of service due to the disparity in leverage and sophistication that normally exists between guardian and institution. These potential conflicts should always be recognized. Consent by proxy to experimentation should be viewed with suspicion, and should not be accepted as valid and legally adequate until it has been critically reviewed to assure that it serves its original purpose: the protection of the best interests of the individual subject.

THERAPEUTIC EXPERIMENTATION

Biomedical Procedures

There is little statutory or case law dealing specifically with experimentation on institutionalized mental patients. Therefore, it is necessary to analogize to the factors involved in the nonexperimental situation. While this is worthwhile, it is also potentially dangerous. Theoretically at least, in the therapeutic situation the physician's only concern is the patient's well-being. With an experiment, not only are there usually more uncertainties and greater risks, but the physician-scientist who contemplates the procedure is motivated in part or entirely by using a protocol to test a hypothesis in search of scientific knowledge of general applicability. The physician-patient relationship is radically altered by the broadened objectives of the physician-researcher, who may be more interested in conducting an objective study than in the best interests of the subjects . Thus, with informed consent, the law is stricter and more protective of the subject's rights in its analysis of the experimental situation. Some therapeutic procedures may be given separate and different consideration by the law. For example, it is possible that the sterilization of an incompetent individual may be deemed to be "therapeutic," or in his best interests. For those incompetent persons who do not have the requisite mental capacity to adequately use alternative forms of birth control, sterilization may be the only viable option for preventing pregnancy. There may be medical reasons preventing the use of other birth control options, as well as social and psychological information which contraindicate these methods.

Regardless of this, the court in *Relf v. Weinberger*[25] decided that the consent of a representative of a mentally incompetent individual cannot impute voluntariness to a person actually undergoing irreversible sterilization. This finding was based on the determination that, at least when important human rights are at stake, there is a requirement that "the individual have at his disposal the information necessary to make his decision and the mental competence to appreciate the significance of that information." Therefore, since the federal statute under consideration only permitted federally assisted family planning sterilizations on a voluntary basis, the court held that they cannot be performed on any person incompetent to personally consent to the procedure. Thus, proxy consent to sterilization was found not to be voluntary consent, seemingly regardless of whether the particular sterilization was considered therapeutic or not.

In *Wyatt v. Aderholt*,[26] a three-judge federal court declared the Alabama involuntary sterilization statute unconstitutional. In addition, it promulgated guidelines for the voluntary sterilization of institutionalized mental patients. Initially, the court determined that, not only must the sterilization be in the "best interest" of the resident, but it also may not be performed without the consent of the person to be sterilized if he is competent to consent. If the individual is incompetent the court does not allow guardian/proxy consent, even though the procedure must be, according to the guidelines, in the best interest of the ward, and therefore traditionally within the scope of authority of a guardian. Instead, the court held that sterilization may not be performed under these circumstances unless it is approved by the director of the institution, a review committee, *and* a court of competent jurisdiction. This principle of protecting the incompetent person's interests by requiring court review was followed by the court in *In re Anderson*,[27] a case involving a father's petition to authorize the sterilization of his mentally retarded daughter.

Thus, there is legal authority for the performance of serious therapeutic medical procedures upon a mental patient without the patient's consent. There is also legal authority for the proposition that irreversible sterilization is, by its very nature, so important and intrusive that either proxy consent will not be found valid at all, or it will only be allowed in the context of stringent procedural safeguards, including judicial review. It is unclear exactly which other procedures would be included in this category. The more drastic the procedure and its possible effect upon the patient and the exercise of his rights, however, the more likely that the stricter standards will apply.

Behavior Modification

The problem of consent is even more complex with behavior modification procedures. The term behavior modification has come to mean *all* of the ways in which human behavior is modified, changed, or influenced. Thus, behavior modification may include milieu therapy, psychotherapy, positive reinforcement, token economy programs, aversive conditioning, as well as electroconvulsive therapy, injection of psychoactive drugs, and psychosurgery.[28] In this sense, behavior modification refers only to the end product of the process—a change in behavior. As always, we should begin with the assumption that a mental patient has the power to consent or withhold consent to behavior modification, unless he is adjudicated legally incompetent, in which case a guardian can consent to those procedures which are for his benefit.

Thus, in *Winters v. Miller*[29] an involuntarily committed mental patient alleged that her rights had been violated due to the imposition of forced medication, mostly in the form of tranquilizers. The court based its decision on First Amendment grounds, because the patient was a Christian Scientist who was refusing to consent to the treatment on religious grounds. Nonetheless, the court emphasized that although Winters was involuntarily committed, she had never been legally determined to be incompetent, and therefore retained the ability to make her own choice concerning consent to treatment. Similarly, the court in *Belger v. Arnot*[30] found that the consent of the husband to the care and treatment of his wife's mental condition was not valid and did not bar an assault and battery action against the treating physician. Since the woman had never been declared legally incompetent, her own consent was essential.

The court in *Wyatt v. Stickney* recognized that certain behavior modification procedures may be deemed so offensive, frightening, or risky that their use should be restricted by requiring the patient's informed consent.[31] Although there are some provisions for proxy consent, the court added layers of protection by requiring the opportunity for outside, independent consultation, as well as the involvement of a Human Rights Committee.

In deciding whether a particular procedure is so intrusive or coercive as to require these added protections, one commentator has suggested the following guidelines:

1. The extent and duration of changes in behavior patterns and mental activity effected by the therapy—the degree of change in personality.
2. The side effects associated with the therapy.
3. The extent to which the therapy requires physical intrusion into the inmate's body.
4. The degree of pain, if any, associated with the therapy.
5. The extent to which an uncooperative inmate can avoid the effects of the therapy.[32]

Informed consent requirements will vary depending on the nature of the procedure for which it is requested. The more potentially harmful, intrusive, or experimental the procedure, the stricter and more numerous must be the safeguards to protect the individual. There is precedent for the scrutiny of potentially hazardous or intrusive "treatments" and for an attempt to delimit the conditions under which informed consent is obtained. Since each state has differing statutes and case law concerning the use of behavioral techniques, it is impossible to generalize as to the limitations, but the trend is toward more, rather than less regulation and protection.

In general, in order to protect a patient involved in an experimental procedure, there should be a review of the patient's consent. This would help ensure the competent and voluntary character of the consent. The closer that the institution comes to meeting the constitutional minimum standard of *Wyatt v. Stickney*, the more likely it is that, as concerns the effect of institutionalization on a patient's competence and voluntariness, the consent will be valid. For an adjudicatively incompetent patient, the best interest determination made by his guardian in his proxy consent decision would be reviewed. For the practically incompetent patient, an original determination of the best interest of the patient would be made.

These review mechanisms may take different forms. The director or superintendent of the facility could perform this function. However, there may be a possible conflict or interest problem here. A problem may also be presented by the possibility of role conflict arising from the entrusting of the notice and explanation of right function to the same agency which undertakes to perform the therapeutic function.[33] Instead, a committee structure could be used, either totally independent of the institution or one composed partly of institutional administration and staff and partly of independent people. The committee could be patterned after the

Human Rights Committee provided for in *Wyatt*. Alternatively, this review could be done by an agency specially created to protect mental patients' rights. Finally, there could be court review of the adequacy of these procedures. However, this last procedure might prove costly and cumbersome, and it might be best to reserve it for those cases in which particularly coercive or intrusive experiments are being considered.

THE PRISON SETTING

The U.S. Supreme Court approved the involuntary use of behavior modifying drugs in the prison system for the control of potentially disruptive behavior in the 1990 case of *Harper v. Washington*.[34] In terms of substantive rights, the Court agreed that prisoners have a right "to avoid the unwanted administration of antipsychotic drugs." Nonetheless, this right must yield if the state can demonstrate that forcing treatment is "reasonably related to legitimate penological interests." Such interests do not include the use of drugs as punishment, but do include the use of drugs for treatment and to maintain order in the prison environment. This latter interest in "prison safety and security," the Court noted, is especially important because prisons are "by definition" made up of "persons with a demonstrated proclivity for antisocial criminal, and often violent, conduct."

In deciding whether a forced drug treatment regulation meets the "reasonably related to a legitimate penological interest" standard, a reviewing court should examine three factors: (1) the connection between the regulation and the state interest it is meant to foster; (2) the impact accommodating the asserted right will have on other prisoners, the guards, "and on the allocation of prison resources generally"; and (3) the absence of ready alternatives. Using these factors, the Court concluded that SOC regulation is constitutional as a matter of substantive due process. The Court found that problems of potential abuse are avoided because:

The fact that the medication must first be prescribed by a psychiatrist ensures that the treatment in question will be ordered only if it is in the prison's medical interests, given the legitimate needs of his institutional confinement.

As to procedural due process, the Court concluded that no judicial review is needed because "an inmate's interests are adequately protected, and perhaps better served, by allowing the decision to medicate to be made by medical professionals rather than a judge." Such professionals, the Court noted, are bound by the Hippocratic Oath, are in the best position to assess the risks of antipsychotic medication, and are best able to make medical judgments. Nor need the prisoner be provided with legal counsel at the hearing. In the Court's words: "It is less than crystal clear why *lawyers* must be available to identify possible errors in *medical* judgment."

What we should do about competent mentally ill persons who refuse drug treatment is an extremely difficult and contentious question. In the free world, however, we continue to give the individual patient, who is presumed to be competent, the legal right to refuse medication, even if the result is the unsatisfactory one of continued confinement to a mental institution (as previously noted, guardians are generally appointed to make decisions for those adjudged incompetent). This general rule now has an exception: competent mentally ill prisoners who are a danger to themselves, others, or property may be forcibly medicated. The psychiatric profession has generally hailed this decision as a victory for themselves and their patients. But their celebration is premature. This decision will likely serve only to make the public in general, and prisoners in particular, more suspicious of psychiatrists who will be seen as agents of an oppressive state rather than as independent physicians pledged to act in the best interests of their patients.

Psychiatrists must decide if they want to be agents of the state, just as other physicians have had to decide if they want to deliver lethal injections for capital punishment. Psychiatrists should take seriously the likelihood that even if they see forced medication as treatment, others will see it simply as a substitute for bars, straight jackets, and more guards. This case permits "treatment" solely for prison management. Although the holding is limited to prisoners, who have already been deprived of their basic right to liberty, the Court's general message cannot be so easily confined. For example, while the flagrant use of lobotomy as punishment, so well captured by Ken Kesey in *One Flew Over the Cuckoo's Nest*, seems constitutionally forbidden by the opinion, the use of ECT for prison management may be permissible.[35]

NONTHERAPEUTIC EXPERIMENTATION

Since nontherapeutic experimentation is, by definition, not for the benefit of the subject, no proxy consent is theoretically permissible. Therefore, unless the particular patient is legally competent to give informed consent, it would seem that·there could be no nontherapeutic experimentation on institutionalized mental patients. In *Frazier v. Levi*,[36] for example, a mother, acting as guardian, sought a sterilization for her adult pregnant daughter who had a mental age of six years, was sexually permissive, and had two retarded illegitimate children. Although the mother maintained that she was no longer financially and emotionally able to support any more of her daughter's children and that the operation would therefore be to everyone's benefit, she admitted that the operation was not medically necessary. The court refused to authorize the procedure and held that the daughter lacked the mental capacity to consent to the operation and that, without consent, she could not be deprived of her legal rights.[37]

Similarly, in *In re Richardson*,[38] an action was brought by the parents of a minor retarded child to permit the donation of one of the child's kidneys for transplantation into the child's older sister. The mother, father, and older sister all consented to the procedure, but the mentally retarded child, having a mental age of a three- or four-year old, was not capable of giving legal consent. The court defined its duties to be the protection and promotion of the ultimate best interest of the child. In particular, it determined that the minor had a right to be free from bodily intrusion to the extent of the loss of an organ unless it was specifically found that the removal of the kidney was in the child's best interest. Rejecting a claim that the child would benefit by a successful operation because, when his mother and father die, his older sister would be able to take care of him, the majority found that the operation would clearly be against the child's best interest, and that therefore neither his parents nor the courts could authorize the surgery.

However, there are circumstances under which nontherapeutic procedures are performed on incompetents. These situations also involve the transplantation of organs and sterilization. In *Strunk v. Strunk*,[39] the mother of an incompetent ward of the state petitioned the court of equity to permit the removal of one of his kidneys for transplantation into his twenty-eight-year-old brother. The potential donor was twenty-seven years of age, but had a mental age of approximately six years and had been previously committed to a state institution for the feeble-minded. All other members of the family and the Department of Mental Health had consented to the operation, but the donor was incompetent to give legally valid consent. A guardian *ad litem* had been appointed to contest the state's authority to allow the operation at every stage of the proceeding.

The court placed controlling emphasis on the psychiatric testimony. A psychiatrist who examined the incompetent determined that, in his opinion, the death of the brother would have "an extremely traumatic effect" on the potential donor. It was also argued that, while mental incompetents have difficulty establishing a sense of identity with other people, they nevertheless have a need for close intimacy, so that the donor's identification with the brother, who was his family tie, made it vital to the incompetent's improvement that his brother survive. Even though the transplant, from the donor's point of view, was physically nonbeneficial, the Kentucky Court of Appeals implicitly and summarily equated benefit in the constitutional sense with a vague showing that possible psychological detriment might be avoided. The court concluded that, while a parent did not have the authority to consent to such an operation, the court did have the ability to do so by exercising its equitable powers under the doctrine of *parens patriae*.[40]

The most notorious case of nontherapeutic experimentation on institutionalized individuals took place in New York's Willowbrook State School.[41] The facility, once the world's largest institution for the mentally retarded, had become infamous for its deplorable conditions.[42] Built in 1941 to house 3,000 residents, Willowbrook at one time had a population of 5,200. At least three-quarters of the patients were classified as "profoundly" or "severely" retarded. The facility was grim, with the stench of sweat, feces, and urine everywhere. For most residents it was a warehouse, providing only shelter and the barest essentials. Conditions have been described as hazardous to the health, safety, and sanity of the residents, contributing to the deterioration of the patients through overcrowded conditions and shortage of adequately trained personnel. Indeed, it has been said that Willowbrook was "an atrocity, a massive regressive machine."[43] Charges had been made that ninety percent of the school's mortalities have as their underlying cause "neglect, deprivation, and malnutrition. You can see it in the kids' skin condition, loss of hair, slow hearing." There were tales of maggot-infested wounds, assembly-line bathing, inadequate medical care, inadequate clothing, and cruel and inappropriate use of restraints. In addition, infectious diseases, such as hepatitis and shigella were rampant in the wards.

The crowding and unsanitary conditions of the facility, coupled with the poor personal hygiene of residents, caused an epidemic of fecally borne infectious hepatitis.[44] Hepatitis is frequently protracted and debilitating and sometimes fatal to the victim. Nearly everyone at the school was infected, so that new arrivals would probably have contracted the virus within six months. Beginning in 1956, physicians at the institution worked at finding a vaccine for this particular strain of infectious hepatitis. They isolated strains of the virus, and, with parental consent, infected several retarded children newly admitted to the school. Many of the children became quite ill. All of them risked serious illness. As a result of these efforts, a vaccine for hepatitis virus was developed. The physicians involved in the study felt it was justified for numerous reasons. First, the residents were bound to be exposed to the same strains under the natural conditions in the facility anyway. Second, if they were admitted to the special hepatitis unit, they would be protected from the risk of multiple infections which were prevalent in the remainder of the institution. Finally, it was hoped that the residents, following a subclinical infection, would have an immunity to the particular hepatitis virus. In addition, it was maintained that, under circumstances in which active individuals are in constant contact with each other and with fecal matter, a vaccine is the best protection against infectious disease.

There were many criticisms of the Willowbrook experiment. For example, children were infected with the virus, but had protective doses of gamma-globulin withheld from them. This was at a time when other residents not in the study, as well as staff were receiving the gamma-globulin. By entering the study, a child increased his risk of contracting chronic liver disease while at the institution. Moreover, it was argued that the research money should have been spent on cleaning up the institution as a better means of eradication this disease. It is documented that the incidence of clinical hepatitis is very low at small, well-staffed institutions. Highly susceptible Down syndrome children admitted to small, private institutions showed only a 1½ percent prevalence of a particular strain of hepatitis, while similar children admitted to large state institutions showed a thirty percent rate.

A particularly controversial aspect of the program involved the means by which parental consent was obtained. In late 1964, Willowbrook was closed to all new admissions because of overcrowding. Parents seeking admission for their children were informed of this and placed on a waiting list. However, the hepatitis study, occupying its own space in the institution, continued to admit new patients. For a period of time, letters were sent to the parents informing them that there were a few vacancies in the hepatitis research unit if they cared to consider volunteering their child for that.[45] Many parents felt that this situation coerced them into consenting to have their children participate.

The hepatitis experiment was one of the factors which, combined with the general horrible conditions of the facility, led to the filing of the suit in *New York State Association for Retarded Children v. Carey*.[46] The court found that voluntarily institutionalized mentally retarded individuals have a constitutional right to protection from harm. Appropriately, the court approved a detailed consent decree which set up standards and procedures, similar to those in *Wyatt*, which would serve to ensure the recognition of the residents' right to protection from harm. This was felt to be necessary because "harm can result not only from neglect but from conditions which cause regression or which prevent development of an individual's capabilities." Significantly, the decree absolutely forbade medical experimentation. In addition, it created three boards with important functions. The Review Panel will oversee the implementation of standards and procedures mandated in the consent decree, the Consumer Advisory Board will evaluate alleged dehumanizing practices and violations of individual and legal rights, and a Professional Advisory Board will give advice on professional programs and plans, budget requests, and objectives, as well as investigate alleged violations. Ultimately, Willowbrook was closed.

CONCLUSIONS

Because of the special problems of competence and voluntariness, and the history of abuse and neglect of the institutionalized mental patients, this group should not be used as subjects of experiments at all unless the experiment involves important scientific work that can only be done on this population, and very stringent guidelines, to protect the rights and welfare of subjects are followed. The law has generally scrutinized consent given by the institutionalized mental patient with special care, but has usually permitted either personal or proxy consent to reasonable medical interventions after making sure that responsible efforts have been undertaken to insure capacity to consent and voluntariness, and to protect subject welfare.

In the institutional setting it is especially important to distinguish between minimal risk therapeutic experimentation (which holds out a reasonable prospect of benefit), and nontherapeutic experimentation which holds no prospect for benefit and some prospect of harm. Competent individuals should be able to consent to participate in the former, but participation in the latter should either be excluded altogether or limited to minimal risk research that can be done in no other population, and holds great promise to gain important knowledge that can be of benefit to the institutionalized mentally disabled population.

In 1978 the U.S. Department of Health, Education and Welfare, (now Health and Human Services) proposed regulations that delineated "Additional Protections Pertaining to Biomedical and Behavioral Research Involving as Subjects Individuals Institutionalized as Mentally Disabled" (Appendix B). This is the only set of regulations to protect the subjects of human experimentation that were proposed by the National Commission for the Protection of Subjects of Biomedical and Behavioral Research that were not adopted by the Department following publication in the *Federal Register*.

It is time to reconsider the adoption of federal regulations designed to protect the rights and welfare of mentally disabled persons in research studies. It is not reasonable, however, to simply recommend the adoption of the 1978 proposals for two reasons: (1) the nature and content of medical research on the mentally disabled has changed substantially that as deinstitutionalization has proceeded, most research is now done outside of institutions; and (2) some provisions of the 1978 proposal were written in the alternative, and a decision must be made concerning which, if any, alternative to adopt. Specifically, we believe that special regulations for experimentation on the mentally disabled should be adopted, and should contain the following provisions:

1. *The regulations should be expanded to include all mentally disabled patients, with special provision for the institutionalized patient.* We believe the regulations should be expanded to include all mentally disabled persons, although we recognize that this will require a definition that is likely to be much less precise than either age (children's regulations) or institutional status (prisoner regulations). One possibility is to include only individuals who have a "severe mental illness," using a definition similar to that employed in Massachusetts for involuntary commitment purposes ("a substantial disorder of thought, mood, perception, orientation, or memory which grossly impairs judgment, behavior, capacity to recognize reality or ability to meet the ordinary demands of life, but shall not include alcoholism"). The primary concern in addressing this issue is to balance measures to protect the welfare of individuals against stigmatizing them further.

2. *The role of a "consent auditor" should be clarified.* The proposed regulations [sec. 46.505(3) and 46.506(4)(iii)] include alternative language that requires a "consent auditor" to witness the informed consent process and make an assessment of whether or not the subject had the capacity to consent and did in fact consent. This requirement was suggested because of the likelihood that a researcher might overestimate a prospective subject's capacity to consent if the subject agreed to participate in research. The proposed regulations require that this person not be "involved with the research, nor should they be employed by or otherwise associated with the institution conducting or sponsoring the research, or with the institution in which the subject resides." Although the concept is reasonable, the proposal may have so restricted the persons who can function in this role as to make the proposal impracticable. In this regard it seems reasonable to either permit a designated employee of the institution (such as a nurse) to function in this role, or to combine this role with that of an "advocate" (discussed below).

3. *The role of an "advocate" should be clarified.* The proposed regulations [sec. 46.507(a)(4)(c)] include alternative language requiring an advocate to act as a "guardian ad litem" in the "best interests" of the subject in monitoring and approving consent to research in situations in which research involves "greater than minimal risk and no prospect of direct benefit" to the subject. The problem with this formulation is that a guardian *must* act in the best interests of his ward, and it is *never* in the best interests of the subject to participate in this class of research. Thus, as formulated, the only role the advocate can have is to refuse to have the potential subject participate in nonbeneficial research that involves *any* risk.

Although the formulation in the proposed regulations does not work, it makes sense to insure that every mentally disabled person in a research project involving more than minimal risk be assigned a personal physician who is kept fully informed of the research, and whose only loyalty and obligation is to the patient and the patient's welfare. This would ensure that there is at least one person whose sole function is to advocate for the patient's good, unimpeded with any contamination by the "requirements" of the research protocol.

4. *Certain types of research should not be performed on mentally disabled persons.* Research that could not be done includes research that is unrelated to the subject's condition, or to the class of patients suffering from the mental condition of the subject. We also believe that more than minimal risk research should never be

performed on severely mentally disabled persons who are incapable of giving their informed consent (as we read the proposed regulations, such research is not permitted as a practical matter, but we think it would be better to explicitly state this prohibition).

5. *A member of the IRB should always be appointed to argue against using the mentally disabled as a research population for any study in which this is proposed.* The proposed federal regulations, like existing regulations, put very heavy emphasis and responsibility of the IRB to protect human subjects. Nonetheless, the IRB may be unaware of the particular problems associated with the use of mentally disabled persons, and may inadequately consider problems not only of rights (consent), but welfare (protection). In this regard, whether the study should be conducted at all on mentally disabled persons becomes the central issue. The IRB is unlikely to give this matter full consideration unless special provision is made to assign a member of the IRB to present the case *against* using this population. This role would be similar to that of a guardian ad litem assigned the role of presenting all the arguments against a proposed treatment plan to a judge who is asked to authorize treatment that is not or cannot be consented to by an individual (such as involuntary sterilization or kidney donation by a mentally disabled person). The concept behind this proposal is that a reasonable decision on this issue can only be reached by the IRB if *both* sides of the argument are presented and considered. This procedure, combined with a substantive rule prohibiting use of the mentally disabled as research subjects in research unrelated to their condition, could significantly protect this population from exploitation.

APPENDIX A

The Nuremberg Code

1. The voluntary consent of the human subject is absolutely essential. This means that the person involved should have legal capacity to give consent; should be so situated as to be able to exercise free power of choice, without the intervention of any element of force, fraud, deceit, duress, overreaching, or other ulterior form of constraint or coercion; and should have sufficient knowledge and comprehension of the elements of the subject matter involved as to enable him to make an understanding and enlightened decision. This latter element requires that before the acceptance of an affirmative decision by the experimental subject there should be made known to him the nature, duration, and purpose of the experiment; the method and means by which it is to be conducted; all inconveniences and hazards reasonably to be expected; and the effects upon his health or person which may possibly come from his participation in the experiment.

The duty and responsibility for ascertaining the quality of the consent rests upon each individual who initiates, directs or engages in the experiment. It is a personal duty and responsibility which may not be delegated to another with impunity.

2. The experiment should be such as to yield fruitful results for the good of society, unprocurable by other methods or means of study, and not random and unnecessary in nature.

3. The experiment should be so designed and based on the results of animal experimentation and a knowledge of the natural history of the disease or other problem under study that the anticipated results will justify the performance of the experiment.

4. The experiment should be so conducted as to avoid all unnecessary physical and mental suffering and injury.

5. No experiment should be conducted where there is an *a priori* reason to believe that death or disabling injury will occur; except, perhaps, in those experiments where the experimental physicians also serve as subjects.

6. The degree of risk to be taken should never exceed that determined by the humanitarian importance of the problem to be solved by the experiment.

7. Proper preparations should be made and adequate facilities provided to protect the experimental subject against even remote possibilities of injury, disability, or death.

8. The experiment should be conducted only by scientifically qualified persons. The highest degree of skill and care should be required through all stages of the experiment of those who conduct or engage in the experiment.

9. During the course of the experiment the human subject should be at liberty to bring the experiment to an end if he has reached the physical or mental state where continuation of the experiment seems to him to be impossible.

10. During the course of the experiment the scientist in charge must be prepared to terminate the experiment at any stage, if he has probable cause to believe, in the exercise of the good faith, superior skill, and careful judgment required of him, that a continuation of the experiment is likely to result in injury, disability, or death to the experimental subject.

APPENDIX B

Subpart E—Additional Protections Pertaining to Biomedical and Behavioral Research Involving as Subjects Individuals Institutionalized as Mentally Disabled

Sec.
46.501   Applicability.
46.502   Purpose.
46.503   Definitions.
46.504   Additional duties of the Institutional Review Boards where individuals institutionalized as mentally disabled are involved.
46.505   Research not involving greater than minimal risk.
46.506   Research involving greater than minimal risk but present the prospect of direct benefit to the individual.
46.507   Research involving greater than minimal risk and no direct benefit to individual subjects, but likely to yield generalizable knowledge about the subjects' disorder or condition.
46.508   Research not otherwise approvable which presents an opportunity to understand, prevent, or alleviate a serious problem affecting the health or welfare of individuals institutionalized as mentally disabled.
AUTHORITY: 5 U.S.C. 301.

Subpart E—Additional Protections Pertaining to Biomedical and Behavioral Research Involving as Subjects Individuals Institutionalized as Mentally Disabled.

§ 46.501   Applicability.

(a) The regulations in this subpart are applicable to all biomedical and behavioral research conducted or supported by the Department of Health, Education, and Welfare involving as subjects individuals institutionalized as mentally disabled.

(b) Nothing in this subpart shall be construed as indicating that compliance with the procedures set forth herein will in any way render inapplicable pertinent State or local laws bearing upon activities covered by this subpart.

(c) The requirement of this subpart are in addition to those imposed under the other subparts of this part.

§ 46.502   Purpose.

Individuals institutionalized as mentally disabled are confined in institutional settings in which their freedom and rights are potentially subject to limitation. In addition, because of their impairment they may be unable to comprehend sufficient information to give a truly informed consent. Also, in some cases they may be legally incompetent to consent to their own participation in research.

At the same time, so little is known about the factors that cause mental disability that efforts to prevent and treat such disabilities are in the primitive stages. There is widespread uncertainty regarding the nature of the disabilities, the proper identification of persons who are disabled, the appropriate treatment of such persons, and the best approaches to their daily care. The need for research is clearly manifest. It is the purpose of this subpart to permit the conduct of responsible investigations while providing additional safeguards for those institutionalized as mentally disabled.

§ 46.503   Definitions.

As used in this subpart:

(a) "Secretary" means the Secretary of Health, Education, and Welfare and any other officer or employee of the Department of Health, Education, and Welfare to whom authority has been delegated.

(b) "DHEW" means the Department of Health, Education, and Welfare.

(c) "Mentally disabled" individuals includes those who are mentally ill, mentally retarded, emotionally disturbed, psychotic or senile, regardless of their legal status or the reason for their being institutionalized.

(d) "Individuals institutionalized as mentally disabled" means individuals residing, whether by voluntary admission or involuntary confinement, in institutions for the care and treatment of the mentally disabled. Such individuals include but are not limited to patients in public or private mental hospitals, psychiatric patients in general hospitals, inpatients of community mental health centers, and mentally disabled individuals who reside in half-way houses or nursing homes.

(e) "Children" are persons who have not attained the legal age of consent to general medical care as determined under the applicable law of the jurisdiction in which the research will be conducted.

(f) "Parent" means a child's natural or adoptive parent.

(g) "Legally authorized representative" means an individual or judicial or other body authorized under applicable law to consent on behalf of a prospective subject to such subject's participation in the particular activity or procedure. An official serving in an institutional capacity may not be considered a legally authorized representative for purposes of this subpart.

(h) "Minimal risk" is the probability and magnitude of physical or psychological harm or discomfort that is normally encountered in the daily lives, or in the routine medical or psychological examination, of normal individuals.

(i) "Assent" means a prospective subject's affirmative agreement to participate in research. Mere failure to object shall not, absent affirmative agreement, be construed as assent. Assent can only be given following an explanation, based on the types of information specified in § 46.103(c), appropriate to the level of understanding of the subject, in accordance with procedures established by the Institutional Review Board.

(j) "Consent auditor" means a person appointed by the Institutional Review Board to ensure the adequacy of the consent process, particularly when there is a substantial question about the ability of a subject to consent or assent or when there is a significant degree of risk involved. Consent auditors are responsible only to the Board and should not be involved with the research, nor should they be employed by or otherwise associated with the institution conducting or sponsoring the research, or with the institution in which the subject resides. They should be persons familiar with the physical, psychological, and social needs of the class of prospective subjects as well as with their legal status.

(k) "Advocate" means an individual appointed by the Institutional Review Board to act in the best interests of the subject. The advocate will, although he or she is not appointed by a court, be construed to carry the fiduciary responsibilities of a guardian ad litem toward the person whose interests the advocate represents. No individual may serve as an advocate if the individual has any financial interest in, or other association with, the institution conducting or sponsoring the research, nor with the institution in which this research is conducted; nor, where the subject is the ward of a State or other agency, institution, or entity, may the advocate have any financial interest in, or other association with, that State, agency, institution, or entity. An advocate must be familiar with the physical, psychological, and social needs and the legal status of the class of individuals institutionalized as mentally disabled in the institution in which the research is conducted. [This definition will be retained in the final regulations if duties are assigned to "advocates."]

§ 46.504   Additional duties of the institutional review boards where individuals institutionalized as mentally disabled are involved.

(a) In addition to all other responsibilities prescribed for Institutional Review Boards under this part, the Board shall review research covered by this subpart and approve such research only if it finds that:

(1) The research methods are appropriate to the objectives of the research;

(2) The competence of the investigator(s) and the quality of the research facility are sufficient for the conduct of the research;

(3) Appropriate studies in nonhuman systems have been conducted prior to the involvement of human subjects;

(4) There are good reasons to involve institutionalized individuals as subjects of the research. In reviewing proposals to involve institutionalized persons in research, the Board should evaluate the appropriateness of involving alternative, noninstitutionalized populations in the study instead of, or along with, the institu-

tionalized individuals. Sometimes, the participation of alternative populations will not be possible or relevant, as when the research is designed to study problems or functions that have no parallel in free-living persons, (e.g., studies of the effects of institutionalization or studies related to persons, such as the profoundly retarded or severely handicapped, who are almost always found in residential facilities.)

(5) Risk of harm or discomfort is minimized by using the safest procedures consistent with sound research design and by using procedures performed for the diagnosis or treatment of the particular subject whenever possible;

(6) Adequate provisions are made to protect the privacy of the subjects and to maintain confidentiality of data. For example, data may be disclosed to authorized personnel and used for authorized purposes only; data should be collected only if they are relevant and necessary for the purposes of the research and analysis; data should be maintained only as long as they are necessary to the research or to benefit the subjects; and all data should be maintained in accordance with fair information practices;

(7) Selection of subjects among those institutionalized as mentally disabled will be equitable. Subjects in an institution should be selected so that the burdens of research do not fall disproportionately on those who are least able to consent or assent, nor should one group of patients be offered opportunities to participate in research from which they may derive benefit to the unfair exclusion of other equally suitable groups of patients.

(8) Adequate provisions are made to assure that no prospective subject will be approached to participate in the research unless the health care professional who is responsible for the health care of the subject has determined that the invitation to participate in the research and the participation itself will not interfere with the health care of the subject;

(9) The Board shall appoint a consent auditor to ensure the adequacy of the consent procedures when, in the opinion of the Board, such a person is considered necessary, e.g., when there is a substantial question about the ability of a subject to consent or to assent or when there is a significant degree of risk involved; and

[In the event the Department decides that there should be consent auditors for all projects, the above paragraph will be appropriately modified.]

(10) The conditions of all applicable subsequent sections of this subpart are met.

(b) The Board shall carry out such other duties as may be assigned by the Secretary.

(c) The institution shall certify to the Secretary, in such manner as the Secretary may require, that the duties of the Board under this subpart have been fulfilled.

§ 46.505   Research not involving greater than minimal risk.

Biomedical or behavioral research that does not involve greater than minimal risk to subjects who are institutionalized as mentally disabled may be conducted or supported by DHEW provided the Institutional Review Board has determined that:

(a) The conditions of § 46.504 are met; and

(b) Adequate provisions are made to assure that no subject will participate in the research unless:

(1) The subject gives informed consent to participation;

(2) If the subject lacks the capacity to give informed consent, the research is

relevant to the subject's condition, the subject assents or does not object to participation, and the subject's legally authorized representative consents to the subject's participation; or

(3) If a subject, who lacks the capacity to give informed consent, objects to participation: (i) The research includes an intervention that holds out the prospect of direct benefit to the subject, or includes a monitoring procedure required for the well-being of the subject, (ii) the subject's legally authorized representative consents to the subject's participation, and (iii) the subject's participation is authorized by a court of competent jurisdiction.

[Consideration is being given to mandating that, in addition to the above requirements: (1) A "consent auditor" be appointed by the Institutional Review Board to ensure the adequacy of the consent process and determine whether each subject consents, or is incapable of consent but assents, or objects to participation, and (2) whenever the consent auditor determines that a subject is incapable of consenting, the subject may not participate without the authorization of an "advocate."]

§46.506   Research involving greater than minimal risk but presenting the prospect of direct benefit to the individual subjects.

(a) Biomedical or behavioral research in which more than minimal risk to subjects who are institutionalized as mentally disabled is presented by an intervention that holds out the prospect of direct benefit for the individual subjects, or by a monitoring procedure likely to contribute to the well-being of the subjects, may be conducted or supported provided the Institutional Review Board has determined that:

(1) The conditions of section 46.504 are met;

(2) The risk is justified by the prospect of benefit to the subjects;

(3) The relation of the risk to anticipated benefit to subjects is at least as favorable as that presented by available alternative approaches;

(4) Adequate provisions are made to assure that no adult will participate in the research unless:

(i) The subject gives informed consent to participation;

(ii) If the subject lacks the capacity to give informed consent, the subject assents to participation, and the subject's legally authorized representative consents to the subject's participation; or

(iii) If a subject who lacks the capacity to give informed consent, does not assent, or objects to participation: (A) The intervention or monitoring procedure is only available in the context of the research, (B) The subject's legally authorized representative consents to the subject's participation, and (C) the subject's participation is authorized by a court of competent jurisdiction.

[Consideration is being given to mandating that, in addition to the above requirements: (1) A "consent auditor" be appointed by the Institutional Review Board to ensure the adequacy of the consent process and determine whether each subject consents, or is incapable of consent but assents, or objects to participation, and (2) whenever the consent auditor determines that a subject is incapable of consenting, the subject may not participate without the authorization of an "advocate."]

(5) Adequate provisions are made to assure that no child will participate in the research unless:

(i) The subject assents (if capable) and the subject's parent(s) or guardian(s) give permission, as provided in section 46.409 of this part; or

(ii) If the subject objects to participation, the intervention or monitoring procedure is available only in the context of the research, the subject's parent(s) or guardian(s) give permission, and the subject's participation is authorized by a court of competent jurisdiction.

(b) Where appropriate, the Institutional Review Board shall appoint a consent auditor to ensure the adequacy of the consent process and determine whether each subject consents, or is incapable of consent but assents, or objects to participation. [This paragraph will be deleted if a consent auditor is required in all cases.]

§ 46.507   Research involving greater than minimal risk and no prospect of direct benefit to individual subjects, but likely to yield generalizable knowledge about the subject's disorder or condition.

(a) Biomedical or behavioral research in which more than minimal risk to subjects who are institutionalized as mentally disabled is presented by an intervention that does not hold out the prospect of direct benefit for the individual subjects, or by a monitoring procedure that is not likely to contribute to the well-being of the subjects, may be conducted or supported provided an Institutional Review Board has determined that:

(1) The conditions of section 46.504 are met;

(2) The risk represents a minor increase over minimal risk:

(3) The anticipated knowledge (i) is of vital importance for the understanding or amelioration of the type of disorder or condition of the subjects, or (ii) may reasonably be expected to benefit the subjects in the future;

(4) Adequate provisions are made to assure that no adult will participate in the research unless the following conditions are met:

(i) The subject gives informed consent to participation;

(ii) If the subject lacks the capacity to give informed consent, the subject assents to participation, and the subject's legally authorized representative consents to the subject's participation; or

(iii) If the subject lacks the capacity to assent but does not object, the subject's legally authorized representative and a court of competent jurisdiction consent to the subject's participation.

[The Department is considering the following additions to the above provisions:

In § 46.507(a)(4)(B), with respect to subjects capable of consenting: (i) Adding the requirement that inclusion of each subject be approved by the Secretary based upon the advice of a panel of experts, or (ii) requiring the approval of an "advocate."

In § 46.507(a)(4)(C), with respect to subjects incapable of assenting: (i) Prohibiting use of such subjects on the theory that there is no research which can be performed only with these subjects, (ii) requiring approval by the Secretary based upon the advice of a panel of experts, or (iii) requiring the approval of an "advocate."]

(5) If the subject is a child, the requirements of §§ 46.407 and 409 of subpart D (relating to research involving children) are satisfied.

(b) No subject may be involved in the research over his or her objection.

(c) The Institutional Review Board shall appoint a consent auditor to ensure the adequacy of the consent process and determine whether each subject consents, or is incapable of consenting but assents, or is incapable of assenting but does not object, or objects to participation. [This paragraph will be deleted if a consent auditor is required for all research covered by this subpart.]

§ 46.508   Research not otherwise approvable which presents an opportunity to understand, prevent, or alleviate a serious problem affecting the health or welfare of individuals institutionalized as mentally disabled.

Biomedical or behavioral research that the Institutional Review Board does not believe meets the requirements of §§ 46.505, 46.506, or 46.507 may nevertheless be conducted or supported by DHEW provided:

(a) The Institutional Review Board has determined the following:

(1) The conditions of § 46.504 are met; and

(2) The research presents a reasonable opportunity to further the understanding, prevention, or alleviation of a serious problem affecting the health or welfare of individuals institutionalized as mentally disabled; and

(b) The Secretary, after consultation with a panel of experts in pertinent disciplines (e.g., science, medicine, education, ethics, law) and following opportunity for public review and comment, has determined either (1) that the research in fact satisfies the conditions of §§ 46.505, 46.506, or § 46.507, as applicable, or (2) the following:

(i) The research presents a reasonable opportunity to further the understanding, prevention, or alleviation of a serious problem affecting the health or welfare of individuals institutionalized as mentally disabled;

(ii) The conduct of the research will be in accord with basic ethical principles of beneficence, justice, and respect for persons, that should underlie the conduct of research involving human subjects; and

(iii) Adequate provisions are made for obtaining consent of those subjects capable of giving fully informed consent, the assent of other subjects and the consent of their legally authorized representatives, and, where appropriate, the authorization of a court of competent jurisdiction [and if §§ 46.505, 46.506, 46.507 require an advocate, the authorization of that advocate].

[FR Doc. 78-31822 Filed 11-16-78; 8:45 am]

NOTES

1. Portions of this paper are adapted from a background paper on consent to research by institutionalized mentally ill patients prepared by the authors for The National Commission for the Protection of Human Subjects of Biomedical and Behavioral Research (1976) which was published by the Commission and also by the authors in Annas, G.J., Glantz, L.H. and Katz B. *Informed Consent to Human Experimentation: The Subject's Dilemma*, Ballinger Books, Cambridge, Massachusetts, 1977. It is concerned with the problem of informed consent to experimentation by those institutionalized individuals considered mentally infirm, including the mentally ill, the mentally retarded, the emotionally disturbed, the psychotic, and the senile. However, the important issues are common to all of these categories, and since most of the law in this area deals specifically with the mentally ill and/or the mentally retarded, we refer to these particular groups in the paper.

2. See President's Committee on Mental Retardation, *Silent Minority*, DHEW Pub. No. (OHD) 74-21002, at 10.

3. See, e.g., Dybwad, Challenges in Mental Retardation, 21 (1964); Herr, Civil Rights, Uncivil Asylums

and the Retarded, 43 *Cinn. L. Rev.* 679 (1974); Comment, Civil Restraint, Mental Illness, and the Right to Treatment, 77 *Yale L.J.* 87, 88 (1967), Harter, Mental Age, IQ and Motivational Factors in the Discrimination Learning Set Performance of Normal and Retarded Children, 5 J. *Experimental Child Psych.* 123 (1967); Iscoe and McCann, The Perception of an Emotional Continuum by Older and Younger Mental Retardates, 1 J. *Personality Social Psych.* 383 (1965); Lyle, The Effect of an Institutional Environment upon the Verbal Development of Imbecile Children, 3 J. *Mental Deficiency Research* 122 (1059); Bloomberg, A Proposal for a Community Based Hospital as a Branch of a State Hospital, 116 *Am. J. Psych.* 814 (1960); Blatt and Kaplan, *Christmas in Purgatory: A Photographic Essay on Mental Retardation* (1966).

4. Comment, Behavior Modification and Other Legal Imbroglios of Human Experimentation, 52 J. *Urban L.* 155, 157 (1974).

5. *See* Ritts, A Physician's View of Informed Consent in Human Experimentation, 36 *Fordham L. Rev.* 631 (1968).

6. Schloendorff v. Society of N.Y. Hosp., 211 N.Y. 125, 105 N.E. 92, 93 (1914).

7. E.g., Natanson v. Kline, 186 Kan. 393, 350 P.2d 1093 (1960); Bang v. Charles T. Miller Hosp., 251 Minn. 427, 88 N.W.2d 186 (1958).

8. See Annas, G.J. *The Rights of Patients*, 2d ed, Southern Illinois University Press, Carbondale, IL, 1989.

9. See Annas, G. J. and Grodin, M., eds. *The Nazi Doctors and the Nuremberg Code: Human Rights in Human Experimentation*, Oxford University Press, New York, 1992.

10. Shapiro, Legislating the Control of Behavior Control; Autonomy and the Coercive Use of Organ Therapies, 47 *So. Calif. L. Rev.* 237, 308–309 (1974).

11. Schoenfeld, Human Rights for the Mentally Retarded: Their Recognition by the Providers of Service, 4 *Human Rights* 31 (1974), in which he discusses the 1971 United Nations Declaration of General and Special Rights of the Mentally Retarded, in which the international community recognized the principle that the mentally retarded have the same rights as other citizens of the same country and age.

12. Morris, Institutionalizing the Rights of Mental Patients: Committing the Legislature, 62 *Calif. L. Rev.* 957, 967 (1974). See, generally Allen, Ferster and Weihoffen, *Mental Impairment and Legal Incompetency* (1968), and Brooks, The Right to Refuse Antispychotic Medications: Law and Policy, 39 *Rutgers L. Rev.* 339 (1987).

13. Rennie v. Klien, 462 F. Supp. 1131, 1145 (1978). See, also Wyatt v. Stickney, 344 F. Supp. 373, 387, 399 (M.D. Ala. 1972), *aff'd sub nom.* Wyatt v. Aderholt, 503 F.2d 1305 (5th Cir. 1974): "No person shall be presumed mentally incompetent soley by reason of his admission or commitment to the institution." See, also Davis v. Watkins, 384 F. Supp. 1196, 1206 (N.D. Ohio 1974).

14. Rogers v. Okin, 478 F. Supp. 1342 (1979); Rogers v. Okin, 821 F. 2d 22 (1987).

15. See also McAuliffe v. Carlson, 377 F. Supp. 896 (D. Conn. 1974), supplemental decision, 386 F. Supp. 1245 (D. Conn. 1975) held that a hearing was required before finding a mental patient incompetent. This case was later reversed by the Second Circuit, 520 F.2d 1305 (2d Cir. 1975), but for reasons which do not conflict with the analysis of the lower court as to the effect of institutionalization on the status of mental patients. This principle is applicable to the ability of an institutionalized patient to give or withhold consent to medical treatment. In the case of In re Yetter, 62 Penn Dist. & Cty. Rpts. 2d 619 (1972) a sixty-year-old involuntarily committed mental patient declined to consent to a recommended surgical breast biopsy. Her fears were based on the death of an aunt following such surgery (although the court was presented evidence indicating that the aunt had died fifteen years following the surgery from unrelated causes), as well as the concern that the operation would interfere with her genital system, affecting her ability to have babies, and would prohibit a movie career. Although her reasoning was becoming somewhat delusional, the court found that at the time the patient made her initial decision not to have the surgery, she was lucid, rational, and had the ability to understand the recommended procedure and the possible consequences of her refusal, including the risk of death. Even though it indicated that the patient's decision in this situation might be "irrational and foolish," the court nevertheless determined that Yetter was competent to reach this conclusion, and therefore declined to appoint a guardian for her for the purpose of consenting to the surgery. The court stated that the mere commitment of an individual to a state facility does not destroy the person's competency nor require the appointment of a guardian. Other courts are in accord, Brooks, *supra* note 12, and 74 A.L.R. 4th 1099 (Nonconsensual Treatment of Mentally Ill Persons with Neuroleptic or Antipsychotic Drugs as Violative of State Constitutional Guarantee) (1993)

16. Goffman, *Asylums*, (1962). *See also*, Kramer, The Subtle Subversion of Patients' Rights by Hospital Staff Members, 25 *Hosp. Community Psychiat.* 475 (1974).
17. E.g., Kaimowitz v. Dept. of Mental Health, Civ. No. 73-19434-AW (Cir. Ct. Wayne County, Mich., July 10, 1973); Annas and Glantz, Psychosurgery: The Law's Response, 54 *B.U.L. Rev.* 249, 263 (1974).
18. Annas, G.J. & Densberger J., Competence to Refuse Treatment: Autonomy vs. Paternalism, 15 *Toledo L. Rev.* 84 (1974).
19. Katz, The Right to Treatment—An Enchanting Legal Fiction, 36 *U. Chi. L. Rev.* 755, 773 (1069).
20. Note, Civil Commitment of the Mentally Ill: Theories and Procedures, *Harv. L. Rev.* 1288 (1966).
21. See, e.g., Postel, Civil Commitment: A Functional Analysis, 38 *Brooklyn L. Rev.* 1 (1971), and Annas & Densberger, *supra* note 18.
22. From Roman Law comes the idea that in some circumstances the state should relate to the citizen as the parent to his child. Known as the doctrine of *parens patriae*, this concept is firmly recognized in Anglo-American Law. It gives the sovereign both the right and the duty to protect the persons and property of those who are unable to care for themselves because of minority or mental illness. Ross, Commitment of the Mentally Ill: Problems of Law and Policy, 57 *Mich. L. Rev.* 945, 956–957 (1959).
23. See Allen, *Legal Rights of the Disabled and Disadvantaged* 23 (1969).
24. Allen, The Retarded Citizen: Victim of Legal and Mental Deficiency, 2 *U. Md. L.Rev.* 4 (1971). Cf., Lewis, McCollum, Schwartz and Grunt, Informed Consent in Pediatric Research, 16 *Children* 143, 144–145 (1969). This was, in fact, the case in the most important case to reach the U.S. Supreme Court, O'Connor v. Donaldson, 422 U.S. 563 (1975).
25. 372 F. Supp. 1196 (D.D.C. 1974). In a further development in this case, the court in *Relf v. Mathews* rejected proposed modifications of its previous judgment. The court noted that it intended to implement its decision that federal funds be available for sterilization's only for persons having the necessary capacity to decide voluntarily and free of coercion, and that the modifications were designed to substitute a universal federal standard of voluntariness which would permit sterilizations of persons eighteen years of age and older even when such persons were otherwise incompetent in fact because of age or mental condition under state standards.
26. 368 F. Supp. 1382 (M.D. Ala. 1973)
27. Dane Cty. Ct., Branch I, Wis. (Nov. 1974). Cf., Wade v. Bethesda Hosp., 337 F. Supp. 671 (S.D. Ohio 1971); In re M.K.R., 515 S.W.2d 467 (Mo. 1974); In re Kemp, 118 Cal. Rep. 64 (Ct. App. 1974); Holmes v. Powers, 439 S.W. 2d 579 (Ky. 1969).
28. See Krasner & Ullmann, *Case Studies in Behavior Modification* 1–2 (1965), Ayllon, Behavior Modification in Institutional Settings, 17 *Ariz. L. Rev.* 3 (1975); Kassirer, Behavior Modification for Patients and Prisoners: Constitutional Ramifications of Enforced Therapy, 2 *J. Psychiatry L.* 245 (1974).
29. 306 F. Supp 1158 (E.D.N.Y. 1969), *rev'd on other grounds*, 446 F.2d (2d Cir. 1971), *cert. denied*, 404 U.S. 985 (1971).
30. 344 Mass. 679, 183 N.E.2d 866 (1962) (*dicta*).
31. 344 F. Supp. 373 and 387 (M.D. Ala. 1972), *aff'd sub nom.* Wyattt v. Aderholt, 503 F. 2d 1305 (5th Cir. 1974). See generally Wexler, Token and Taboo: Behavior Modification, Token Economies, and the Law, 61 *Calif. L. Rev.* 81 (1973).
32. Note, Conditioning and Other Technologies Used to "Treat?" "Rehabilitate?" "Demolish?" Prisoners and Mental Health Patients, 45 *So. Calif. L. Rev.* 616, 659 (1972). See Note, Advances in Mental Health: A case for the Right to Refuse Treatment, 48 *Temple L. Q.* 354, 363 (1975); Friedman, Legal Regulation of Applied Behavior Analysis in Mental Institutions and Prisons, 17 *Ariz. L. Rev.* 39, 90 (1975).
33. Thorn v. Superior Court, 1 Cal. 3d 666, 675, 462 P.2d 56, 62, 83 Cal. Rptr. 600, 606 (1970).
34. Washington v. Harper, 494 U.S. 210 (1990).
35. See generally Annas, G.J., *Standard of Care: The Law of American Bioethics*, Oxford University Press, New York, 1993, 28–34.
36. 440 S.W.2d 393 (Tex. Ct. Civ. App. 1969); *Accord* In re M.K.R., 515 S.W.2d 467 (Mo. 1974). Cf., Wade v. Bethesda Hosp., 337 F. Supp. 671 (S.D. Ohio 1971).
37. The case of A.L. v. G.R.H. involved a petition by a mother to receive authorization to order the sterilization of her son. The fifteen-year-old child had suffered brain damage as the result of being struck by an automobile during early childhood, and presently had an I.Q. of 83, which is considered to be within the "dull" or "borderline" area. However, the boy appeared capable of further improvement. Indeed, it seemed that he would eventually be capable of earning his own living

either in specially supervised work or in entry-level jobs in the general marketplace. His mental disability would not be transmittable to his offspring, nor did he exhibit a propensity to force his attentions on others, although he had shown interest in social activity with girls. Expert testimony indicated that he was sufficiently intelligent to understand what was involved in sterilization and to participate in the decision-making process. The court decided that the facts of the case did not bring it within the legal principles which either permit parents to consent on behalf of their child to necessary medical services, or allow state intervention over the parents' wishes to rescue the child from parental neglect or to save his life. It maintained that the desirability of sterilization did not emanate from any life-saving necessities, but rather had as its sole purpose the prevention of the capability of fathering children. Thus, the court held that parents do not have the power to consent to sterilization of their children even though they believe that it would benefit the child, and accordingly affirmed the trial court's denial of the request. Accord In re Kemp's Estate, 43 Cal. App. 3d 758, 118 Cal. Rptr. 64 (1974). See Holmes v. Powers, 439 S. W.2d 579 (Ky. 1968).

38. 284 So. 2d 185 (La. App. 1973). *Accord* In re Pescinski, 67 Wis. 2d 4, 226 N.W.2d 180 (1975).
39. 445 S.W.2d 145 (Ky. 1969); See Baron, Botsford & Cole, Live Organ and Tissue Transplants from Minor Donors in Massachusetts, 55 *B.U.L Rev.* 159, 179 (1975); Comment, Spare Parts From Incompetents: A Problem of Consent, 9 *J. Family L.* 309 (1969).
40. Another case following this mold is Howard v. Fulton-Dekalb Hospital Authority, 42 USLW 2322 (GA. Sup.Ct., Nov. 29, 1975). In this case, a mother was suffering from chronic renal disease and the only person medically suitable for transplant purposes was her fifteen-year-old "moderately retarded" daughter. Both mother and daughter consented to the operation. However, the court found that, due to her minority and mental retardation, the daughter's consent was not legally valid. It also recognized the duty of the court, through its function as *parens patriae*, to independently review the circumstances of the case to assure that the best interests of the child were being protected, regardless of the existence of the mother's consent. However, this court also paid special attention to the psychiatric testimony, and decided that the kidney transplant should be allowed to proceed so as to protect the daughter from "the physical deprivation and emotional shock" which would result from the loss of her mother. However, the factors involved in this type of situation, including the fact that a specific life will be saved in exchange for the imposition of a minimal risk on the incompetent donor, as well as the concept of family unity in making determinations of this type make this line of cases somewhat inapplicable to other instances of nontherapeutic procedures on incompetent persons.
41. Krugman, Giles and Jacobs, Infectious Hepatitis: Studies on the Effect of Gamma Globulin on the Incidence of Inapparent Infection, 174 *JAMA* 823 (1960); Krugman, Giles and Hammond, Viral Hepatitis, Type B (MS-2 Strain), 218 *JAMA* 1665 (1971); Krugman and Giles, Viral Hepatitis, Type B (MS-2-Strain), 288 *JAMA* 755 (1973).
42. For some description of conditions at Willowbrook, see, Rivera, *Willowbrook: A Report on How It Is and Why It Doesn't Have to Be That Way* (1972); Ramsey, *The Patient as Person* 47–58 (1970); N.Y.U. Medical Center, Proceedings of the the Symposium on Ethical Issues in Human Experimentation: The Case of Willowbrook State Hospital Research (Urban Health Affairs Program, N.Y.U. Medical Center, May 4, 1972).
43. Bedlam in 1972—Retarded Care at Willowbrook, *Med. World News*, Jan. 28, 1972, at 15.
44. Ratnoff, Who Shall Decide When Doctors Disagree? A Review of the Legal Development of Informed Consent and the Implications of Proposed Lay Review of Human Experimentation, 25 *Case Western Res. L. Rev.* 472, 489 (1975).
45. Studies with Children Backed on Medical Ethical Grounds, 8 *Med. Tribune* 1, 23 (1967). See Krugman and Giles, Viral Hepatitis—New Light on an Old Disease, 212 *JAMA* 1019, 1020 (1970).
46. 393 F. Supp. 715 (E.D.N.Y. 1975), 357 F. Supp. 752 (E.D.N.Y. 1973).

Government Oversight

Robert A. Destro

The Catholic University of America, Columbus School of Law, Washington, DC 20064

A well-developed ethical theory provides a framework of principles within which an agent can determine morally appropriate actions.[1]

o'ver sight n superintendence, watchful care, supervision[2]

INTRODUCTION: FORMS OF OVERSIGHT

There are several levels at which the government "supervises" the activities of biomedical and behavioral researchers.[3] The most pervasive and familiar form of regulatory oversight is the complex web of institutional, State and federal regulations governing research in all of its forms, including biomedical and behavioral human subjects research. Designed to assure that research is evaluated and performed in an ethical, competent and professional manner, their viewpoint is "prospective"; that is, their goal is to *avoid* problems by providing mechanisms which will flag inappropriate practices, behaviors, and research designs before damage is done.

At the next level is tort law. Operating under standards of care defined by State law, its focus is "retrospective"; that is, it is designed to compensate for harms which result from the breach of a legal duty. The threat of malpractice liability is a significant one for all professionals, especially medical practitioners and researchers. As a result, the States have undertaken the difficult task of striking a delicate balance between the need to compensate individuals in a fair and timely manner for harms done, and the inherent uncertainties of professional practice.

At the foundational level is the realm of morality, ethics and professional responsibility. It is here that we find the basis for the entire system of government oversight; for law and regulation are merely the means by which important individual and social interests, including morality, are protected. In graphic form, the system can be represented as an inverted triangle (Figure 1).

This paper has two main parts. The first, designated "Part Two" will address briefly the forms of oversight which are common to all forms of biomedical research. The second, designated "Part Three" will discuss oversight issues related to vulnerable subjects; and the paper will conclude with a short discussion of the central question which faces those concerned with the ethics of neurobiological research involving vulnerable human subjects: What kind of "professional rela-

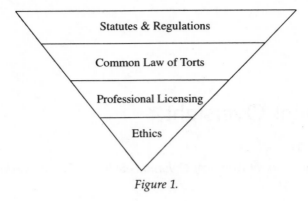

Figure 1.

tionship" exists between subject and researcher when the nature of the subjects' condition is such that there are grave questions concerning their mental facilities?

OVERSIGHT ISSUES COMMON TO ALL BIOMEDICAL AND BEHAVIORAL RESEARCH

Reviewing Issues at the "Core" of Law and Regulation

Before addressing specific "oversight" questions, it will be worthwhile to review several of the core ethical concepts relevant to biomedical and behavioral human subjects research. The laws, administrative regulations, and codes of professional ethics which govern the field are rooted in specific understandings related to the following[4]:

1. "Duty"
2. Respect for the Dignity of Human Persons, including
   2.1.   Nonmalificence
   2.2.   Beneficence, and
   2.3.   Justice
   2.4.   Autonomy
3. Truthfulness
4. Fidelity

"*Duty*" is defined by Webster as "1. that which a person is bound, by any moral or legal obligation, to pay, do or perform...; 2. the moral or legal obligation to follow a certain line of conduct, or to do a certain thing;...".[5] The concept lies at the heart of codes of morality, professional ethics and tort law.[6]

Respect for the Dignity of Human Persons covers a wide range of territory, including the key principles of nonmaleficence (do not inflict harm), beneficence (the prevention and removal of evil or harm, as well as the doing or promotion of good), justice ("giving to each his due") and autonomy (i.e., respect for individuals as moral agents).[7]

Though it is certainly possible to include *truthfulness* as a virtue which is required in any relationship with another human person,[8] it is particularly relevant in the context of professional relationships, where the specialized knowledge or experience of the professional is the basis for the relationship between the parties.[9]

The virtue of *fidelity* is included in this list because it is a key concern of every professional code of ethics.[10] Because it is the foundation of the duty of confidentiality, the duty to avoid conflicting interests, and the duty to serve as an advocate for client or patient interests, the voluntary assumption of a duty of loyalty marks the starting point for most professional relationships.[11]

Regulatory Controls Relating to the Integrity of the Research Process

There are a host of government regulations which relate to the general integrity of the research process. Among these are institutional and federal regulations which prohibit financial fraud and abuse,[12] including conflicts of interest,[13] and institutional, administrative, and other rules which prohibit research and other forms of academic fraud.[14]

Regulations Governing Human Subjects Research

Professor Jesse Goldner has noted that in order to "understand the scheme which currently controls the circumstances under which biomedical and behavioral research involving human subjects can take place, there is a need to have some familiarity with (1) the recent history of formal experimentation with human subjects and (2) the ethical principles which have now been identified as necessarily underlying the conduct of such research."[15] It is important to underscore that observation here. Laws and regulations are, in a sense, the tip of the iceberg; for they are the *result* of bitter experience both at home and abroad,[16] extensive debate in Congress leading to the passage of the National Research Act[17] and the submission of *The Belmont Report*,[18] and continuing oversight by federal and state agencies charged with oversight of human subjects research. It is impossible to debate modifications in either the scope or substantive content of these rules without first acquiring a keen sense for the evils the regulatory edifice was intended to eliminate.

Jurisdictional Issues in the Protection of Human Subjects Involved in Biomedical and Behavioral Research. Oversight of human subjects research takes place at both federal and state levels. At the federal level, the most comprehensive regulations have been issued by the agencies most directly involved in massive programs involving research with human subjects: the Department of Health and Human Services (DHHS) and the Food and Drug Administration (FDA). Other agencies doing or funding human subjects research, such as the Environmental Protection Agency, and the Departments of Defense and Education, have regulations on the books as well.

Federal agencies are responsible for oversight of activities within their jurisdiction, but there are key "cross-cutting" sources of additional federal oversight as well, including the supervisory roles of the United States Department of Justice[19] and the National Institutes of Health's Office for Protection From Research Risks of the Division of Human Subject Protections.[20] There is also "retrospective" oversight in the form of private suits for damages under either applicable civil rights laws, or the Federal Tort Claims Act,[21] and concurrent regulation under the terms of applicable State or foreign law.[22]

At present, seven states, California,[23] Wisconsin,[24] New York,[25] Delaware,[26] Montana,[27] Florida,[28] and Virginia[29] have statutes which govern human subjects research, and a bill which would "require the informed consent of human subjects as a condition of performing research involving the Commonwealth's facilities, services or funds" was submitted to the Joint Committee on Health Care of the Massachusetts House on February 28, 1994.[30] Other States have more limited laws which apply to specific kinds of human subjects research.[31] Regardless of scope or substantive content, however, all State policies governing human subjects research explicitly reference the federal guidelines, and recognize them as setting the minimum standards for acceptable research behavior.

Institutional Review Boards: Composition and Oversight Responsibilities.[32] The main "oversight" mechanism in human subjects research is the local Institutional Review Board (IRB), and the concerns which motivated the National Commission to recommend "systematic, non-arbitrary analysis of risks and benefits"[33] are clearly discernible from the federal regulations which govern their composition and function.

In summary form, the federal "assurance" requirement[34] is that IRBs be groups of highly competent professional and lay persons, who are free of direct conflicts of interest (save for those which arise from institutional loyalties), well-qualified by reference either to professional training, experience in their respective fields of endeavor, personal background, and position in the community *and* who are sensitive to the moral, cultural, legal, professional, and ethical questions which arise whenever human subjects are used in biomedical and behavioral research.[35] They must be provided with adequate staff and resources to support their review and recordkeeping activities, and with a set of written procedural guidelines for conducting their initial and continuing review of research and for reporting their findings, for keeping abreast of changes in any research activity subject to approval, for catching any unanticipated problems involving risks to subjects or others or any serious or continuing noncompliance with federal regulations or IRB policy, and for suspending or terminating approval.[36] As an additional precaution, they must keep extensive records which will enable evaluators outside the IRB, whether within the institution itself or outside authority, to review the details of the IRB's deliberations.[37]

Their basic oversight duties are to assure that:

1. risks are reasonable in relation to anticipated benefits, if any, to the subjects of the research, and the importance of the knowledge to be gained, *and* that such risks be minimized to the extent consistent with sound research design and clinical practice;
2. that selection of subjects is equitable and that vulnerable populations such as children (born and unborn), prisoners, pregnant women, mentally disabled persons, or persons subject to economic or educational disadvantage, are neither unfairly targeted, nor subjected to coercion or undue influence;[38]

3. that the autonomy and privacy of all research subjects is protected by scrupulous attention to informed consent and confidentiality guidelines; and
4. that research plans include, "when appropriate...adequate provision for the monitoring the data collected to ensure the safety of subjects"[39]

Though there are specific "additional protections" which are mandated for research activities which involve the most obvious of the members of the population most vulnerable to coercion or undue influence [children and unborn children (fetuses), pregnant women, human *in vitro* fertilization, and prisoners], the basic regulations also indicate that persons with mental disabilities, or who are subject to economic or social disadvantage are also considered "vulnerable." Notably absent from the Code of Federal Regulations, however, are any specific "additional protections" designed to protect these groups.[40] Several States have taken steps to fill a small part of this void.[41]

Informed Consent

The instructions for applicants seeking Public Health Service Grants for research involving human subjects states that "[r]esearch investigators are entrusted with an essential role in assuring the adequate protection of human subjects. In activities they conduct, or which are conducted under their direction, they have a direct and continuing responsibility to safeguard the rights and the welfare of the individuals who are or may become subjects of their research."[42] The relationship between researcher and subject is, in fact and in law, the ultimate focus of the oversight process.

The detail of federal and state regulations which govern the informed consent process in research settings, the recordkeeping required of IRB in the course of their oversight responsibilities, and the tendency of lawyers and medical professionals to "cover" themselves with an appropriate "disclosure" form has led "[m]any people [to] think of consent to treatment as a form..., the document through which patients agree to procedures their physician believes are advisable or necessary. Such a definition is incorrect and misleading, and in some instances, can be dangerous."[43]

From a legal perspective, such dangers arise whenever form is elevated over substance. The concept of "informed consent" is one which reflects the balancing of right and duty which lies at the heart of both the common law of torts, and any robust understanding of professional ethics. It refers, not to forms or notations in charts, but rather to "the dialogue between the patient and the provider of services in which both parties exchange information and questions culminating in the patient's agreeing to a specific" plan of medical, surgical or other professional (including legal) intervention.[44]

Consent is necessary because the law protects both bodily integrity and human dignity (of which autonomy is but one constituent part).[45] Under the common law, any unconsented touching is a battery, while placing person in fear of being touched is an assault.[46] Where consent has been obtained on the basis of inadequate information, the common law of negligence (malpractice) provides the remedy, and in cases where material information is *intentionally* misrepresented, withheld, or provided in a misleading manner, the consent is invalid (thus invoking the law of battery),[47] and an action for fraud or deceit may be appropriate as well.

At least insofar as medical *treatment* is concerned, the law presumes that, subject to certain constraints not relevant at this point, "[e]very human being of adult years and sound mind has a right to determine what shall be done with his own body."[48] What this means in practice is that reasonable physicians and patients are free to choose, or reject, any legally and professionally acceptable therapy, even if the course of action chooses poses considerable risk to the life or the health of the patient.[49]

When the issue is biomedical or behavioral *experimentation*, however, the calculation is different, even where the experimental treatment or procedure is thought to be therapeutic. The reason is straightforward: experiments involve the unknown, and thus impose upon the treating physician or researcher additional duties of care and disclosure beyond those which exist in the normal, or even the "innovative," treatment setting.[50] The dialogue between researcher and subject must therefore be as open and as accurate as possible; for any information which might cause a reasonable *volunteer*—not "patient"—to grant or withhold consent will be deemed material.[51]

It is with this background in mind that we return to the regulations governing informed consent. Under the federal regulatory scheme, the dialogue between researcher and subject is broken into elements which include the information which must be provided to the subject:

1. A statement that the study involves research, an explanation of the purposes of the research and the expected duration of the subject's participation, a description of the procedures to be followed, and identification of any procedures which are experimental.
2. A description of any foreseeable risks or discomforts to the subject.
3. A description of the possible benefits to the subject or to others which may be expected from the research.
4. A disclosure of appropriate alternative procedures or courses of treatment, if any, that might be advantageous to the subject.
5. A statement describing the extent, if any, to which confidentiality of records identifying the subject will be maintained and, in the case of FDA research, the possibility that the FDA may inspect the records.
6. For research involving more than minimal risk,[52] an explanation as to whether any compensation and an explanation as to whether any medical treatments are available if injury occurs and, if so, what they consist of, or where further information may be obtained.
7. An explanation of whom to contact for answers to pertinent questions about the research and research subject's rights, and whom to contact in the event of a research-related injury to the subject.
8. A statement that participation is voluntary, that refusal to participate will involve no penalty or loss of benefits to which the subject is otherwise entitled, and that the subject may discontinue participation at any time without penalty or loss of benefits to which the subject is otherwise entitled.[53]

That these disclosures relate to "material" facts a "reasonable volunteer" (or even a "reasonable person") would want to know before deciding to participate should be obvious. Equally important are the facts contained in a list of additional elements to be disclosed to each subject "when appropriate."[54] This information includes:

1. A statement that the particular treatment or procedure may involve risks to the subject (or embryo or fetus, if the subject is or may become pregnant) which are currently unforeseeable.
2. Anticipated circumstances under which the subject's participation may be terminated by the investigator without regard to the subject's consent.
3. Any additional costs to the subject that may result from participation in the research.
4. The consequences of a subject's decision to withdraw from the research and procedures for orderly termination of participation by the subject.
5. A statement that significant new findings developed during the course of the research which may relate to the subject's willingness to continue participation will be provided to the subject.
6. The approximate number of subjects involved in the study.[55]

One need only look at the sheer volume and complexity of the information to be disclosed to appreciate the great difficulties inherent in attempts to obtain a truly "informed" consent under the best of circumstances. Even healthy, competent persons without any disabilities or other factors in their personal history which would make them vulnerable may have difficulties understanding the information disseminated to them. When investigators play "hide the ball," there may be no consent at all.[56]

OVERSIGHT AND VULNERABLE POPULATIONS

It is against this backdrop that we must consider the rules to be applied to vulnerable populations. Though the basic ethical and legal issues are the same for all research involving human subjects, the addition of *any* "vulnerability factor" whatever raises very serious issues which strike at the moral foundation of the entire regulatory enterprise. Dr. Jay Katz, for example, has written that "the celebrated *Belmont Report* on principles and guidelines for human research is confusing by not making clear distinctions between the principles that should govern research with competent and incompetent persons."[57] He notes that the *Report*'s reliance on the three basic principles of "respect for persons, beneficence, and justice" as the basis for a balancing of benefits and burdens gloss over our understanding that "the point of Kantian principles is precisely to say that certain things cannot be 'balanced out,' *i.e.*, if certain actions are unjust or disrespectful of persons then they are wrong and therefore simply should not be done."[58] The trick, of course, is to agree on a set of rules, the violation or compromise of which will be branded as unethical *per se*.[59]

I will assume, for present purposes, that all parties will agree that the requirement that "informed consent" be obtained prior to commencing research on a human subject is, regardless of its philosophical grounding, just such a rule. We must, therefore, consider the manner in which various vulnerability factors affect this critical requirement.

Some of these factors are relatively straightforward ones that affect the ability of the individual to understand the information given, and hence limit that person's ability to give an *informed* consent. Such factors can range from simple language differences which can be overcome by retaining competent translators, to far more serious questions involving the use of language which may, due to its technical or specialized nature, be inaccessible even to well-educated persons who are not medical specialists.[60] Even more difficult are problems involving persons who may *never* understand either the substance or the significance what they are being told because of capacity problems stemming from age (children or elderly persons), mental or developmental disability, and the total incompetence of those in comas or persistent vegetative states.[61] Other vulnerability issues arise because the population at issue may be uniquely susceptible to coercion, undue influence, or conflict of interest. In these cases, the question is not whether the consent is "informed," but the far more basic question of whether there is, or can be, any "consent" at all.

Oversight of "Capacity-Related" Issues

As noted in the discussion of the federal and State regulatory framework, the bulk of "oversight," insofar as it relates to vulnerable populations, occurs at the *state* level. At the most basic level are issues of "competence" which go to the ability of an individual to make legally binding decisions *at all*. "Competency, like dangerousness, is an elusive concept[, and] applying generally accepted criteria will [often] produce inappropriate results."[62] Commonly accepted factors include an individual's ability to appreciate his condition and the nature and consequences of important choices; the ability to understand information; knowledge of place, self and time; and the ability to engage in coherent, lucid discussions "are sometimes seen as the benchmarks of competency."[63] Adults are generally presumed competent unless the contrary is shown by "clear and convincing" evidence. Due to their immaturity, minors are presumed to be "legally incompetent" to make certain decisions,[64] and both State and federal regulations governing research on human subjects contain special rules which govern children as research subjects.

Closely related to issues of competence are issues of "capacity"; that is, an individual's *ability* to make specific kinds of informed decisions. Due to a medical condition or the administration of mind-altering drugs, an adult who is otherwise competent might lack capacity to make treatment or other kinds of decisions, and an individual who seems rarely, if ever, lucid might well have the capacity, on occasion, to understand the nature and consequences of the decisions he or she is being called upon to make.[65] By the same reasoning, state law also has come to recognize the capacity of certain minors, usually those considered "mature" or "emancipated", to make certain kinds of decisions concerning health care.[66]

The laws of every State provide procedures for the determination of competence and capacity, and authorize the appointment of full or temporary guardians who can make decisions for those found, in an appropriate judicial proceeding, to lack capacity. In the medical setting, however, most such determinations are made, for adults at least, "at the bedside, not in a court room."[67] For children, the issue is necessarily more complicated, given not only the parent's guardianship rights and responsibilities, but also the constitutional right of parents to the care, custody and control of the education and upbringing of their children.[68]

Where capacity is limited by a mental or developmental disability, the issues are identical to those discussed above, but some States have gone farther, and have enacted laws which mandate that special care be taken to assure both the rights and the ethical treatment of vulnerable populations. Wisconsin, for example, has enacted its own "Alcohol, Drug Abuse, Developmental Disabilities and Mental Health Act". That law provides, among other things, that all "patients"[69] must be informed, upon admission, that they

Have a right not to be subjected to experimental research without the express and informed consent of the patient and of the patient's guardian after consultation with independent specialists and the patient's legal counsel. Such proposed research shall first be reviewed and approved by the institution's research and human rights committee created under sub. (4) and by the department before such consent may be sought. Prior to such approval, the committee and the department shall determine that research complies with the principles of the statement on the use of human subjects for research adopted by the American Association on Mental Deficiency, and with the regulations for research involving human subjects required by the U.S. department of health and human services for projects supported by that agency.[70]

Virginia takes this reasoning one step farther, and restricts the ability of persons lacking capacity to participate in non-therapeutic research at all. Virginia follows the general rule that informed consent[71] be obtained from the person who is to be the subject. If the subject is competent, consent must be "subscribed to in writing by the person and witnessed."[72] If the subject is not competent, then consent shall be "subscribed to in writing by the person's legally authorized representative and witnessed."[73] When the subject is a minor who is otherwise capable of consent, then consent shall be "subscribed to in writing by both the minor and his legally authorized representative."[74] Nevertheless, a legally authorized representative is restricted from giving consent to nontherapeutic research[75] unless the human research committee determines that the nontherapeutic research proposed presents no more than a "minor increase over minimal risk[76] to the human subject."[77]

State law protections such as these are particularly important since they supplement, and are not preempted by,[78] existing federal regulations. Because the federal regulatory framework is largely silent on the topic of mental disability and presumes the propriety of nontherapeutic research on children and other vulnerable persons that may lack capacity to act as "reasonable volunteers", it is important to scrutinize the *entire corpus* of State laws that deal with vulnerable populations, including the elderly and people with specific illnesses, such as HIV. Though State law may not address the specific research ethics question directly, other policies such as standards governing the reasonability of parental, guardianship or surrogate decisions in the health care field, will be influential in the cases where they provide the most appropriate analogy to the issue at hand.

Oversight of Issues Relating to the Voluntariness of the "Consent" Obtained

Precisely the same calculus applies to situations in which there is a substantial question concerning the voluntariness of the consent obtained, whether because of coercion, undue influence, or conflict of interest on the part of the guardian or surrogate decision-maker. Perhaps the most obvious examples of situations where coercion and undue influence are very real possibilities are those involving persons who live in rigorously controlled environments such as the military, prisons, or mental health facilities. It is relatively easy in these situations to envision how either the nature of the environment, the forms of control exercised over those who live there, or the manner in which either the promise of a benefit, or express or implied threats to make life difficult can make an otherwise unwilling person seem to be a "reasonable volunteer."

The possibility of undue influence, rather than direct or indirect coercion, is the concern which motivates the special rules regulating experimentation and informed consent involving persons with diminished capacity. These categories include populations with diminished capacity, including children and persons with mental or developmental disabilities (including the vulnerable elderly), and any population whose need for services is so acute that they may be inordinately willing to subject themselves to greater than normal research risks to achieve the hoped-for benefit. (Some of the controversies over the availability of certain experimental AIDS treatments come to mind here.)

In this setting, the vulnerability arises from the susceptibility of the population to suggestion, manipulation, misunderstanding, or, in the last case, to wishful thinking which may cloud rational judgment. Where the populations are vulnerable because of age or disabling condition, the regulations require the involvement of a guardian who is obligated to act in the best interests of the person under his or her care (usually, but not always, a family member). In these cases, oversight is available not only in the due course of the research protocol, but also by recourse to the procedures provided by state law for the supervision of parent and guardian decision-making. Where the vulnerability arises because of either physical or other need, such as economic or educational deprivation,[79] there is no readily discernible form of oversight other than that of the sensitivities of the membership of the IRB, the sensibilities of the research team, and the possibility of pressure from outside activists and politicians.

OVERSIGHT OF PROBLEMS RELATING TO THE ETHICS OF RESEARCHERS

The final subject which must be addressed is the nature of the professional relationship between researcher and subject. At the formal level, this is the realm of licensing and professional discipline, and the case law is sparse; for there must usually be some sort of egregious lapse before licensing authorities will intervene with even as much as an investigation.[80] Insofar as certain vulnerable populations are subjects of research, there may be civil rights concerns raised as well. This is particularly the case where the research subject is a person with a mental or developmental disability or belongs to a class of persons who have historically been subjected to discriminatory treatment as a result of societal perceptions that they (or their problems) are not worth the effort (e.g., racial minorities and women).

By their nature, these are inherently controversial topics because they raise issues which go far beyond professionalism to the issue of personal morality itself.[81] In general, laws and regulations are concerned about the tangible impact that unethical behavior can have on the research subject, and are framed in a manner which is calculated to prevent it. They are thus the means by which the law assures at least minimal standards of behavior.

Because rules are tangible and provide relatively clear guidance, it is easy to confuse compliance with the rules with the ethical behavior they are intended to foster. This is a tendency to be avoided at all costs, for it exalts form over substance, and leads to an undue focus on the content of rules and forms[82] rather than the *behavior* of the individuals seeking consent, or granting it on behalf of others.

The preamble to the federal regulations governing prisoners,[83] for example, recognizes that special constraints on researchers are appropriate "[i]nasmuch as prisoners may be under constraints because of their incarceration which could affect their ability to make a truly voluntary and uncoerced decision whether or not to participate as subjects in research."[84] But what is it that makes special rules appropriate? The mere fact that the potential subjects may, for one reason or another, be limited in their ability to make a "truly voluntary and uncoerced decision," or the potential that researchers will allow their own self-interest, prejudice concerning the research subjects, or perception of the appropriate balance to be struck between the "public interest" and the interests of various vulnerable populations to affect their professional behavior?

That the concern is for the ethics of the researchers is apparent from the face of the entire regulatory edifice. The task of IRBs is to provide reasonably detached, multi-disciplinary "outsider" oversight of all aspects of the research, and many of the general problems with the structure and function of IRBs relate to issues of professional detachment.[85]

The impact of conflicting interests on the integrity of professional relationships is a common theme of codes of ethics, and everyone recognizes that conflicts of interest are generally to be avoided, no matter what the profession. The most subtle conflict of interest questions, in fact, are those which are not generally perceived to be conflicts of interest at all. Professionals, like everyone else, are occasionally guilty of self-deception.

The best illustrations of the subtle interplay between the ethics of professionalism and the language of regulation are those in which professional or personal interest manifests itself in behavior which is so clearly designed to foster the needs or goals of the professional that the client's interests are effectively ignored. A prime example is the prospective waiver of liability.

Both federal and Virginia regulations governing research on human subjects provide that informed consent forms may not contain "any language through which the person who is to be the human subject waives or appears to waive any of his legal rights, including release of any individual, institution, or agency or any agents thereof from liability for negligence."[86]

It is possible simply to accept the rule as a given, but that would be to miss the point: Why *should* researchers be barred from including prospective waivers from their informed consent forms? One could argue that, to the extent the consent process was "truly informed" and the prospective research subject understood the nature of the risks to be undertaken, it would not be unjust to *ask* for such an agreement, even if, in the end, the prospective subject were to refuse to sign it. After all, respect for the principle of autonomy would seem to indicate that the prospective subject should be able to undertake such risks voluntarily, would it not?

On deeper analysis, however, such requests are clearly inappropriate. The effect of such a waiver is to place *all* of the risks on the subject, including those which may result from breaches of professional standards by those providing services. Is this an appropriate outcome? It is certainly a *possible* one under existing law should the subject be injured, yet fail or decline to enforce his or her rights. And yet, the ability of the subject to consent to a waiver of legal rights, either in exchange for a settlement or because of a lack of inclination to pursue them, does not address the professional ethics question of whether or not a professional should *seek* such waivers in the first instance.

It has been said that ethics is taking responsibility for one's actions. So, the ethical question in this context is not whether it is ethical to allocate risks and benefits in this fashion, but whether the behavior at issue—asking a potential research subject for a prospective waiver of rights—is professionally appropriate given the nature of the proposed relationship? The focus of this question is on the moral and professional duties that a professional owes to a client or patient. Because these are duties which arise because of the nature of the relationship between the professional and client or patient,[87] the consent of the subject is irrelevant. The ethical lapse is complete upon solicitation.

And what is that lapse? It is an attempt by a professional evade his or her responsibility to act in a professionally competent manner. As such, the specific ethical question has little, if anything, to do with research or the kinds of questions which are central to discussions of biomedical research ethics. Such behavior would be no more legitimate for an accountant, architect or lawyer. Under most codes of professional ethics, solicitations of "blanket consents" of this sort are "unjust or disrespectful of persons" who are or will be involved in professional relationships. They are, for that reason, "wrong and therefore simply should not be done."[88]

The same analysis holds true, but is far more subtle, when the ethical lapse is attributable to a perception of the patient or client which does not take full account of their dignity as a human person. Though the Nazi experiments and the Tuskegee syphilis study are often held up as stark reminders of the abuses which can take place when a professional "objectivizes" the human being who provides the occasion for the professional's exercise of skill and judgment to the point of dehumanizing them,[89] those cases are far too extreme to resonate with most practitioners and researchers. But the difference is really only one of degree, and the medical profession is far from alone in this tendency. The lawyers in Dickens' *Bleak House*, a tale of endless litigation in which the clients were forgotten until the fortune over which the lawyers were sparring was consumed, provide a less searing, and far more common, example from another profession where abstract interests are often given precedence over the immediate and practical interests of the persons the professionals are sworn to serve.

Ethics, in this view, is the right ordering of such relationships. When a patient or client is viewed by the professional as a condition, case, or problem—that is, as an object, rather than a person, a "professional" relationship no longer exists (save, perhaps, in the mind of the hapless subject). As a result, the only effective "oversight" which will avoid problems such as these is the insight which comes from increased personal and institutional awareness that professional relationships presuppose a *relationship* between persons.

The rules exist to protect and foster that relationship. Problems arise when there is greater attention given to compliance with rules and abstract principles than to the human needs of our patients, clients or research subjects. Michael Ignatieff has written eloquently of the fallacy of an approach which highlights a legalistic "respect" for the autonomy of others, but which, in reality, springs, from a rather abstract conception of humanity. In *The Needs of Strangers* he cautions us:

Woe betide any man who depends on the abstract humanity of another for his food and protection. Woe betide any man who has no state, no family, no neighbourhood, no community that can stand behind to enforce his claim of need. [King] Lear learns too late that it is power and violence that rule the heath, not obligation.[90]

NOTES

1. Tom L. Beauchamp and James F. Childress, *Principles of Biomedical Ethics*, Oxford University Press, 3d ed. 1989. p. 25.
2. *Webster's New Universal Unabridged Dictionary*, Dorset & Baber, 2d ed. 1983. 1277. The alternative definition is, ironically, "an overlooking, failure to see or notice."
3. For a comprehensive treatment of the legal and regulatory context of patients who are asked to serve as research subjects, see Jesse A. Goldner, "An Overview of Legal Controls on Human Experimentation and the Regulatory Implications of Taking Professor Katz Seriously," in *Health Law Symposium*. Legal and Ethical Controls on Biomedical Research: Seeking Consent, Avoiding Condescension, 38 *St. Louis L.J.* 3, 63 (1993) [hereinafter, Goldner, An Overview of Legal Controls on Human Experimentation]. I am indebted to Professor Goldner, not only for much of the organizational structure of this background paper, but also for his exhaustive background work. I draw on it liberally throughout the course of this paper and wish to give him credit at the outset. It is well worth reading in its entirety, not only as background, but also as a contribution to the ongoing debate over the ethics of biomedical research on competent subjects, which is the subject of the symposium in which it appears.
4. Though there are numerous other concepts, principles, and virtues which could be added to this list, the following have been selected because they form the foundation for the law and regulation which is the subject of this paper.
5. *Webster's New Universal Unabridged Dictionary*, Dorset & Baber, 2d ed. 1983. 566.
6. *Black's Law Dictionary* (6th ed., WESTLAW) indicates that "a 'tort' is broadly defined as 'a private or civil wrong or injury, including action for bad faith breach of contract, for which the court will provide a remedy in the form of an action for damages.... There must always be a violation of some duty owing to plaintiff, and generally such duty must arise by operation of law and not by mere agreement of the parties. A legal wrong committed upon the person or property independent of contract. It may be either (1) a direct invasion of some legal right of the individual; (2) the infraction of some public duty by which special damage accrues to the individual; (3) the violation of some private obligation by which like damage accrues to the individual.'"
7. These concepts are elaborated in Beauchamp and Childress, *Principles of Biomedical Ethics, supra* note 1, at p. 120–93; 194–255; 256–306; and 67–119, respectively.
8. See Sissela Bok, *Lying: Moral Choice in Public and Private Life*, 1979.
9. See Thomas L. Shaffer and Robert F. Cochran, Jr., *Lawyers, Clients and Moral Responsibility* 82–91 (*West*, 1993).
10. See generally Rena A. Gorlin, ed., *Codes of Professional Responsibility*, (Bureau of National Affairs, 2d ed., 1990).

11. See, e.g., American Medical Association, *Principles of Medical Ethics and Current Opinions of the council on Ethical and Judicial Affairs—1989,* Preamble, Principle IV (noting that, except in emergencies, a physician is free to choose whom to serve, with whom to associate, and the environment in which to provide service), *quoted in* Rena A. Gorlin, ed., *Codes of Professional Responsibility, supra* note 10 at 191 [hereafter "Gorlin"]. The caveat, "except in emergencies," reflects the special nature of the physician's training and the demands justice. A similar principle constrains the formation of the lawyer-client relationship. Compare American Bar Association, *Model Rules of Professional Responsibility,* Rule 1.2 (Scope of Representation) with Rule 6.2 (Accepting Appointments), in Gorlin, *op. cit.* at 339, 377.

12. 45 C.F.R. § 76 (1994)

13. See, e.g., sources cited in Goldner, *supra* note 3 at note 161: Marion J. Finkel, Should Informed Consent Include Information on How Research is Funded?, 13 *IRB* 1 (Sept.–Oct. 1991); Roger J. Porter and Thomas E. Malone, *Biomedical Research: Collaboration and Conflict of Interest,* 121–50, 163–84 (1992); Marc A. Rodwin, *Medicine, Money and Morals,* 212–16 (1993). See also Moore v. Regents of the University of California, 51 Cal. 3d 120, 793 P.2d 479 (Cal. 1990); Thomas H. Murray, Who Owns the Body? On the Ethics of Using Human Tissue for Commercial Purposes, 8 *IRB* 1 (Jan.–Feb. 1986).

14. 42 C.F.R. § 50.102 defines research misconduct as "fabrication, falsification, plagiarism, or other practices that seriously deviate from those that are commonly accepted within the scientific community for proposing conducting or reporting research."

15. Goldner, *supra* note 3 at 90.

16. See Office of the Judge Advocate General, Dept. of the Army, *Trials of War Criminals Before the Nuremberg Military Tribunals, Vol. I and II* at 181–82 (Doctors' trial; Nuremberg Code); United States v. Stanley, 483 U.S. 669, 1987 (recounting LSD experiments by the United States Army on the late James Stanley). See generally George J. Annas and Michael A. Grodin, eds., *The Nazi Doctors and The Nuremberg Code,* 1992; Robert J. Lifton, *The Nazi Doctors: Medical Killing and the Psychology of Genocide,* Basic Books, 1986; United States Public Health Service, *Tuskegee Syphilis Study Ad Hoc Advisory Panel: Final Report,* Washington, D.C., April 1973.

17. The National Research Act amended the Public Health Service Act. Pub.L. No. 93-348, 88 Stat. 342, codified as amended at 42 U.S.C. §§ 201 to 300aaa-13 (1994).

18. The National Commission for the Protection of Human Subjects of Biomedical and Behavioral Research, U.S. Department of Health, Education and Welfare, *The Belmont Report: Ethical Principles and Guidelines for the Protection of Human Subjects of Research,* Pub. No. (OS) 78-0012, 1978 [hereinafter *Belmont Report*].

19. See 28 C.F.R. § 46.101 (1995).

20. See 28 C.F.R. § 46.103 (1995); 45 C.F.R. §46.103 (1995).

21. 28 U.S.C. §§1346(b), 1402,

22. See 28 C.F.R §§46.101(f)(g)(h) (indicating that federal rules do not preempt applicable state and foreign law which provide additional protections for human subjects).

23. California has a wide range of statutory protections in place. See *Cal. Health & Safety Code* §§24170 to 24177 (Deering 1994) ("Human Experimentation); *Cal. Health & Safety Code* §26668.6 (requires that prior to administering or prescribing an experimental drug, consent must be obtained according to the guidelines set forth in the Chapter on Human Experimentation); *Cal. Penal Code* §3521 (Deering 1994) (experiments involving prisoners).

24. *Wis. Stat.* §51.61 (1993).

25. *McKinney's Consol. New York Pub. Health Law* §§ 2441–2445 (1994).

26. *Del. Code Ann.* tit. 16 §§5171–75 (1994).

27. *Mont. Code Ann.* §§ 53-21-104, 53-21-147 (1993).

28. Florida Biomedical and Social Research Act, *Fla. Stat.* ch. 402.105 (1993).

29. *Va. Code Ann.* §§ 32.1-162.16, 32.1-162.19, 32.1-162.20 (Michie 1994).

30. 1994 Mass. H.B. 4609, 179th General Court. To date, no action has been taken on the Bill.

31. See, e.g., *D.C. Code* § 6-1969 (1994) (rights of mentally retarded citizens); 20 *Ill. Comp. Stats* 301/30-5 (Michie 1994) (creating a Patient's Bill of Rights for individuals in alcoholism treatment, and requiring that "[b]efore a patient is placed in an experimental research or medical procedure, the provider must first obtain his informed written consent or otherwise comply with the federal requirements regarding the protection of human subjects contained in 45 C.F.R. Part 4"); Montana

Code Ann. §53-20-147 (1994) (assuring the right of residents of residential facilities "not to be subjected to experimental research without the express and informed consent of" either the resident or their authorized guardian).

32. Because there is an extensive literature concerning IRBs, the review which follows will be limited to highlighting only those issues which are of particular relevance to the subject of this conference.
33. *The Belmont Report, supra* note 18 at 16.
34. 45 C.F.R § 46.103 (1994). Assurances are provided to the Office for Protection from Research Risks (OPRR) at the National Institutes of Health. See 45 C.F.R. § 46.101(i). See United States Department of Health & Human Services, Application for Public Health Service Grant (PHS 398; OMB No. 0925-001), Section E(1)(a), p. 25.
35. 21 C.F.R. §56.107 (1994); 45 C.F.R § 46.107 (1994). According to the FDA and DHHS regulations,

(a) Each IRB shall have at least five members, with varying backgrounds to promote complete and adequate review of research activities commonly conducted by the institution. The IRB shall be sufficiently qualified through the experience and expertise of its members, and the diversity of the members' backgrounds including consideration of the racial and cultural backgrounds of members and sensitivity to such issues as community attitudes, to promote respect for its advice and counsel in safeguarding the rights and welfare of human subjects. In addition to possessing the professional competence necessary to renew specific research activities, the IRB shall be able to ascertain the acceptability of proposed research in terms of institutional commitments and regulations, applicable law, and standards of professional conduct and practice. The IRB shall therefore include persons knowledgeable in these areas. If an IRB regularly reviews research that involves a vulnerable category of subjects, such as children, prisoners, pregnant women, or handicapped or mentally disabled persons, consideration shall be given to the inclusion of one or more individuals who are knowledgeable about the experienced in working with those subjects.

(b) Every nondiscriminatory effort will be made to ensure that no IRB consists entirely of men or entirely of women, including the institution's consideration of qualified persons of both sexes, so long as no selection is made to the IRB on the basis of gender. No IRB may consist entirely of members of one profession.

(c) Each IRB shall include at least one member whose primary concerns are in scientific areas and at least one member whose primary concerns are in nonscientific areas.

(d) Each IRB shall include at least one member who is not otherwise affiliated with the institution and who is not part of the immediate family of a person who is affiliated with the institution.

(e) No IRB may have a member participate in the IRB's initial or continuing review of any project in which the member has a conflicting interest, except to provide information requested by the IRB.

(f) An IRB may, in its discretion, invite individuals with competence in special areas to assist in the review of issues which require expertise beyond or in addition to that available on the IRB. These individuals may not vote with the IRB.

36. Id., 45 C.F.R §§46.103(b)(4, 5) (1994).
37. 45 C.F.R §46.115 (1994). See also 45 C.F.R §§46.112 (internal review), 46.114 (cooperative research oversight).
38. See also 45 C.F.R §§ 46.201-211 (1992) (pregnant women, unborn children (fetuses) and human *in vitro* fertilization); 46.301-306 (prisoners); 46.401-409 (children); 45 C.F.R §46. 111(b) (economic or educationally disadvantaged persons). There are no specific federal regulations provisions which address the specific concerns of persons with mental disabilities.
39. 45 C.F.R §46.111 (1994).
40. Federal regulations which would have covered institutionalized persons with mental disabilities were proposed in 1978, 43 Fed Reg. 53, 9570, but no final rules were adopted.
41. See, e.g., *Cal. Welf. & Inst. Code* §4514 (1994) (informed consent requirements for persons with developmental disabilities); *Montana Code* §53-20-104 (1993) (Mental Health Board of Visitors to oversee "all plans for experimental research or hazardous treatment procedures involving persons admitted to a residential facility"); Wisconsin State Alcohol, Drug Abuse, Developmental Disabilities and Mental Health Act, *Wisconsin Stats.* §51.61 (1994) (rights of mental patients, regardless of locale of treatment).
42. United States Department of Health & Human Services, Application for Public Health Service Grant (PHS 398; OMB No. 0925-001), Section E(1)(a), p. 26.

43. Fay A. Rozovsky, J.D., M.P.H., *Consent to Treatment: A Practical Guide*, Little, Brown & Co. 2d ed., 1990. p. 3 (hereafter Rozovsky, *Consent to Treatment*).

44. Id. In the lawyer-client setting, Tom Shaffer and Bob Cochran refer to this discourse as a "moral conversation" See Shaffer & Cochran, *supra* note 9 at 1 ("Law office conversations are almost always moral conversations. This is so because they involve law, and law is a claim that people make on one another. The moral content is often implicit, but it is always there.... If it is possible for a serious conversation between a lawyer and a client in a law office to be without moral content, we cannot think of an example.")

45. See text accompanying notes 4 to 9, *supra*.

46. W. L. Prosser, *Law of Torts* § 10 (West, 4th ed., 1971).

47. See Rozovsky, *Consent to Treatment*, *supra* note 43 at 10, citing Krueger v. San Francisco Forty-Niners, 234 Cal. Rptr. 579, 584 (Cal. App. 1987) (in its zeal to keep Mr. Krueger playing, the team "consciously failed to make full, meaningful disclosure to him respecting the magnitude of the risk he took in continuing to play a violent contact sport with a profoundly damaged left knee.")

48. Schloendorff v. Society of New York Hosp., 211 N.Y. 125, 129–30, 105 N.E. 92, 93 (1914) (Cardozo, J.).

49. The literature on refusal, withdrawing and withholding treatment is immense, and will not be discussed here except to the extent that it casts some light on the reasons *why* treatment can (or should) be refused withdrawn or withheld.

50. Professor Goldner's article, An Overview of Legal Controls on Human Experimentation, *supra* note 3 at pp. 70–88, contains an excellent discussion of the development of legal standards distinguishing experimentation from standard therapeutic treatments.

51. The National Commission proposed that the standard for medical care (a "reasonable person" making information concerning medical care) was "insufficient since the research subject, being in essence a volunteer, may wish to know considerably more about risks gratuitously undertaken than do patients who deliver themselves into the hand of a clinician for needed care." It therefore proposed "a standard of 'the reasonable volunteer'...: because

    the extent and nature of information should be such that persons, knowing that the procedure is neither necessary for their care nor perhaps fully understood, can decide whether they wish to participate in the furthering of knowledge. Even when some direct benefit to them is anticipated, the subjects should understand clearly the range of risk and the voluntary nature of participation.

    The Belmont Report, *supra* note 18, at 11, quoted in Goldner, *supra* note 3 at 97–98

52. " 'Minimal risk' means that the probability and magnitude of harm or discomfort anticipated in the research are not greater in and of themselves than those ordinarily encountered in daily life or during the performance of routine physical or psychological examinations or tests," 21 C.F.R. §50.3 (1994); 45 C.F.R. §46.102(i).

53. 21 C.F.R. §50.25 (1994) (FDA); 45 C.F.R. § 46.116.

54. Id.

55. Id.

56. See, e.g., Halushka v. University of Saskatchewan, 52 W.W.R. 608, 617 (Sask. 1965) (subject told that test was "safe", even though no test had ever been run. The court treated the statement as a "non-disclosure" of the material fact that the researcher did not know whether the test was safe or not.)

57. Jay Katz, M.D., Human Experimentation and Human Rights, in *Health Law Symposium*, Legal and Ethical Controls on Biomedical Research: Seeking Consent, Avoiding Condescension, 38 *St. Louis L.J.* 3, 7, 30 n. 71 (1993).

58. Id., quoting Ernest Marshall, Does the Moral Philosophy of the *Belmont Report* Rest on a Mistake?, 8 *IRB* 5–6 (Nov.–Dec. 1986) at p. 6.

59. See Beauchamp and Childress, *supra* note 1 at 44–47 (discussing the importance of selecting an ethical theory, and summarizing the differences between consequentialist and deontological theories.) They go on to note, in language which is particularly relevant here that

    Theoretical differences can, of course, eventuate in practical disagreements and in different general policies. For example, we shall see throughout this volume that utilitarians tend to support various types of research involving human subjects on grounds of the social benefits of the research for future patients. Deontonlogists, by contrast, tend to be skeptical of some of this research on grounds of its actual or potential violation of individual rights, which are ground in principles of respect for autonomy and protection against harm.

60. Even though the general requirements for informed consent require that "The information that is given to the subject or the representative shall be in language understandable to the subject or the representative," 45 C.F.R § 46.116 (1994), this is not an uncommon problem in many informed consent settings, including those which are not normally viewed as requiring "informed consent" at all—such as the sale of corporate securities. See, e.g., United States Securities and Exchange Commission, Rule 10b-5 (requiring that disclosures be made in language which is sufficiently complete so as not to be misleading.) The regulations involving prisoners make it clear that any information be "presented in language which is understandable to the subject population." 45 C.F.R. § 46.305(a)(5) (1994).

61. The UCLA research study involving schizophrenia research raises many of these questions. The NIH Office for Protection from Research Risks, Division of Human Subject Protection found that neither the informed consent documents nor the extended oral informed consent process complied with "the requirements of HHS regulations.... " What is interesting about the specific findings is that they target *both* the investigators themselves (who are ultimately responsible under both state and federal law for the quality and content of the informed consent process) *and the UCLA School of Medicine IRB.*

62. Rozovsky, *Consent to Treatment, supra* note 43 at p. 674–675.

63. Id.

64. *See* id., at pp. 257–259,

65. See, e.g., Lane v. Candura, 6 Mass. App. Ct. 477, 376 N.E.2d 1232 (1978) (elderly woman who made medically irrational decisions, who had a distorted concept of time, whose mind wandered, and who was confused about details was nevertheless competent because, during her lucid periods, she was fully aware of the nature and consequences of her treatment choices).

66. See generally Rozovsky, *Consent to Treatment, supra* note 43 at pp. 259–360.

67. The New York State Task Force on Life and the Law, *When Others Must Choose: Deciding for Patients Without Capacity,* 1992, at p. 28.

68. See generally Jonathan O. Hafen, Book Note, *The Transformation of Family Law: State, Law, and Family in the United States and Western Europe,* by Mary Ann Glendon (Chicago and London: The University of Chicago Press, 1989) 1991 *B.Y.U. L. Rev.* 719; Bruce C. Hafen, The Constitutional Status of Marriage, Kinship, and Sexual Privacy; Balancing The Individual and Social Interests, 81 *Mich. L. Rev.* 463 (1983); Robert H. Mnookin, *In the Interest of Children: Advocacy, Law Reform, and Public Policy,* New York: W.H. Freeman & Co., 1985

69. Wis. Star. 51.61 (a) provides that the term "patient" includes

   any individual who is receiving services for mental illness, developmental disabilities, alcoholism or drug dependency, including any individual who is admitted to a treatment facility in accordance with this chapter or ch. 55 or who is detained, committed or placed under this chapter or ch. 55, 971, 975 or 980, or who is transferred to a treatment facility under §51.35(3) or 51.37 or who is receiving care or treatment for those conditions through the department or a county department under s. 51.42 or 51.437 or in a private treatment facility. "Patient" does not include persons committed under ch. 975 who are transferred to or residing in any state prison listed under § 302.01. In private hospitals and in public general hospitals, "patient" includes any individual who is admitted for the primary purpose of treatment of mental illness, developmental disability, alcoholism or drug abuse but does not include an individual who receives treatment in a hospital emergency room nor an individual who receives treatment on an outpatient basis at those hospitals, unless the individual is otherwise covered under this subsection.

70. *Wis. Stats.* 51.61(j) (1994).

71. Under Virginia law, " 'Informed consent' means the knowing and voluntary agreement, without undue inducement or any clement of force, fraud, deceit, duress, or other form of constraint or coercion of a person who is capable of exercising free power of choice. For the purposes of human research, the basic elements of information necessary to such consent shall include: (1) A reasonable and comprehensible explanation to the person of the proposed procedures or protocols to be followed, their purposes, including descriptions of any attendant discomforts, and risks and benefits reasonably to be expected; (2) A disclosure of the any appropriate procedures or therapies that might be advantageous for the person; (3) An instruction that the person may withdraw his consent and discontinue participation in the human research at any time without prejudice to him;

(4) An explanation of any costs or compensation which may accrue to the person and, if applicable, the availability of third party reimbursement for the proposed procedures or protocols; and (5) An offer to answer and answers to any inquiries by the person concerning the procedures an protocols." *Va. Code Ann.* § 32.1-162.16.

72. *Va. Code Ann.* §. 32.1-162.18 (1994).

73. Id.

74. Id.

75. " 'Nontherapeutic research' means human research in which there is no reasonable expectation of direct benefit to the physical or mental condition of the human subject." *Va. Code Ann.* § 32.1-162.16 (1994).

76. *Va. Code Ann.* § 32.1-162.18(B). Like most States, Virginia borrows key definitions from extant federal regulations governing the same topic. Thus, for purposes of Virginia law, the term "minimal risk" has the same meaning as in the FDA and DHHS regulations. Compare *Va. Code Ann.* § 32.1-162.16 with 45 C.F.R. §46.102(i) (1994).

77. *Va. Code Ann.* §32.1-162.18. This is not unlike the federal rules governing research on children, 45 C.F.R. §46.406 (1994), but is generalized for all informed consent obtained from a legally authorized representative.

There is also a provision regarding the use of force, which states, "Notwithstanding consent by a legally authorized representative, no person who is otherwise capable of rendering informed consent shall be forced to participate in any human research. In the case of persons suffering from organic brain diseases causing progressive deterioration of cognition for which there is no known cure or medically accepted treatment, the implementation of experimental courses of therapeutic treatment to which a legally authorized representative has given informed consent shall not constitute the use of force." Id.

78. See 45 C.F.R. § 46.116(e) ("The informed consent requirements in this policy are not intended to preempt any applicable Federal, State, or local laws which require additional information to be disclosed for informed consent to be legally effective.")

79. 45 C.F.R. §46.111(b) (economic or educationally disadvantaged persons).

80. See Goldner, *supra* note 3 (collecting sources).

81. Some of the same issues arise in the highly-charged area of medical care discrimination, and the related, but conceptually distinct, areas of health care rationing and the ethics of withholding and withdrawing medical treatment.

82. See text at note 43 *supra*.

83. A "prisoner" is anyone who is incarcerated involuntarily in a penal institution, regardless of the nature of the authorizing legislation, or the status of the proceeding. See 45 C.F.R. § 46.303 (1994).

84. 45 C.F.R. §46.302 (1994).

85. See generally Goldner, *supra* note 3, at pp. 102–112 and nn. 242–310 (collecting sources).

86. 45 C.F.R. §46.116 (1994); *Va. Code Ann.* § 32.1-162.18.

87. Beauchamp & Childress are of the view that '[m]ost obligations of positive beneficence in health care rest on fidelity-generating contracts and role relations." Beauchamp and Childress, *supra* note 1 at p. 341. They reject attempts "to capture relationships between health-care professionals and patients in any single metaphor or mold such as contractors, partners, parents, friends, or technicians" because "[n]o single metaphor or model adequately expresses the complexity of health care or the moral principles and rules that should govern such relationships." Id. at p. 341–42.

88. See notes 57–58, *supra*. Though it has been argued by some philosophers that the autonomy principle must be taken to mean that "consent cures" in many situations which might otherwise be defined by others as unethical or immoral, see, e.g., H. Tristam Engelhardt, Death By Free Choice: Variations on an Antique Theme, in Baruch Brody, ed., *Suicide and Euthanasia*, Dordrecht: Kluwer Academic Publishers, 1989, 251, 264–65, and some professions will permit conflicts of interest to continue if the client has consented after "full disclosure" of the limitations and risks caused by the conflict, compare, e.g. American Bar Assn., *Model Rules of Professional Responsibility*, Rules 1.7 to 1.9 (1994), with, e.g., American Association for Counseling and Development, *Ethical Standards*, Section B ¶¶ 12–14 in Gorelin, *Codes of Professional Responsibility*, *supra* note 10, at p. 228 (permitting certain conflicts only where other alternatives are unavailable, and those which might impair objectivity or professional judgment "must be avoided and/or the counseling relationship terminated through referral to another competent professional."), "prospective" or "prophylactic" oversight can also be

justified on informed consent grounds. In order to secure an "informed consent" to such a waiver, the professional would necessarily need to disclose not only his or her professional qualifications, but also a relatively detailed summary of the kinds of negligence which might reasonably be expected to occur. Since, in the normal professional setting, a *reasonable* expectation is that *no* negligence will occur, it follows that a competent professional will rarely, if ever, discuss the possibility during the solicitation or informed consent process.

89. See Jay Katz, *supra* note 57, at 30, quoting Edmund D. Pellegrino, Beneficence, Scientific Autonomy and Self-Interest: Ethical Dilemmas in Clinical Research, *Geo. Med.* 21 (1991).
90. Michael Ignatieff, *The Needs of Strangers An Essay on Privacy, Solidarity, and the Politics of Being Human,* New York: Viking Press, 1984; London: Chatto & Windus, The Hogarth Press, 1984, at 53.

Institutional Review Boards and Research on Individuals with Mental Disorders

Sue Keir Hoppe

Department of Psychiatry, The University of Texas Health Science Center at San Antonio, 7703 Floyd Curl Drive, San Antonio, Texas 78284-7792

Exclusion of "those institutionalized as mentally infirm" from special regulatory protection by Institutional Review Boards (IRBs) has been a widely-discussed issue since the recommendations of the National Commission for the Protection of Human Subjects of Biomedical and Behavioral Research were published in 1978. Just when IRBs had begun to "adjust" to the exclusion by developing their own policies and procedures pertaining to research on individuals with mental disorders, a series of reports of unethical research on such individuals appeared in the media.

One of the most sensational reports involved a longitudinal study of "developmental processes" in schizophrenia at The University of California at Los Angeles (UCLA), funded by the National Institute of Mental Health (NIMH), beginning in 1983. Reports of the study challenged some of the assumptions that many IRBs had been making about the adequacy of their review of studies of individuals with mental disorders. In most major universities and research centers, the media reports generated considerable interest and set off a process of self-examination. Public outcry and IRB initiatives in response to reports of the schizophrenia study confirmed that our current ethical approach to human subject research was "born in scandal and reared in protectionism" and will probably continue to evolve in this manner (C. Levine, 1991:79).

The schizophrenia study was conducted in two phases. In the first phase, individuals with a diagnosis of schizophrenia received biweekly injections of the antipsychotic medication Prolixin and were monitored for psychosocial stressors and physiological states to identify predictors of relapse. In the second phase, subjects were randomly assigned to 12 weeks on Prolixin, then 12 weeks on placebo, or vice versa. The number of subjects who relapsed in the second phase of the study is not consistent across available accounts, but it appears that almost one-half experienced a "return to a state of active and severe psychotic symptoms," including "hallucinations, unusual thought content, bizarreness, self-neglect, hostility, depressive mood and suicidality." As apparently stated in the NIMH proposal, the

This article will also appear in the journal: *Accountability in Research*, Volume 4, pp. 187–195 (1996).

risk of relapse was "expected to occur in some ... subjects, probably in most" (Horowitz, 1994).

Following the suicide of a former subject, questions were raised by his family about a program that "emphasized research over treatment of mental illness." Another family with a relative who had participated in the study complained about an inability to reach investigators when the relative experienced hallucinations and threatened homicide. The Office for Protection from Research Risks (OPRR) at the National Institutes of Health (NIH) investigated the allegations and ultimately required changes in the consent forms to include more apt descriptions of risks in phase two of the study. Equally important, but not the object of OPRR censure, are aspects of the study design that produced harm in many subjects who took part.

The schizophrenia study has been a catalyst for evaluating progress and pitfalls in the ethical protections IRBs afford research subjects with mental disorders. This paper provides an overview of some of the issues to be considered in the evaluation. Although there is a large body of literature on the evolution of IRBs, regulatory developments are briefly reviewed first to establish the social context of the discussion that follows. Then, differing viewpoints on the need for additional regulation of research on individuals with mental disorders are presented. Finally, themes and trends in IRB review of research on individuals with mental disorders are explored.

EVOLUTION OF THE IRB

In 1953, the NIH opened a Clinical Center for human subject research by staff scientists. The first Federal document requiring committee review of research ("Group Consideration for Clinical Research Procedures Deviating from Accepted Medical Practice or Involving Unusual Hazard") was issued, its jurisdiction limited to the Clinical Center. In 1966, the Surgeon General of the Public Health Service (PHS) ordered that no extramural research grant (i.e., a grant to an institution outside the PHS or NIH) involving human subjects be awarded unless a committee of the grantee institution had reviewed the ethics of the proposed work. The committee was to assess the adequacy of measures to protect the rights and welfare of potential research subjects, including procedures for obtaining informed consent and the potential risks and benefits of the research.

Subsequent revisions of the Surgeon General's policy statement allowed grantee institutions to give "assurance" of prior committee review on an institution-wide basis, rather than for individual projects, and required committees to consider local laws and norms in their deliberations. The Institutional Relations Branch of the NIH Division of Research Grants was established to carry out the Surgeon General's order. The Branch later was upgraded and the name was changed to, and remains, the Office for Protection from Research Risks (OPRR).

In 1971, the Food and Drug Administration (FDA) promulgated regulations requiring peer review of all new and unapproved drugs and devices used in institutions such as universities, hospitals, and research centers. In 1974, the Department of Health, Education, and Welfare (DHEW) issued the first Federal regulations on IRBs, committees to be established in institutions to carry out the peer review required by the NIH in the 1966 Surgeon General's order. In the same year, Congress passed the National Research Act, which mandated IRBs and called for the creation of the National Commission for the Protection of Human Subjects of Biomedical and Behavioral Research.

The charge to the National Commission was to identify problems in carrying out government obligations for the ethical conduct of research utilizing human research subjects and to suggest means for solving the problems. The National Commission deliberated for four years and in 1978 issued the Belmont Report (National Commission for the Protection of Human Subjects of Biomedical and Behavioral Research, 1978), named for a conference center near Baltimore at which the Commission met. The Belmont Report is a succinct summary of the basic principles that should be used to evaluate the ethicality of human subject research: 1) respect for persons, which includes requirements to treat individuals as autonomous and to protect those who are not autonomous; 2) beneficience, which involves maximizing benefit and minimizing harm; and 3) justice, which implies fairness in the distribution of both benefits and harms of research among individuals and/or groups.

In 1981, major revisions of FDA and DHEW (which had by then become the Department of Health and Human Services, DHHS) regulations on the protection of human research subjects were made. The revisions built upon the ethical principles outlined in the Belmont Report and set out fairly specific procedures to be followed by IRBs. For example, the regulations specified the minimum size and composition of the membership of an IRB; categories of research, primarily based on potential risks to subjects, for which there are different levels of review; and basic elements of informed consent and conditions under which consent is to be obtained. The regulations also defined special, "vulnerable" populations such as children and prisoners and stipulated additional precautions pertaining to their participation in research. As noted earlier, no special protections were afforded those institutionalized as mentally infirm. The regulations have been since revised twice—once in 1983 to incorporate changes after experience had been gained in implementing them, and then in 1991 to expand their authority beyond the NIH to other governmental entities as well.

Despite the relative specificity of the current regulations governing human subject research, there is variability among IRBs in the ways in which they carry out their mandate. To some extent, the variability is by design: IRBs were created as agents or committees of their own institutions rather than as arms of the Federal government. In the process of creating IRBs, it was anticipated that they would need "sufficient flexibility to bring about adequate resolutions to particular problems as they present[ed] themselves in peculiarly local contexts" (R.J. Levine, 1986:342). In the context of this decentralized system, most IRBs act "thoughtfully and responsibly" and are usually seen as "bulwarks" of human subject protection.

However, when there are allegations of unethical research, negative views of IRBs attract public attention, even if the views represent minority opinion. Recently, IRBs have been portrayed as "rubber stamps" and the IRB system has been described as "patchwork policing" (Anderson and Binstein, 1994). Rubber-stamp IRBs are seen as approving proposed research with little or no dissent, discussion, or judgment. Bulwark IRBs and rubber-stamp IRBs are at opposite ends of a continuum signifying a hodgepodge of approaches to human subject protection. These characterizations of IRBs suggest that variability in their review is problematic and they imply at least two methods of achieving greater standardization. One method is increased system-wide guidance through special Federal regulations for the protection of subjects with mental disorders. Another method involves strengthening the commitment of institutions to the importance of the IRB and to high standards in the review of proposed human subject research.

FEDERAL REGULATION AND INSTITUTIONAL RESPONSIBILITY

DHHS justified the exclusion of those institutionalized as mentally infirm from special regulatory protection for two reasons: first, DHHS asserted that the basic regulations provided adequate protection for such individuals; and, second, that there was a "lack of consensus" on special regulations as evidenced in the report of the National Commission and in the proposed regulations drafted by DHHS (National Commission for the Protection of Human Subjects of Biomedical and Behavioral Research, 1978; R.J. Levine, 1986:269–270). Proponents of increased Federal oversight of research on individuals with mental disorders argue that the basic regulations do not adequately protect such individuals and that additional regulations should be added to existing ones. Failure of DHHS to mandate special protections for those institutionalized as mentally infirm was the result of "pressure from a few important research psychiatrists, in the absence of an effective advocacy group, at the time, on behalf of the mentally ill" (Shamoo, 1994). Psychiatrists feared that excessive bureaucracy would impede their research efforts, and consumer advocates feared that it would "decrease funding for much needed research, frighten families into discouraging their ill relatives from participating in research, and dissuade psychiatrists from conducting neurobiological research" (Hassner, 1994). It can be argued that a lack of consensus on regulations to protect those institutionalized as mentally infirm reflected the extremely complicated ethical issues surrounding such research—a potentially compelling reason to provide direction through rather than abandon regulation.

Several special issues or needs in the protection of research subjects with mental disorders have been identified by proponents of increased regulation (Shamoo and Irving, 1993). These include, for example, whether and/or under what circumstances research that involves more than minimal risk and no potential direct benefit to a subject and research that is unrelated to a subject's mental disorder should be permitted. The extent to which professionals involved in the treatment of potential subjects with mental disorders should be allowed to obtain consent has also been questioned. Proponents of increased regulation have raised issues about whether assent of subjects who are incapable of consenting to research should be required, as well as the use of consent auditors in research involving more than minimal risk. Representation on IRBs of individuals who are "knowledgeable about and experienced in" working with individuals with mental disorders is also a concern.

As a result of technological developments, increased attempts to understand the causes and course of major mental disorders, and heightened consumer advocacy, new ethical issues have arisen since the regulations were implemented. Proponents of increased regulation have suggested that a new national commission of broadly-based, committed, knowledgeable individuals is needed to reevaluate the need for special regulations for research subjects with mental disorders.

Several other regulatory mechanisms have been suggested. One is the establishment of a national research review or ethical advisory board (Levine and Caplan, 1986; Caplan, 1994) with several possible functions. A national board could review studies that involve more than minimal risk and no probability of direct benefit to subjects. It could also consider projects with broad national implications and the long-term social consequences of research, a task which local IRBs "cannot and perhaps should not undertake." Random audits or site visits by Federal authorities of studies involving individuals with mental disorders have also been proposed (Caplan, 1994).

Those who question the need for additional Federal regulation of research on individuals with mental disorders justify their position in three ways. First, there is concern that increased Federal regulation may promote conformity to paperwork at the expense of the "spirit" of human subject protection and encourage "a high level of patient distrust and investigator defensiveness," fostering "timid researchers concerned with self-protection" (Wirshing, 1994). The tendency for some investigators and some IRBs to focus more on legal than ethical issues in the development and review of research gives substance to this concern. Second, the "excessive bureaucracy" that additional Federal regulation might involve is seen as reflecting a preoccupation with "back ward" populations which are not representative of most individuals with mental disorders. Thus, additional regulation, albeit undertaken to increase research protection, might have unintended social consequences, further stigmatizing the mentally ill (Reatig, 1981; Stanley and Stanley, 1982).

Finally, the belief in institutional responsibility for the conduct of research is strong and rooted in the importance of integrating scientific investigation as part of the broader culture (Parsons, 1951:335–348). Formal institutional responsibility for ethical review of research existed as early as 1966 when the Surgeon General of the PHS ordered local committee review before a study could be funded. This precedent has been carried through the subsequent evolution of research regulation. The model was adopted because, until very recently, scientists predominantly were affiliated with and closely tied to universities. A variety of social changes, the availability of industry- and government-supported research funding primary among them, have weakened the ties of scientists to universities and diminished the role of universities in the pursuit of scientific knowledge. The extent to which it is possible to maintain IRBs as agents of institutions depends on the continued willingness of universities to assume responsibility for human subject protection and their ability to effectively carry out the responsibility.

IRB REVIEW OF RESEARCH ON INDIVIDUALS WITH MENTAL DISORDERS

IRBs employ two basic criteria in the review of research. One criterion involves "norms protecting the rights of subjects"; the other, "norms protecting their welfare" (Williams, 1984). The first criterion involves safeguarding certain values, such as autonomy, irrespective of the probable benefit or harm to a subject associated with participation in research. The second criterion involves minimization of risks to subjects and appraisal that risks are "reasonable" in relation to anticipated benefits. There is consensus that IRBs attend to and are more effective in applying the first criterion than the second.

The imbalance in the application of the two criteria can be attributed in part to the fact that there are fairly specific regulations and a large and sophisticated literature on informed consent, while risk-benefit assessment is "notoriously ambiguous" (Williams, 1984). The Belmont Report stated that risk-benefit assessment must be "balanced" and shown to be "in a favorable ratio." The "metaphorical character" of these terms exemplifies the difficulties IRBs confront in making precise or reliable judgments about risks and benefits (Meslin, 1990). Further, the regulations only define minimal risk, leaving other fundamental terms such as harm and benefit open to wider interpretation.

In addition to these concrete issues, broader social norms, personal attitudes, and institutional forces account for the relative rarity of "serious" risk-benefit assessments by IRBs (Williams, 1984). Social norms dictate an emphasis on subjects' rights over their welfare. IRB disapproval of a study on the basis on apparent risks could be seen as paternalistic, robbing subjects of the freedom to exercise their own choice in deciding to participate in a study. In terms of personal attitudes, some individuals assign greater value than others to science *per se* and IRB members are likely to be overly-represented in this regard. Taken together, these pro-individual, pro-research biases decrease the likelihood that subject welfare will be used by IRBs to evaluate proposed studies (as well as the probability that an IRB will disapprove *any* proposed study).

This is ironic in that the issue of whether IRBs should ever consider the scientific design of studies was a matter of intense debate after the regulations were first implemented. Currently, most IRBs acknowledge that "science" is within the scope of their review, including whether the knowledge that is being sought is worth acquiring and whether the way in which the study is designed will yield that knowledge without unreasonable risks to subjects. However, risk-benefit assessment necessarily involves judging the competence of the investigator, a conflictual decision for IRB members when the investigator is a colleague. If risks are deemed by the IRB to be excessive in relation to anticipated benefits, this often requires redesigning an entire study, whereas consent forms can be fairly easily rewritten. Thus, the extent to which scientific review, in the context of risk assessment, can be accomplished by IRBs depends on the willingness of its scientific members to transcend "entrenched professional biases" and "restrained criticism of colleagues" (Frankel, 1989).

It has been asserted that "Most current psychiatric research projects do not place research subjects at significant risk of harm" (Winslade and Douard, 1992). The UCLA schizophrenia study counters this assertion, as do other studies of individuals with mental disorders that involve use of wash-out periods or placebos. Risks to subjects in the schizophrenia study were clearly substantial. On the other hand, the benefits to future schizophrenic patients would be great if it were discovered that medication is not always essential for the control of positive symptoms and that periodic discontinuation of medication can prevent serious, sometimes irreversible side effects. In light of the fact that schizophrenia has been described as the "worst disease affecting mankind, even AIDS not excepted" (Carpenter and Buchanan, 1994), the goals of the investigators seem worthwhile.

However, the IRB reviewing the schizophrenia study also had to consider the question "should any experimenter be permitted to allow patients to get worse in the hope of learning more about this illness?" (Hilts, 1994). Some ethicists say "no." One likened the schizophrenia study to taking "kidney transplant patients off their immunosuppression drugs so you could watch them reject their new kidneys." But, as the same ethicist noted, it might be ethical to stop medications in some circumstances, if investigators were "extremely careful not to let people become severely ill" (Hilts, 1994). Similarly, another ethicist believes that "You cannot use people—or you should not use people—as means for other's ends and for ends that might ultimately even be good" (Horowitz, 1994). On the other hand, some ethicists venture that research such as the schizophrenia study is proper "as long as patients [are] informed of the risks they [are] being asked to take" (Hilts, 1994). In Williams' (1984) terminology, the first opinions give priority to subject welfare; the latter, to subject rights.

Over time, the substantive emphases of IRBs in the areas of welfare and rights have shifted (C. Levine, 1991:80–81). In the early to mid-1980s, IRBs emphasized the protection of subjects from harm (nonmaleficence) and informed consent as a mechanism of ensuring voluntariness of participation in research. Today, IRBs tend to emphasize the promotion of subject welfare (beneficence) and personal control or empowerment aspects of autonomy (i.e., freedom from coercion and the right to full information about risks and benefits) in the decision to consent to research participation. The shifts reflect cumulative IRB experience, as well as broader social changes that have taken place.

Many questions about details of the IRB review of the schizophrenia study, not mentioned in the secondary sources that were consulted in the preparation of this paper, would be based on these paradigm shifts. In the initial review of the study, how much IRB discussion centered on subjects' rights and how much of subjects' welfare? What was the IRB's assessment of the level of risk? Was someone knowledgeable about and experienced in working with individuals with mental disorders present at the meeting, or was the consultation of such an individual sought? Were consumers or consumer advocates involved in the design of the study? In addition to the issue of disclosure of risks in the consent process (lack of full disclosure was faulted by OPRR), was attention given to the potential conflict between investigators' roles as researchers and treatment providers (Appelbaum *et al.*, 1987)? What mechanisms were used to ensure that subjects had the capacity to understand and make a choice about participating in the study at the time of enrollment and at intervals throughout its conduct? Were provisions made for changes in capacity to consent over time? Were consent auditors appointed? Equally if not more important than these questions which are typically raised at initial review, what issues were considered by the IRB at each annual (continuing) review?

CONCLUSION

The UCLA schizophrenia study well illustrates the dilemmas that IRBs face in reviewing studies of individuals with mental disorders that involve more than minimal risk and little anticipated direct benefit. As noted, there is a general tendency for IRBs to attend more to subjects' rights than their welfare because the former involves more highly codified principles to guide decision making and is congruent with powerful social norms that emphasize individual autonomy. However, in the review of research on individuals with mental disorders, whose capacity to exercise their rights may be permanently or temporarily diminished, IRBs need to concentrate as much or more on issues of welfare. Clearly, additional Federal regulations governing research on individuals with mental disorders would be of assistance in this regard. Such regulations do not exclude institutional responsibility for human subject research and should not be viewed as a threat to local IRB review, but as an important supplement to it.

REFERENCES

Anderson, J., and Binstein, M. (1994) Would-be assassin turns witness. *Washington Post*, May 23.

Appelbaum, P.S., Roth, L.H., Lidz, C.W., Benson, P., and Winslade, W. (1987) False hopes and best data: consent to research and the therapeutic misconception. *Hastings Center Report* 17:20–24.

Caplan, A.L. (1994) Are existing safeguards adequate? *Journal of the California Alliance for the Mentally Ill* 5:36–38.

Carpenter, W.T., and Buchanan, R.W. (1994) Schizophrenia. *New England Journal of Medicine* 330: 681–690.

Frankel, M.S. (1989) Professional codes: why, how, and with what impact? *Journal of Business Ethics* 8:109–115.

Hassner, V. (1994) What is ethical? What is not? Where do you draw the line? *Journal of the California Alliance for the Mentally Ill* 5:4–5.

Hilts, P.J. (1994) Agency faults a U.C.L.A. study for suffering of mental patients. *New York Times*, March 10.

Horowitz, J. (1994) For the sake of science. *Los Angeles Times Magazine*, September 11.

Levine, C. (1991) AIDS and the ethics of human subject research. In F.G. Reamer (ed.), *AIDS & Ethics*. New York: Columbia University Press.

Levine, C., and Caplan, A.L. (1986) Beyond localism: a proposal for a national research review board. *IRB: A Review of Human Subjects Research* 8:7–9.

Levine, R.J. (1986) *Ethics and Regulation of Clinical Research*. Baltimore, MD: Urban & Schwarzenberg.

Meslin, E.M. (1990) Protecting human subjects from harm through improved risk judgments. *IRB: A Review of Human Subjects Research* 12:7–10.

National Commission for the Protection of Human Subjects of Biomedical and Behavioral Research (1978) *The Belmont Report: Ethical Principles and Guidelines for the Protection of Human Subjects of Research*. Washington, D.C.: U.S. Government Printing Office, DHEW Publication No. (OS) 78-0012.

Parsons, T. (1951) *The Social System*. Glencoe, IL: Free Press.

Reatig, N. (1981) Government regulations affecting psychopharmacology research in the United States: implications for the future. In G.D. Burrows and J.S. Werry (eds.). *Human Psychopharmacology: Research and General Practice, Volume II*. Greenwich, CT: JAI Press.

Shamoo, A.E. (1994) Our responsibilities toward persons with mental illness as human subjects of research. *Journal of the California Alliance for the Mentally Ill* 5:14–16.

Shamoo, A.E. and Irving, D.N. (1993) Accountability in research using persons with mental illness. *Accountability in Research* 3:1–13.

Stanley, B.H., and Stanley, M. (1982) Testing competency in psychiatric patients. *IRB: A Review of Human Subjects Research* 4:1–6.

Williams, P.C. (1984) Why IRBs falter in reviewing risks and benefits. *IRB: A Review of Human Subjects Research* 6:1–4.

Winslade, W.J. and Douard, J.W. (1992) Ethical issues in psychiatric research. In L.K. George Hsu and M. Hersen (eds.), *Research in Psychiatry: Issues, Strategies, and Methods.* New York: Plenum Medical Book Company.

An Urgent Academic Challenge—Treatment of Refractory Patients

Rami Kaminski, M.D.

Mount Sinai School of Medicine, Department of Psychiatry, 1 Gustave Levy Place, New York, NY 10128

We are now midway through the decade of the brain. Amazing fields are opening in molecular biology, brain imaging, neuropsychology, neurophysiology; the list is long. Justifiably, those of us who work with the severely mentally ill, are thrilled by the promise of those new frontiers. It is equally justified, however, to be disappointed by the fact that prognosis and rate of treatment refractoriness have improved very little during the neuroleptic era. As the century is ending, it is time for pause and introspection. Are we really doing everything possible for the so-called treatment nonresponders? Have we, in the rush for new and better treatments, neglected the scores of patients currently wasting away in long-term facilities? Have we offered them more than a fleeting and perennial promise that the breakthrough is around the corner?

I would like to make the following argument; The quest for *optimizing* current available treatments is as important as the quest for newer ones. The academic community has to lead the drive for optimizing treatments by way of research and education. The breakthrough may be around the corner, but it could be too late for the hundreds of thousands who suffer now.

OPTIMIZING CURRENT AVAILABLE TREATMENTS

First, why is this topic not widely researched and taught? Our system of academic rewards and promotions is based on grants and publications. In most cases, those who do not respond to standard treatments, need individually tailored ones to succeed. The number of variables that can account for refractoriness to treatment is dazzling. The variety of confounding factors is a great obstacle in reaching statistical conclusions. Those who work in tertiary centers for nonresponders, can write a book of single case reports that are not readily reproducible in other patients. Moreover, often the solution for medication resistance is very simple and almost unworthy of mention in our sophisticated academic literature. As a result few academicians can "afford" to engage in such an endeavor. A good example of what could be done, is a classic study by Dr. Arthur Rifkin.[1] In a city hospital he

111

conducted a definitive study on required doses of haldol. Thanks to him one can refer now to a solid body of data in justifying the administration of low doses of neuroleptic. One can only wonder how many patients were saved from overshooting the therapeutic window hence become "treatment refractory." Studies of this nature are all too few.

THE CHALLENGE

In many nonacademic settings, psychiatrists follow very rigid dosing schedules. Since most of the published works seeking improvement of treatment concentrate on augmentation strategies, refractory patients often receive "cocktails" of neuroleptics, anti-seizure medication, lithium, alprazolam, and antidepressants. There is little interest, time, or knowledge in fine tuning the treatments individually. Here, I believe, lies the problem.

Luckily, as in other medical fields, most patients respond to universal guidelines of treatment. It is the refractory patient, who has some unique attributes which render him a nonresponder, who is the exception. Those attributes could involve a spectrum of causes, ranging from individual metabolic differences to co-morbid factors such as another brain disorder or a neuroendocrine one. Therefore, it behooves the attending psychiatrist to pay individual attention to the refractory patient, to engage in skillful differential diagnosis and embark on a careful trial-and-error drug administration protocols. This is rarely the case currently. In fact, most treatment refractory patients are consigned to long-term hospitals where the chance of getting such a work up is small.

Change is quite possible. In working as a psychopharmacology consultant in three large state hospitals I have discovered that all necessary ingredients are present. In my experience, after an initial period of understandable reticence, psychiatrists in long-term facilities engage enthusiastically in such endeavors. The intellectual and humane challenges can be invigorating to this segment of our colleagues who feel helpless and believe they are nonparticipants in research. They cannot, however, engage in conceptual and theoretical research, simply because the chronic care system is not even remotely geared toward that objective. This is the duty of the academic community.

The task is admittedly daunting. Since I speak from my personal experience I would like to make a disclaimer: Nothing I am about to say, is news to most academic psychiatrists. This may be the problem. It appears so obvious that the temptation to study and publish it is small. Yet, our rush for the future, noble and intriguing as it is, does not obviate the responsibility to the patients of the present. I will try to summarize the most poignant problems although as I said they may be as many as there are refractory patients.

Pharmacokinetics and Other Pharmacological Issues

A consultant to long-term facilities quickly becomes aware of the inattention to basic pharmacokinetic considerations. Often even blatant, commonplace drug interactions are not addressed. Issues of drug metabolism, half-life, peak of activity, and steady-state, are ignored. Endocrine aspects, basal metabolic state, and other patient-specific factors that can contribute to refractoriness are similarly unaddressed. To do justice to our colleagues one has to acknowledge the large caseload that psychiatrists carry in a custodial rather than a research-oriented environment. The introduction of clozapine helped to sharpen the problem. Despite being considered the primary candidate for the treatment of refractory patients, it has not really been adopted in long-term hospitals. Patients who did receive clozapine were arbitrarily given 600 mg/d. Those who displayed any kind of clozapine side effect were discontinued on the drug with undue haste. In close to a hundred settings I visited, no one could give a good answer to the question: Why not try changing the dose?

One should not assume that the psychiatrists prescribing treatment are not well-versed in the literature. Most can recite the dopamine/serotonin theories, the efficacy of clozapine *vis à vis* negative symptoms, etc. It is simply not in the culture of many prescribers to experiment with the particular properties of a drug and make it appropriate to the patient. When a patient is deemed a nonresponder, medications are switched around, usually in arbitrary fashion. Sooner or later, some augmentation drug is added. Often the patient is maintained on the last medication(s) administered, when his doctors gave up and stopped further modification. It is not uncommon to see a patient on a large dose of neuroleptics, with lithium and carbamazepine, for ten-year periods or longer! All meanwhile the patient may be constantly hallucinating *and* suffering from a myriad of side effects.

There is an urgent need to develop protocols and the collection of data for refractory patients that will concentrate not on *what* to administer but on *how*. We can afford some pharmacological inattentiveness with those who readily respond to treatment. For those who do not, one has to wonder whether they really suffer from iatrogenic refractoriness brought upon by a standard rather than creative approach. Long-term facilities provide the treating psychiatrist with a rare commodity: Time. Often the same psychiatrist takes care of the same patients for years. That reality enables him, even in a managed care era, to investigate without pressure. Should the academic world provide the frontline psychiatrists with both knowledge and education, I submit that many nonresponders will prove to be nonrefractory.

Treat the Symptoms and Not the Diagnosis

A prevalent problem requiring attention and education, is the blurring of distinction between psychosis and schizophrenia. The fact that psychosis is merely a nonspecific symptom(s) that can be present in at least 30 other conditions unrelated to schizophrenia, is often disregarded. As a result, two separate problems ensue: One is the paucity and infrequency of medical work up done to exclude any other etiology in a young psychotic patient. It is probably unprecedented in medicine that a severe, incapacitating condition gets such poor diagnostic (especially differential diagnosis) attention. A young person presenting to the psychiatric emergency room displaying delusions and hallucinations, often gets the diagnosis of schizophrenia at the ER level! The more considerate diagnostician would call it schizophreniform. The likelihood of other underlying causes for psychosis is seldom investigated. In many medical conditions when a patient presents with a set of symptoms that could be the result of many etiologies, the diagnosis with no objective measurable indices would be a diagnosis of exclusion. In young psychotic patients we make this descriptive diagnosis before investigating many other diagnosable and *treatable* conditions.

Admittedly, schizophrenia is probably the most prevalent cause for adolescent psychosis, street drugs excluded. Since we do not have definitive or preventive treatment for schizophrenia, by default, this may be the right practice. However, this philosophy cannot range over to treatment nonresponders. What good is our endeavor to identify more sophisticated high-tech tools when most of our patients do not even get EEGs? Most of the research into the diagnosis of schizophrenia is similarly targeted to the illness itself with little regard to the other causes of psychosis. As long as psychiatrists arel not being taught the importance of a differential diagnosis of psychosis, we will be left with the gnawing suspicion that some of our refractory patients may suffer from a treatable condition other than schizophrenia.

A related problem is, the confusion between antipsychotic and "antischizophrenic" treatment, and the undeserved powers attributed to neuroleptics. It is as if the fact that neuroleptics are merely nonspecific antipsychotic agents is forgotten.

This phenomenon is partially due to the confusion created by the academic community. The leap of faith from neuroleptics and their Parkinsonian side effects to explaining schizophrenia by dopamine disturbance has done a lot to confuse the lines between schizophrenia, psychosis, and the mechanism of action of neuroleptics. This may lead prescribers to think that neuroleptics are targeted at the pathophysiology or even etiology of schizophrenia. In my experience many patients are deemed nonresponder because their *schizophrenia* "fails" to respond to neuroleptics. It is akin to a condition in which a psychotic patient with SLE-cerebritis will be considered a nonresponder to haldol because he still has renal impairment. Moreover, the notion of *what exactly* is considered response is in itself not clear. Every setting has a different goal of treatment. A patient that responds to neuroleptics by relief of psychosis but is still not interactive with others, may be viewed as a nonresponder in a day treatment/rehabilitation setting. Conversely, in a long term hospital a withdrawn, quietly psychotic patient may be seen as a responder, especially should this subdued behavior be the consequence of Parkinsonian side effect and oversedation—which is believed to correlate the administration of neuroleptics with "improvement."

Another common problem, is the questionable practice by which neuroleptics are administered to target nonpsychotic symptoms displayed by schizophrenic patients. It is not rare to see anxiety, OC symptoms, and even depression treated by neuroleptics. The wrong, yet common, notion that our drugs are etiology-targeted, rather than merely symptomatic, can result in diagnosis-oriented treatments. That can and often *does* result in undue restriction of the potentially helpful medical arsenal.

SUGGESTIONS

The aforementioned practices are based, in my opinion, on misguided treatment principles and are too common to be ignored. I suspect that a second look at many refractory patients would reveal that more can be done for them and a large number may even respond to an individually tailored regimen.

The problem is not insurmountable and I would like to make some proposals based on my own experience. I have no pretense that my suggestions are definitive or exclusive. I hope they serve as a contribution to an ongoing discussion about this urgent problem.

The most important suggestion, is the need to change the culture regarding the treatment of nonresponders. An educational drive delineating nonresponders as a separate group of patients is of the essence. That is not to separate them by their misfortune and simply incarcerate them until better treatments become available. Rather, it should be made clear that the difference between nonresponders and responders is that the former group has some special attributes that need to be addressed to improve their symptoms. Granted, some nonresponders may have a different condition that is not amenable to current available treatments regardless of what we do. However, we must be sure we do *everything* before we label them. As is the current practice, patients may be perceived as not amenable to treatment, merely because they were given treatment protocols aimed at responders. True refractory patients are those who do not respond to *specialized* treatments aimed at *nonresponders*. Should we apply them to the current refractory population, we can probably reduce their numbers.

(This incidently, could contribute to clinical research. In our everlasting quest to define "pure" patient population, we are constantly left with the uncomfortable feeling that some of those who partake in refractory studies are not really refractory but iatrogenically so. Once we can establish new guidelines for treatment of refractory patients, we will probably be left with much more clearly definable patient population.)

In other words, we should seek to divide the refractory patients into two groups: True nonresponders and iatrogenic ones. This conceptual shift will help to eradicate the unnecessary hopelessness one feels when a patient does not respond readily to our standard treatments. To do so, some concepts need to be sharpened. A few examples:

1. *Treatment Goals.* We have to come up with clear, universal and practical goals for our treatments. Many patients are currently subjected to erratic assessment of what qualifies as response. For some patients merely "failing" rehabilitative efforts, or getting readmitted often, is an incentive for the staff to start thinking about long term admission.

In collaboration with psychiatric rehabilitation experts we can devise an array of functional tests that would be universally agreed upon. That, and understanding of what outpatient settings offer, will prevent our patients from being assigned to unattainable goals, "failure," and being branded nonresponders. It may sound simple, but patients can become nonresponders merely because we do not match our response criteria to their abilities. The stakes are high for the patients, since treatment failure carries with it the risk of long-term admission where at least for now, the possibility of a" second chance" is slim.

2. Special Pharmacological Considerations. A great deal of research and education is needed to alert the prescribing community to individual differences in drug response. We can learn a lesson from primary care physicians. When the SSRIs came to the market, drug companies launched an extensive campaign to uproot two misguided notions held by this group of colleagues. One was that if a depressive disorder is secondary or reactive or exogenous there is no need to use antidepressants. The other one was that nonpsychiatrists tended to administer antidepressants in a half-hearted manner, thereby not achieving therapeutic doses and response. Although drug companies did this for their own commercial reasons, this campaign has been successful in changing the culture of treatment of depression among primary care physicians. Similar work can be done with those who treat refractory patients.

Seven years ago my department at Mount Sinai "adopted" a large state hospital. Every week, several consultants conduct case conferences there, with all of the treatment team present. Coordinators at the long-term hospital facilitate and assure follow up of the recommendations made by consultants. Having been a part in this interesting pilot model from its inception, I can testify to the enthusiasm and, more importantly, the change of treatment culture among the treatment teams. Many patients, long considered hopeless, got another chance. Their cases were reviewed anew. Consequently, some got discharged and most had a marked improvement in their side effect and quality of life. But the most important change was the renewed spirit of hope among staff, patients, and families. The sense that something can be done for almost every patient has lifted the morale and made the treating physicians much more tenacious in the effort to improve the patients' conditions. A system of consultants from academic hospitals to chronic care ones, could be mutually beneficial and contribute to the abolition of the sense that nonresponders are hopeless cases.

3. Psychosis as Independent from Schizophrenia. The clinical reality, until proven otherwise, supports the notion of psychosis as a common pathophysiological pathway, rather than diagnostically specific. There exists a need for more studies of psychosis across the diagnostic boundaries. Hopefully, this could better illuminate the mechanism of action of neuroleptics (which in reality function as a symptomatic, nonspecific treatment for psychosis). One may extrapolate from this notion to other areas of schizophrenia: A rising direction in neuroscience, views clusters of symptoms as reflection of central functional mechanisms rather than diagnosis-specific. In that vein, we should not ignore the appearance of schizophrenia-"specific" symptoms in seemingly unrelated neurodegenerative disorders. A case in point is the indistinguishable characteristics of bradyphrenia in Parkinson's disease from negative symptoms of schizophrenia. Beyond being a clinical reality, it may help to put schizophrenia back into the realm of severe (neurodegenerative?) brain disorders rather than an entity all by its own. Conceptually, it will promote the notion that young persons presenting with a myriad of CNS symptoms, deserve the same diagnostic rigor commonly lent to any other severe brain disorder. Perhaps than, the uncomfortable notion shared by many about the heterogeneity of schizophrenia would get further credence, or equally important, be disputed.

CONCLUSION

It is a sad reality that the field of schizophrenia is subjected to unusual dichotomy: On one hand there exists a supreme, high-tech, sophisticated research into the underpinning of the disease. On the other hand, regrettably, most patients are still diagnosed and treated in the same impressionistic, inaccurate ways, prevalent years ago, and not befitting the decade of the brain. It is a duty of those who guide to present to the community the limitations of our knowledge and understanding. Otherwise, when dealing with the treatment refractory patients, many psychiatrists can conclude that their treatment is beyond the reach of our current knowledge. That leads to the cessation of further attempts to explore different, additional strategies. By emphasizing the limitations and uncertainties involving psychotropics, one could help psychiatrists to be more creative and intellectually daring in their regimen. Giving dictums about dosage and the purpose of psychotropics can lead to rigid adherence to protocols rather than individual tailoring of medication. It may suffice for those who are prone to respond (perhaps because their therapeutic window is wider) but be detrimental to nonresponders.

I would like to use this forum of Ethics in Neurobiological Research to call for the formation of an association of clinicians and academicians dedicated to research and education in practical guidelines for the management of treatment refractory patients. Those who are now in long-term hospitals cannot wait until we come up with better treatments. Their time is now.

REFERENCE

1. *Archives of General Psychiatry* 1991, 48(2):166–70

Washouts/Relapses in Patients Participating in Neurobiological Research Studies in Schizophrenia

Dianne N. Irving[†] and Adil E. Shamoo[‡]

*[†]Department of Philosophy, De Sales School of Theology, Washington, D.C.
[‡]Center for Biomedical Ethics, and the Department of Biochemistry, School of Medicine, University of Maryland at Baltimore, 108 Greene St., Baltimore, Maryland 21201*

INTRODUCTION

Neural biological disorders (NBD; mental illnesses) are common widespread afflictions of brain functions manifested in modifications of behavior, ranging from socially unacceptable behavior to psychosis and delusions. Schizophrenia, one of the most serious forms of NBD, afflicts over two million persons in the United States (Andreasen and Carpenter, 1993)[1] and tens of millions of people worldwide. About 10% of these people commit suicide. Since the 1950's pharmacologic agents (e.g., neuroleptics) have become the primary tool in treating schizophrenia (Kane and Marder, 1993).[2] However, the pharmacological treatment for schizophrenia is accompanied by large numbers of serious side effects (such as tardive dyskinesia) and large variations of patients' responses to drug treatments. Further, schizophrenic patients can have periods of "wellness" even without any medications before a relapse occurs.

This situation has encouraged the scientific community to explore research on new drugs, using patients with schizophrenia as human research subjects, with increasing success. In reviewing the literature, some of these studies included the withdrawal of medications (washouts) for periods ranging from one day to over one year. Some studies involved the use of placebos or double-blind protocols; others involved the use of "challenge" doses of drugs, e.g., amphetamines or L-dopa (to induce relapse-like symptoms in patients). Many of these studies resulted in patient relapses (considered high-risk consequences to the patients).

The purpose of this paper is to familiarize those inside and outside the field with a brief background on: (1) medical research studies in schizophrenia in which washouts/relapses occurred; and, (2) some of the ethical issues which have been raised concerning such research studies.

WASHOUTS/RELAPSES IN PATIENTS PARTICIPATING IN NEUROBIOLOGICAL RESEARCH STUDIES

Very few systematic studies on the incidence of relapses in patients with schizophrenia participating in neurobiological research protocols have been reported in the scientific literature. The issue of whether or not relapse may cause irreversible brain damage is controversial (Wyatt, 1991; Miller, 1989),[3,4] as is the issue of whether or not relapse may cause severe worsening of the patients' conditions (Wistedt, 1981; Johnson et al., 1983; Hass, Keshavan, and Sweeney, 1994).[5-7]

The most exhaustive study on the incidence of relapses in research studies is by Shamoo and Keay.[8] They conducted a literature search on the National Library of Medicine MEDLINE for the years 1966–1993 (August), using the subject index "schizophrenia," "relapse," "human," and "English language." 142 studies were identified, 141 of which were obtained. In these 141 studies, only 66 studies used patients directly (see notes 9–78). Of these, 41 were of U.S. origin, and 25 were of non-U.S. origin. Shamoo and Keay proceeded to analyze the U.S.-only studies according to several scientific and ethical parameters (the ethical parameters are addressed below).

Some of the important results of this literature study indicate a need to address the incidence of washouts and relapses experienced by the patients who participated in these research protocols. For example, in the 41 U.S.-only studies (involving 2471 patients), 936 patients (37.9%) relapsed; and in the 25 non-U.S. studies (involving 1259 patients), 46% relapsed. Another important result concerned the dropping out of patients from the research protocols. 245 patients in the U.S.-only studies dropped out (9.9%), with no known or noted whereabouts of 98.3% of these patients. 95 patients in the non-U.S. studies dropped out (7.6%), with no known or noted whereabouts of 97.4% of these patients. Of further concern to the authors is the use of placebos, double-blind protocols and "challenge" drugs in some of these studies, which also result in relapses in the patients participating in these protocols.

ETHICAL ISSUES IN NEUROBIOLOGICAL RESEARCH STUDIES RESULTING IN WASHOUTS/RELAPSES

A review of the bioethics literature by Irving and Shamoo on the ethical issues concerning research in schizophrenia was conducted on BIOETHICSLINE (Kennedy Institute of Ethics Library, Georgetown University, Washington, D.C.), generated by MEDLARS II, National Library of Medicine, for the years 1973–1994 (October). The term searched was "schizophrenia" (all citations). 149 citations were identified. Only five citations were found which specifically addressed the issues of washouts/relapses and the use of placebos and/or double-blind protocols. While two of these citations expressed concern about the worsening of the patients' conditions during these studies (Klerman, 1986; Abrams, 1986),[79,80] the third argued that there were no consistent or important clinical or social differences between the patients in the drug and the control groups after a seven-year follow-up study (Curson et al., 1986).[81] This was the only citation found to systematically address any follow-up studies on these patients in view of their relapse conditions. Additionally, two citations were articles which were published in *The New York Times* (Hilts, 1994; editorial, 1994)[82,83] concerning problems associated with recent UCLA experiments involving patients who had relapsed.

Why so few citations addressing schizophrenia in general, and the issues of washouts/relapses, the use of placebos, double-blind protocols, or "challenge" drugs were identified in this search is unknown. This could be a question of a lack of interest or knowledge within the bioethics or ethics communities, or the range of journals, books, media sources, etc., which are indexed in the BIOETHICSLINE or MEDLAR II computer programs. Perhaps these issues are addressed more often within the context of the strictly scientific literature.

Some other citations independently identified are included here. For example, Shamoo and Irving (1993) addressed their concerns about possible harms caused to relapsing patients in their article on accountability in research using persons with mental illness.[84] Rothman and Michels (Aug. 1994) expressed similar concern for patients with schizophrenia participating in experiments using placebo controls.[85] Hass, Keshavan and Sweeney (Dec. 1994) recently indicated that failure to treat psychosis in patients with schizophrenia may be associated with worse treatment responses and a more severe course of illness.[86] Articles by Davidson *et al.* (1987)[87], and Angrist *et al.* (1973)[88] report the effects of using "challenge" doses of L-dopa on patients with schizophrenia during their research studies. In a recent book by Maris (1986),[89] *Biology of Suicide,* an article by van Praag[90] discusses the importance of the use of "challenge" drugs in experiments in schizophrenia; and Tanney's article[91] in the same book argues for the use of electroconvulsive therapy in patients with schizophrenia who are suicidal.

There were several articles specifically relevant to washouts/relapses in the Spring 1994 issue of *The Journal of the California Alliance for the Mentally Ill.*[92] In that issue, articles by Hassner,[93] Shamoo,[94] Becker,[95] Caplan,[96] Irving,[97] and the *Journal*'s "Postscript"[98] express specific concerns for the harmful effects of relapses, washouts or the use of placebos or double-blind protocols with patients who have schizophrenia because of the harmful effects often experienced by these patients. Also included was the reprint of the *New York Times* editorial (above) identifying some of the harmful effects of washouts and relapses on patients with schizophrenia in a recent UCLA study.[99] In other articles Levine,[100] Smith,[101] Kane,[102] Angrist,[103] Lieberman and Sloan,[104] Goodwin,[105] and Shore *et al.*,[106] acknowledge similar concerns, but argue generally that some research of this type, under appropriate conditions, is required in order to develop new drugs for treatments for patients with schizophrenia. (Articles by Annas,[107] Barchas and Barchas,[108] Lazarus,[109] Rubenstein,[110] Wirshing,[111] Opler,[112] Schell,[113] Ghaemi and Hundert,[114] Destro,[115] Thomasma,[116] Frese,[117] Kaminski,[118] Moline and Aisen,[119] Meiback,[120] Zarin and West[121] address more specifically other ethical issues involved in research using patients with schizophrenia in general.)

In terms of media coverage of the issues of washouts/relapses (aside from the above-mentioned editorial published in *The New York Times*), little has been written. Phil Hilts discussed the problems of the UCLA experiments in an article in *The New York Times*.[122] The *Los Angeles Times Magazine* (1994) published an extensive article on the suicide of one of the patients with schizophrenia who had participated in a UCLA research program.[123] Additionally, a television program in Minneapolis (1994) produced a piece entitled "Fatal Experiment," exploring the circumstances surrounding the suicide of a patient with schizophrenia shortly after taking part in a drug protocol at a local hospital in Minneapolis.[124]

In the medical research literature study mentioned above (Shamoo and Keay, 1994), several ethical issues specifically associated with washouts/relapses were addressed. The focus of their concerns was on ethical issues such as:

- the need to establish the mentally ill as a vulnerable group when using them in medical research
- whether these patients were being used solely as experimental subjects
- the need to evaluate patient recruitment standards
- what criteria were used to determine whether patients were judged to be competent to sign the informed consent forms, and the need for documentation of informed consent in the research papers submitted for publication
- the lack of important safeguards during the consent procedures
- the lack of information in the consent forms about the possibility of relapses or that one of the purposes of the experiment was to study the relapse process
- whether the psychiatrist/researcher should sign the consent forms when a patient was determined incompetent by the same researcher, and the need for an independent psychiatrist to determine competency in such situations if there is a possible conflict of interests or undue dependency of the patient on the psychiatric researcher
- the proper role of the surrogate decisionmaker
- who should make the decision to withdraw the patient from the studies
- the *timely* removal and treatment of research subjects when they express a decision to withdraw from the study, or when the researcher considers such withdrawal necessary for the well-being of the patient
- accurate estimations of the duration of washouts and relapses
- the need to document the location and medical conditions of those patients who had dropped out, and to provide medical treatment for them if needed
- the number of patients who had suffered relapses, especially when foreseeable
- the claim that the FDA mandates relapse-induction studies
- the use of "challenge" doses of amphetamines or of L-dopa
- the safeguards necessary for the use of outpatients in these studies
- the need to readdress the structure and composition of the IRB review procedures

Given the large numbers of incidences of washouts/relapses, there is considerable need to address these and other procedural, scientific, medical and ethical issues when using patients with schizophrenia in neurobiological research.

REFERENCES

1. Andreasen, N.C., and Carpenter, W.T. Diagnosis and classification of schizophrenia. *Schiz. Bull.* 1993; 19:25–40.
2. Kane, J.M., and Marder, S.R. Psychopharmacologic treatment of schizophrenia. *Schiz. Bull.* 1993; 19: 113–128.

3. Wyatt, R.J. Neuroleptics and the natural course of schizophrenia. *Schiz. Bull.* 1991; 17:325–347.
4. Miller, R. Schizophrenia as a progressive disorder: relations to EEG, CT, neuropathological and other evidence. *Prog. Neurobiol.* 1989; 33:17–44.
5. Wistedt, B. A depot neuroleptic withdrawal study: a controlled study of the clinical effects of the withdrawal of depot fluphenazine decanoate and depot flupenthixol decanoate in chronic schizophrenic patients. *Acta Psychiatr. Scand.* 1981; 64:65–84.
6. Johnson, D.A.W., Pasterski, G., Ludlow, J.M., Street, K., and Taylor, R.D.W. The discontinuance of maintenance neuroleptic therapy in chronic schizophrenic patients: drug and social consequences. *Acta Psychiatr. Scand.* 1983; 67:339–352.
7. Hass, G.L., Keshavan, M.S., and Sweeney, J.A. Delay to first medication in schizophrenia: evidence for a possible negative impact of exposure to psychosis. *Abstracts of Panels and Posters* (Poster Session II), 33rd Annual Meeting, American College of Neuropsychopharmacology, San Juan, Puerto Rico, Dec. 12–16, 1994; 171.
8. Shamoo, A.E., and Keay, T.J. Ethical concerns about relapse studies. *Cambridge Quarterly of Health Care Ethics* 1996; 5:373–386.
9. Ravaris, C.L., Weaver, L.A., and Brooks, G.W. Further studies with fluphenazine enanthate: II. Relapse rate in patients deprived of medication. *Am. J. Psychiatry* 1967; 124:248–249.
10. Prien, R.F., Cole, J.O., and Belkin, N.F. Relapse in chronic schizophrenics following abrupt withdrawal of tranquilizing medication. *Br. J. Psychiatry* 1968; 115:679–686.
11. Gallant, D.M., Mielke, D., Bishop, G., Oelsner, T., and Guerrero-Figueroa, R. Pipotiazine palmitate: an evaluation of a new long acting intramuscular antipsychotic agent in severely ill schizophrenic patients. *Dis. Nerv. Sys.* 1975; 36:193–196.
12. Quitkin, R., Rifkin, A., and Kane, J., *et al.* Long-acting oral vs injectable antipsychotic drugs in schizophrenics. A one-year double-blind comparison in multiple episode schizophrenics. *Arch. Gen. Psychiatry* 1978; 35:889–892.
13. Lonowski, D.J., Sterling, F.E., and Kennedy, J.C. Gradual reduction of neuroleptic drugs among chronic schizophrenics: a double blind controlled study. *Acta Psychiatr. Scand.* 1978; 57:97–102.
14. Prange, A.J., Loosen, P.T., Wilson, I.C., *et al.* Behavioral and endoctrine responses of schizophrenic patients to TRH (protirelin). *Arch Gen Psychiatry* 1979; 36:1086–1093.
15. Rifkin, A., Quitkin, F., Kane, J.M., *et al.* The effect of fluphenazine upon social and vocational functioning in remitted schizophrenics. *Biol. Psychiatry* 1979; 14:499–508.
16. Kane, J.M., Rifkin, A., Quitkin, F., *et al.* Low dose fluphenazine decanoate in maintenance treatment of schizophrenia. *Psychiat. Res.* 1979; 1:341–348.
17. Schooler, N.R., Levine, J., Severe, J.B., and the NIMH-PRB Collaborative Fluphenazine Study Group. Depot fluphenazine in the prevention of relapse in schizophrenia: evaluation of a treatment regimen. *Psychopharmacol. Bull.* 1979; 15:44–47.
18. Brown, W.A., and Laughren, T. Low serum prolactin and early relapse following neuroleptic withdrawal. *Am. J. Psychiatry* 1981; 138:237–239.
19. Branchey, M.H., Branchey, L.B., and Richardson M.A. Effects of neuroleptic adjustment on clinical condition and tardive dyskinesia in schizophrenic patients. *Am. J. Psychiatry* 1981; 138:608–612.
20. Shenoy R.S., Sedler A.G., Goldberg, S.C., *et al.* Effects of a six week drug holiday on symptom status, relapse, and tardive dyskinesia in chronic schizophrenics. *J. Clin. Psychopharmacol.* 1981; 1:141–145.
21. Goldberg, S.C., Shenoy, R.S., Sadler, A., *et al.* The effects of a drug holiday on relapse and tardive dyskinesia in chronic schizophrenics. *Psychopharmacol. Bull.* 1981; 17:116–117.
22. Herz, M.I., Szymanski, H.V., and Simon, J. Intermittent medication for stable schizophrenic outpatients: an alternative to maintenance medication. *Am. J. Psychiatry* 1982; 139:918–922.
23. Van Kemmen, D.P., Docherty, J.P., and Bunney, W.E. Prediction of early relapse after pimozide discontinuation by response to d-amphetamine during pimozide treatment. *Biol. Psychiatry* 1982; 17:233–242.
24. Brown, W.A., Laughren, T., Chisholm, E., and Williams, B.W. Low serum neuroleptic levels predict relapse in schizophrenic patients. *Arch. Gen. Psychiatry* 1982; 39:998–1000.
25. Kane, J.M., Rifkin, A., Quitkin, F., *et al.* Fluphenazine vs placebo in patients with remitted, acute first-episode schizophrenia. *Arch. Gen. Psychiatry* 1982; 39:70–73.
26. Mandel, M.R., Severe, J.B., Schooler, M.R., *et al.* Development and prediction of postpsychotic depression in neuroleptic treated schizophrenics. *Arch. Gen. Psychiatry* 1982; 39:197–203.
27. Glazer, W.M., and Sheard, M.H. Relapse in patients with shifting RDC diagnoses treated with lithium alone. *J. Clin. Psychiatry* 1982; 43:134–136.

28. Lehmann, H.E., and Wilson, W.H., Durtsch, M. Minimal maintenance medication: effects of three dose schedules on relapse rates and symptoms in chronic schizophrenic outpatients. *Comp. Psychiatry* 1983; 24:293–303.

29. Kane, J.M., Rifkin, A., Woerner, M., *et al.* Low dose neuroleptic treatment of outpatient schizophrenics. *Arch. Gen. Psychiatry* 1983; 40:893–896.

30. Carpenter, W.T., and Heinrichs, D.W. Early intervention, time limited, targeted pharmacotherapy of schizophrenia. *Schiz. Bull.* 1983; 9:533–542.

31. Lieberman, J.A., Kane, J.M., Gadaleta, D., *et al.* Methylphenidate challenge as predictor of relapse in schizophrenia. *Am. J. Psychiatry* 1984; 141:633–638.

32. Marder, S.R., Van Putten, T., and Mintz, J. Costs and benefits of two doses of fluphenazine. *Arch. Gen. Psychiatry* 1984; 41:1025–1029.

33. Angrist, B., Rubinstein, M., Wolkin, A., and Rotrosen, J. Amphetamine response and relapse risk after depot neuroleptic discontinuation. *Psychopharmacol.* 1985; 85:277–283.

34. Jayaram, G., Coyle, J., and Tune, L. Relapse in chronic schizophrenics treated with fluphenazine decanoate is associated with low serum neuroleptic levels. *J. Clin. Psychiatry* 1986; 47:247–248.

35. McMillan, D.E., Fody, E.P., and Couch, L., *et al.* Drug holidays and serum haloperidol levels in schizophrenic patients. *J. Clin. Psychiatry* 1986; 47:373–374.

36. Faraone, S.V., Curran, J.P., Laughren, R., *et al.* Neuroleptic bioavailability, psychosocial factors, and clinical status: a 1 year study of schizophrenic outpatients after dose reduction. *Psychiatry Res.* 1986; 19:311–322.

37. Nuechterlein, K.H., Snyder, K.S., Dawson, M.E., *et al.* Expressed emotion, fixed dose fluphenazine decanoate maintenance, and relapse in recent onset schizophrenia. *Psychopharmacol. Bull.* 1986; 22: 633–639.

38. Lieberman, J.A., Kane, J.M., Sarantakos, S., *et al.* Influences of pharmacologic and psychosocial factors on the course of schizophrenia: prediction of relapse in schizophrenia. *Psychopharmacol. Bull.* 1986; 22:845–853.

39. Davidson, M., Keefe, R.S.E., Mohs, R.C., *et al.* L-dopa challenge and relapse in schizophrenia. *Am. J. Psychiatry* 1987; 144:934–938.

40. Faraone, S.V., Brown, W.A., and Laughren, T.P. Serum neuroleptic levels, prolactin levels, and relapse: a two year study of schizophrenic outpatients. *J. Clin. Psychiatry* 1987; 48:151–154.

41. Kriesman, D., Blumenthal, R., Borenstein, M., *et al.* Family attitudes and patient social adjustment in a longitudinal study of outpatient schizophrenics receiving low dose neuroleptics: the family's view. *Psychiatry* 1988; 51:3–13.

42. Boza, R.A., Milanes, F., Starkey, T., and Dominguez, F. Relapse rate and low serum neuroleptic levels in schizophrenics treated with fluphenazine: another view. *J. Clin. Psychiatry* 1988; 49:245.

43. Van Kammen, D.P., Peters, J., van Kammen, W.B., *et al.* CSF norepinephrine in schizophrenia is elevated prior to relapse after halperidol withdrawal. *Biol. Psychiatry* 1989; 26:176–188.

44. Herz, M.I., Blazer, W., Mirza, M., *et al.* Treating prodromal episodes to prevent relapse in schizophrenia. *Br. J. Psychiatry* 1989; 155:123–127.

45. Glazer, W.M., Morgenstern, H., Schooler, N., *et al.* Predictors of improvement in tardive dyskinesia following discontinuation of neuroleptic medication. *Br. J. Psychiary* 1990; 157:585–592.

46. Buchanan, R.W., Kirkpatrick, B., Sumerfelt, A., *et al.* Clinical predictors of relapse following neuroleptic withdrawal. *Biol. Psychiatry* 1992; 32:72–78.

47. Steinhauer, S.R., van Kammen, D.P., Colbert, K., *et al.* Pupillary constriction during haloperidol treatment as a predictor of relapse following drug withdrawal in schizophrenic patients. *Psychiatry Res.* 1992; 43:287–298.

48. Ventura, J., Nuechterlein, K.H., Hardesty, J.P., and Gitlin, M. Life events and schizophrenic relapse after withdrawal of medication. *Br. J. Psychiatry* 1992; 161:615–620.

49. Barbee, J.G., Mancuso, D.M., Freed, C.R., and Todorov, A.A. Alprazolam as a neuroleptic adjunct in the emergency treatment of schizophrenia. *Am. J. Psychiatry* 1992; 149:506–510.

50. McKane, J.P., Robinson, A.D.T., Wiles, D.H., McCreadie, R.G., and Stirling, G.S. Haloperidol decanoate v. fluphenazine decanoate as maintenance therapy in chronic schizophrenic in-patients. *Br. J. Psychiatry* 1987; 151:333–336.

51. MacMillan, J.F., Crow, T.J., Johnson, A.L., and Johnstone, E.C., III. Short-term outcome in trial entrants and trial eligible patients. *Br. J. Psychiatry* 1986; 148:128–133.

52. McCreadie, R.G., Dingwall, J.M., Wiles, D.H., and Heykants, J.J.P. Intermittent pimozide versus fluphenazine decanoate as maintenance therapy in chronic schizophrenia. *Br. J. Psychiatry* 1980; 137:510–517.

53. Stevens, B. The social value of fluphenazine decanoate. *Acta Psychiatr. Belg.* 1976; 76:792–804.

54. Johnson, D.A.W. The duration of maintenance therapy in chronic schizophrenia. *Acta Psychiatr. Scand.* 1976; 53:298–301.

55. Johnson, D.A.W., and Breen, M. Weight changes with depot neuroleptic maintenance therapy. *Acta Psychiatr. Scand.* 1979; 59:525–528.

56. Pan, P.C., and Tantam, D. Clinical characteristics, health beliefs and compliance with maintenance treatment: a comparison between regular and irregular attenders at a depot clinic. *Acta Psychiatr. Scand.* 1989; 79:564–570.

57. Cheung, H.K. Schizophrenics fully remitted on neuroleptics for 3–5 years—to stop or continue drugs? *Br. J. Psychiatry* 1981; 138:490–494.

58. Johnson, D.A.W., Ludlow, J.M., Street, K., and Taylor, R.D.W. Double-blind comparison of half-dose and standard-dose flupenthixol decanoate in the maintenance treatment of stabilised outpatients with schizophrenia. *Br. J. Psychiatry* 1987; 151:634–638.

59. Andrews, P., Hall, J.N., and Snaith, R.P. A controlled trial of phenothiazine withdrawal in chronic schizophrenic patients. *Br. J. Psychiatry* 1976; 128:451–5.

60. McCreadie, R., Mackie, M., Morrison, D., and Kidd, J. Once weekly pimozide versus fluphenazine decanoate as maintenance therapy in chronic schizophrenia. *Br. J. Psychiatry* 1982; 140:280–286.

61. Capstick, N. Long-term fluphenazine decanoate maintenance dosage requirements of chronic schizophrenic patients. *Acta Psychiatr. Scand.* 1980; 61:256–262.

62. Odejide, O.A, and Aderounmu, A.F. Double-blind placebo substitution: withdrawal of fluphenazine decanoate in schizophrenic patients. *J. Clin. Psychiatry* 1982; 43(5):195–196.

63. Soni, S.D., Sampath, G., Shah, A., and Kiska, J. Rationalizing neuroleptic polypharmacy in chronic schizophrenics: effects of changing to a single depot preparation. *Acta Psychiatr. Scand.* 1992; 85: 354–359.

64. Nishikawa, T., Tanaka, M., Tsuda, A., Koga, I., and Uchida, Y. Prophylactic effects of neuroleptics in symptom-free schizophrenics: a comparative dose-response study of timiperone and sulpiride. *Biol. Psychiatry* 1989; 25:861–866.

65. Menkes, D.B., Clarkson, H.O., Caradoc-Davies, G., and Mullen, P.E. Anticholinergic equivalents and parkinsonism: a model for predicting side-effects of antipsychotic drugs. *Int. Clin. Psychopharmacol.* 1987; 2:55–67.

66. Fleischbacker, W.W., Stuppack, C., Barnas, C., Unterweger, B., and Hinterhuber, H. Low-dose zotepine in the maintenance treatment of schizophrenia. *Pharmacopsychiat.* 1987; 20:61–63.

67. Dencker, S.J., Malm, U., and Lepp, M. Schizophrenic relapse after drug withdrawal is predictable. *Acta Psychiatr. Scand.* 1986; 73:181–185.

68. Frecska, E., Perenyi, A., Bagdy, G., and Revai, K. CSF dopamine turnover and positive schizophrenic symptoms after withdrawal of long-term neuroleptic treatment. *Psychiatry Res.* 1985; 16: 221–226.

69. Perenyi, A., Kuncz, E., and Bagdy, G. Early relapse after sudden withdrawal or dose reduction of clozapine. *Psychopharmacology* 1985; 86:244.

70. Wistedt, B., and Palmstierna, T. Depressive symptoms in chronic schizophrenic patients after withdrawal of long-acting neuroleptics. *J. Clin. Psychiatry* 1983; 44:369–371.

71. Zander, K.J., Fischer, B., Zimmer, R., and Ackenheil, M. Long-term neuroleptic treatment of chronic schizophrenic patients: clinical and biochemical effects of withdrawal. *Psychopharmacology* 1981; 73:43–47.

72. Dencker, S.J., Lepp, M., and Malm, U. Do schizophrenics well adapted in the community need neuroleptics? *Acta Psychiatr. Scand.* (Suppl) 1980; 279:64–76.

73. Dencker, S.J., Lepp, M., and Malm, U. Clopenthixol and flupenthixol depot preparations in outpatient schizophrenics. I. A one year double-blind study of clopenthixol decanoate and flupenthixol palmitate. *Acta Psychiatr. Scand.* (Suppl) 1980; 279:10–28.

74. Kingstone, E., Grof, P., Furlong, W., Jacques, W., Virc, L., and Daigle, L. Penfluridol, a peroral long-acting neuroleptic, for the maintenance treatment of schizophrenic patients who relapse. *J. Clin. Pharmacology* 1977; 17:252–8.

75. Johnson, D.A.W., Pasterski, G., Ludlow, J.M., Street, K., and Taylor, R.D.W. The discontinuance of maintenance neuroleptic therapy in chronic schizophrenic patients: drug and social consequences. *Acta Psychiatr. Scand.* 1983; 67:339–352.

76. Wistedt, B. A depot neuroleptic withdrawal study: a controlled study of the clincal effects of the withdrawal of depot fluphenazine decanoate and depot flupenthixol decanoate in chronic schizophrenic patients. *Acta Psychiatr. Scand.* 1981; 64:65–84.

77. Inanaga, K., Inoue, K., Tachibana, H., Oshima, M., and Kotorii, T. Effect of L-dopa in schizophrenia. *Folia Psychiatr. Neurolog. Japonica* 1972; 26(2):145–157.

78. Inanaga, K., Nakazawa, Y., Inoue, K., Tachibana, H., Oshima, M., and Kotorii, T. Double-blind controlled study of L-dopa therapy in schizophrenia. *Folia Psychiatr. Neurolog. Japonica* 1975; 29(2): 123–143.

79. Kierman, G.L. Scientific and ethical considerations in the use of placebo controls in clinical trials in psychopharamacology. *Psychopharmacol. Bull.* 1986; 22(1):25–29.

80. Abrams, R.A. Ethical questions raised by a proposed randomized, double-blind study involving placebo, standard drug therapy and experimental antipsychotic drug therapy for patients with schizophrenia. *Clin. Res.* 1986; 34:6–9.

81. Curson, D.A., Hirsch, S.R., Platt, S.D., Bamber, R.W., and Barnes, T.R.E. Does short term placebo treatment of chronic schizophrenia produce long term harm? *Br. Med. J.* 1986; 293:726–728.

82. Hilts, P. Agency faults a U.C.L.A. study for suffering of mental patients. *New York Times*, 1994 Mar 10; A1, B10.

83. Editorial. Medical ethics in the dock. *The New York Times*, 1994 Mar 14.

84. Shamoo, A.E., and Irving, D.N. Accountability in research using persons with mental illness. *Accountability in Research* 1993; 3:1–17.

85. Rothman, K.J., and Michels, K.B. Sounding Board: The continuing unethical use of placebo controls. *N. Engl. J. Med.* 1994; 331:394–398.

86. Haas, G., Keshavan, M.S., and Sweeney, J.A. Delay to first medication in schizophrenia: evidence for a possible negative impact of exposure to psychoses. *Abstracts of Panels and Posters* (Poster Session II), 33rd annual Meeting, American College of Neuropsychopharamacology, San Juan, Puerto Rico, Dec. 12–16, 1994; 171.

87. Davidson, M., *et al.* L-dopa challenge and relapse in schophrenia. *Am. J. Psychiatry* 1987; 144: 934–938.

88. Angrist, B., Sathananthan, G., and Gershon, S. Behavioral effects of L-dopa in schizophrenic patients. *Psychopharmacologia* (Berl.) 1979; 31:1–12.

89. Maris, R. (ed.). *Biology of Suicide*, New York: The Guilford Press, 1986. pp. 21–50.

90. Van Praag, H.M. Affective disorders and aggression disorders: evidence for a common biological mechanism. In Maris 1986; 21–50.

91. Tanney, B.L. Electroconvulsive therapy and suicide. In Maris 1986; 116–140.

92. Weisburd, D. (ed.). *The Journal of the California Alliance for the Mentally Ill* Spring 1994; 5(1).

93. Hassner, V. What is ethical? what is not? where do you draw the line? *Ibid.*, 4–5.

94. Shamoo, A.E. Our responsibilities toward persons with mental illness as human subjects in research. *Ibid.*, 14–16.

95. Becker, J.C. A mother's testimony, *Ibid.*, 17.

96. Caplan, A.L. Are existing safeguards adequate? *ibid.*, 36–38.

97. Irving, D.N. Psychiatric research: reality check. *Ibid.*, 42–44.

98. Editor. A disturbing postscript. *Ibid.*, 69.

99. Editorial. Medical ethics in the dock. *The New York Times* (Mon., Mar. 14, 1994); *Ibid.*, 12.

100. Levine, R.H. A researcher's concern with ethics in human research. *Ibid.*, 6–8.

101. Smith, M.L. Power, advocacy, and informed consent. *Ibid.*, 25–27.

102. Kane, J.M. Enormous ethical challenges. *Ibid.*, 28–29.

103. Angrist, B. Ethical issues regarding prospective studies of amphetamine psychosis. *Ibid.*, 32–35.

104. Lieberman, J.A., and Sloan, J. The moral imperatives of medical research in human subjects. *Ibid.*, 40–42.

105. Goodwin, F.K. Questions and answers. *Ibid.*, 45–47.

106. Shore, D., Berg, K., and Mullican, C. Ethical issues in clinical neurological research. *Ibid.*, 61–62.

107. Annas, G.J. Experimentation and research. *Ibid.*, 9–11.

108. Barchas, J.D, and Barchas, I.D. The imperative for research on severe mental disorders. *Ibid.*, 18–19.

109. Lazarus, J. Critical ethical issues and conflicts. *Ibid.*, 20–21.
110. Rubenstein, L.S. Psychiatric experimentation: the lessons of history. *Ibid.*, 22–24.
111. Wirshing, W.C. In a perfect world none of this would concern us. *Ibid.*, 30.
112. Opler, L.A. Conducting clinical psychiatric research in "non-research" settings. *Ibid.*, 30–31.
113. Schell, B.H. The ominous shadow of the CIA has imprinted itself on the brain research community. *Ibid.*, 38–40.
114. Ghaemi, S.N, and Hundert, E.M. The ethics of research in mental illness. *Ibid.*, 47–49.
115. Destro, R.A. Law, professionalism, and bad attitude. *Ibid.*, 50–53.
116. Thomasma, D.C. Obtaining consent for research in the neurobiologically impaired. *Ibid.*, 54–55.
117. Frese, F.J. Informed consent and the right to refuse or participate. *Ibid.*, 56–57.
118. Kaminski, R. The importance of maintaining patient's continued and informed consent. *Ibid.*, 57–58.
119. Moline, M.L, and Aisen, M.W. Perspectives of protocol reviewers. *Ibid.*, 59–60.
120. Melback, R.C. Why does it take so long now for drugs to get to the marketplace? *Ibid.*, 63–65.
121. Zarin, D.A, and West, J. ACORN: APA clinical outcomes research network. *Ibid.*, 66–67.
122. Hilts, P. Agency faults a U.C.L.A. study for suffering of mental patients. *New York Times* 1994 Mar 10; B10.
123. Horowitz, J. For the sake of science. *Los Angeles Times Magazine* Sept. 11, 1994; 16 (cover story).
124. Leer, R. A fatal experiment. Nov. 3, 1994; Station KSET, Minneapolis, MN (Unit Five Investigation).

Standards of Accountability for Consent in Research

Leonard S. Rubenstein

Bazelon Center for Mental Health Law, 1101 15th Street, N.W., Suite 1212, Washington, D.C. 20005

I.

It is fitting that this conference on ethics in neurobiological research takes place as we open 1995. This year marks the 50th anniversary of the end of the Second World War and thus the threshold of modern concern about the ethics of biomedical research. The first product of this concern, the Nuremburg Code, has remained a benchmark for the regulation of biomedical research since 1949 (*United States v. Brandt*).[1] The Code, like the Declaration of Independence, is a document more talked about than read. Yet, like the Declaration, its first, unequivocal sentence, set the tone for the future: "The voluntary consent of the human subject is absolutely essential." Despite its flaws, the Nuremburg Code compares favorably with many of the most explicit contemporary documents, and is indeed more protective of human subjects in experimentation than many codes adopted since.

Today the principle of informed consent in human subject research has become part of the fabric of regulation of research practices. It would seem that, as a result, the violations of the principle that led to the scandals of the past half-century— Tuskegee, the Jewish Chronic Disease Hospital case, radiation research and the rest—would be behind us. Yet new research scandals involving a faulty consent process in human subject research keep reappearing, most recently in the UCLA drug discontinuation experiments, which brought about a rebuke from the Office of Protection from Research Risks, a critique from Jay Katz (Katz, 1993)[2] a front page story in the *New York Times* (1994)[3] and even an editorial in that newspaper (*New York Times*, 1994).[4]

The difficulties in the process of informed consent are well known and well documented. Some investigators resist a thoroughgoing consent process because they fear it will discourage participation in the research project, lead to unnecessary misunderstandings or interfere with efforts at randomization, all of which impede discovery of new and effective treatments for the afflictions of our day. Appelbaum and Roth, for example, found that of 17 investigators they surveyed

This article will also appear in the journal: *Accountability in Research*, Volume 4, pp. 197–206 (1996).

who worked with human subjects only half believed in the importance of informed consent as an ethical or practical principle, and the others rejected it as an unattainable intrusion on the conduct of the research.[5] Further, as Katz observes, researchers tend to blur the distinction between research and clinical practice, and with it, the differentiated roles of physician and the investigator. This leads to insufficient and sometimes misleading discussion with subjects. In the UCLA case, for example, he identifies manipulation of subjects' consent, failure to distinguish trivial and non-trivial risks, lack of explanation of the incidence or severity of predictable risks, and failure to explain the risks and benefits of nonparticipation (Katz, 1993). Indeed, he argues that inadequate obtaining of consent is the norm, not the exception.

To be sure, today these problems are usually of a different order than the injection of live cancer cells and other grotesque practices that Henry Beecher identified in 1966 (Beecher, 1966).[6] Still, as in Beecher's day, problems of consent persist in the research conducted by prominent researchers at leading institutions. Moreover, they continue in the face of federal regulations establishing rules for informed consent and establishing Institutional Review Boards to enforce them. Some of the problems no doubt derive, as many have argued, from flaws in the regulations and in their implementation, e.g., in-bred IRBs, failure to attend to special risks among subjects whose competence to consent is questionable, and undue focus on consent forms rather than the consent process.

But I believe another factor is at work as well. At least part of the reason problems of consent in research persist is confusion about whose standards of accountability in research apply. Are they the standards of the research field or of the larger society? Is the proper analogy for the accountability of the researcher medical malpractice, where standards traditionally derive from the practices of the community of physicians, or civil rights law, where statutory or judicially created standards of discrimination derive from the larger society? The existence of federal regulations that constitute a body of law tends to mask these questions rather than do away with them. It is my purpose in this paper to identify the conflict and propose a resolution.

II.

I propose to illustrate these questions in an unusual way, through a piece of litigation in the late 1980s in which I was personally involved that pitted two of the country's leading authorities on the history of informed consent in biomedical research against each other, under oath, on the subject of the applicable standard in judging a researcher's conduct. Indeed, each expert was cross-examined at length, one by me.

The case was *Orlikow v. United States*,[7] where nine former patients of Dr. Ewen Cameron of McGill University in Montreal sought damages from the United States for injuries they claimed to have received as a result of Central Intelligence Agency funding of radical experiments performed on them in the 1950s by Dr. Cameron without their informed consent. I was one of the lawyers for the nine plaintiffs. Social historian David Rothman, whose many writings on biomedical ethics include *Strangers at the Bedside* (Rothman, 1991),[8] was our expert. Tom Beauchamp, who, with Ruth Faden, wrote *A History and Theory of Informed Consent*,[9] was an expert for the government.

Cameron, who died in 1967, is a fascinating figure, one of the most renowned psychiatrists of his day and once president of the American Psychiatric Association. He was a pioneer of social psychiatry and an innovator who developed the day hospital. Cameron was one of the psychiatrists commissioned by Nuremburg authorities to interview Rudolph Hess to assess his competence to stand trial. Indeed, he was deeply affected by the experience of Nazism and wanted to find ways to change thought patterns that be believed were destructive either to an individual's mental health or the democratic social fabric. His science, however, was deeply flawed. In the early 1950s, he developed experiments designed to test theories about changing human thought patterns and personality through a technique called "psychic driving," which involved the repetition of a recorded message a few seconds in length thousands of times for up to 20 hours at a time. Using private patients who came to him for treatment for anxiety, depression and other mental health problems, he sought to "implant"—his word—these new ideas and personality characteristics into the subject. He published the first of many articles on the subject, appropriately enough entitled "Psychic Driving," in the *American Journal of Psychiatry* in January, 1956.

He neither disclosed to his patients that he was engaged in research, or indeed in any unconventional form of therapy, nor sought anyone's consent for it. The patients were competent to make decisions about their lives but were never told that experiments, much less radical experiments, were being conducted on them, that Cameron's research interests included clinically untested therapies that could cause harm, nor that they would not be able to stop the research by a subsequent refusal.

In the course of his research, Cameron soon enough discovered that his patients resisted the repeated messages. Some refused to participate; some actually ran away. So Cameron sought to develop techniques to lower both conscious and unconscious resistance to listening to the messages for hours on end. He also looked for new ways to "break down" behavior by reducing the patient to a state of complete confusion and helplessness. The most notorious of these methods was "depatterning," Cameron's term for achieving a state "where the patient has developed an organic brain syndrome with acute confusion, disorientation and interference with his learned habits of eating and bladder and bowel control." Cameron wrote that, at this point, the patient's "conceptual span is limited to a few minutes and to entirely concrete events" and that "he cannot conceptualize where he is, nor does he recognize those who treat him."

To achieve this, Cameron used massive regressive electro-convulsive therapy combined with drugs that sought to reduce the person to an infantile state. He also used barbiturate-induced sleep for periods exceeding 30 days, LSD, sensory deprivation and curare.

Cameron published papers concerning this work in major psychiatric journals and lectured widely about it. In the mid-1950's, his Psychic Driving article came to the attention of the CIA, which was then interested in brainwashing techniques allegedly used by the Chinese in the Korean War. Using a front organization called the Society for the Investigation of Human Ecology, the CIA invited Cameron to submit a research grant proposal. Cameron's proposal included additional psychic driving experiments that included the following elements:

1. Breaking down of ongoing patterns of behavior—depatterning—through intensive electroshocks.
2. Intensive repetition (16 hours a day for 6–7 days) of the messages.
3. Keeping the patient in partial sensory deprivation.
4. Following the psychic driving with 7–10 days of barbiturate-induced sleep.

He also proposed to put LSD into the mix and inactivate the patient through the use of curare. The CIA funded the proposal in 1957.

It was not until the late 1970s, following Senate Intelligence Committee hearings, Freedom of Information Act requests by an enterprising reporter, John Marks,[10] and reports of the CIA documents in the *New York Times*, that the connection between the CIA and Cameron came to light. By then, Cameron had died, but nine of Cameron's former patients sued the United States, claiming that they had been harmed by experiments funded by the CIA. They alleged, among other wrongs, that the CIA had acted negligently in funding experimentation on patients without requiring and assuring their informed consent to participate in it. As the case proceeded through discovery, the CIA witnesses conceded that they had no interest in protecting Cameron's patients, had not considered the question of informed consent, and had made no effort to require any safeguards of the subjects or to require reports of the status of the subjects.[11]

Nevertheless, the government denied liability. Among other defenses, it argued that the CIA's conduct had to be judged by the ethical standards of the time, which it said, did not require informed consent for biomedical research with human subjects. Aside from differences about the appropriate legal standard deriving from relevant cases, the parties differed about what the ethics of research at the time of Cameron's experiments required regarding informed consent. Each side brought in an expert.

Enter Professors Rothman for the plaintiffs and Beauchamp for the CIA. They agreed on some basic facts, particularly that despite the importance of the Nuremburg Code and the principles on which it was based, during the 1950s it garnered little attention from the American research community. Professor Rothman indeed had recently written a retrospective article on Henry Beecher's famous 1966 piece that agreed with Beecher's view that his examples were not anomalies but were fairly representative of practice at the time (Rothman, 1987).[12]

Agreement ended there, however, and I believe their disagreement about the 1950s is relevant now because it demonstrates different standards and approaches for holding researchers accountable for obtaining or failing to obtain consent from research subjects.

Professor Rothman's position was that by the 1950s, the necessity of obtaining the voluntary consent of human subjects to participation in research had become part of the fabric of standards governing that research. The affidavit he submitted in the case traced the history of consent principles in human experimentation from the end of the nineteenth century through Walter Reed's research on yellow fever, the Second World War and into the 1950s. In his words, both before and after the adoption of the Nuremburg Code, the "voluntary consent requirement" was part of the "generally accepted principles applicable to medical experimentation." The affidavit also stated:

[D]uring the 1950's, there was a recognized obligation on the part of entities financing, sponsoring or conducting medical experimentation to adopt ethical standards reflecting the principles set out in the Nuremburg Code, particularly the informed consent requirement; and to make inquiry and to ascertain the competence and prudence in dealing with research subjects of those conducting medical experimentation on their behalf.[13]

Professor Rothman concluded his affidavit by stating that "by the 1950's it was clearly irresponsible for a physician to conduct experiments upon patients without obtaining their voluntary consent to be research subjects."

Professor Rothman agreed that the principle of voluntary consent was frequently violated; indeed, he had just written on that very subject. In his view, however, the many deviations from Nuremburg's principles by researchers did not allow a particular investigator who failed to follow them to escape accountability under those principles. While not discounting contemporary practices in making the decision about accountability, he stressed that extensive violations of a norm do not disprove its existence. Moreover, as stated in his deposition, in his view, the "relevant body by which one judges an ethic is not, in the case at hand, exclusively medical. An ethic guiding human experimentation is not judged only by whether the medical community adheres to it, because there are other communities that are relevant as well." For him, those other communities especially included the larger public from which subjects were selected. Thus, for him, the fact that voluntary consent had not become embedded in the ethos of the research community was not the least dispositive, since the standard derived from a larger community.

Both these points, he said, were apparent in Henry Beecher's 1966 article citing examples of clearly unethical research and the response to it. Professor Rothman explained that Beecher's criticism of unethical research "assumes the existence of the norm" and contended that the public outrage the article generated demonstrated the existence of standards outside the narrow community of the field of medical researchers.

Professor Rothman has made these points in a more elaborate way in *Strangers at the Bedside*, his history of the emergence of biomedical ethics in the modern era. In commenting on the critical responses to Henry Beecher's exposes of unethical research, Rothman states:

The more popular objection (which can still be heard among investigators today) was that he had unfairly assessed 1950s practices in terms of the moral standards of a later era. To these critics, the investigators that Beecher had singled out were pioneers, working before standards were set for human investigation, before it was considered necessary to inform subjects about the research and obtain their formal consent to participation. The enterprise of human investigation was so novel that research ethics had been necessarily primitive and underdeveloped.

However popular—and, on the surface, appealing—that retort is, it not only fails to address the disjuncture between public expectations and researchers' behavior but is woefully short on historical perspective. If the activity was so new and the state of ethics so crude, why did outsiders shudder as they read about the experiments?[14]

Professor Rothman's position was similar to that taken by the Tuskegee Syphilis Study Ad Hoc Panel, which concluded that the failure to obtain consent of the subjects was "ethically unjustified" even in 1932.[15] Similarly, in the well-known Jewish Chronic Disease Hospital case, in the 1960s, a physician was disciplined for violating consent rules despite the absence of the peer review and regulatory requirements concerning consent which now govern experimental research.

Finally, Professor Rothman cited Defense Department policies for funding biomedical research as reinforcing the ethical argument of informed consent. In February, 1953, the Secretary of Defense adopted the Nuremburg Code for atomic, biological and chemical agents research, distributing the memo to the Secretaries of the Army, Air Force and Navy and to the Research and Development Board. The policy required the Secretary of the service conducting the research to approve such experiments. On June 30, 1953, the Chief of Staff of the Army adopted a set of standards distributed to the Chief Chemical Officer and the Surgeon General of the Army governing the use of volunteers in research concerning atomic, biological, and/or chemical warfare defense. The nine-page document recited virtually verbatim the entire Nuremburg Code, including its rules on consent:

The voluntary consent of the human subject is absolutely essential. This means that the person involved should have the legal capacity to give consent, should be so situated as to be able to exercise free power of choice without the intervention of any element of force, fraud, deceit, duress, overreaching, or other ulterior form of constraint or coercion and should have sufficient knowledge and comprehension of the elements of the subject matter as to make an understanding and enlightened decision. This latter element requires that before the acceptance of an affirmative decision by the experimental subject there should be known to him the nature, duration, and purpose of the experiment; the method and means by which it is to be conducted; all inconveniences and hazards reasonably to be expected; and the effects upon his health or person which may possibly come from his participation in the experiment.

The memorandum explained that consent must be in writing and witnessed; that the Secretary of any Service must take responsibility for adherence to consent requirements; that the "duty and responsibility for ascertaining the quality of the consent rests upon the person who initiates, directs or engages in the experiment" and that this duty is a "personal duty and responsibility which may not be delegated to another with impunity." It also states that volunteers must be males under the age of 35, "with no mental or physical diseases." Finally, the memorandum established a series of procedures for implementing these requirements.

Clearly, the CIA did not follow these rules. Recent revelations have provided yet additional examples that these standards were not followed in government-conducted research on the effects of nuclear radiation. The existence of these government-adopted rules, however, reinforced Professor Rothman's view that the CIA could be held accountable for failing to adhere to the principles of informed consent in funding Cameron's work, even if the Defense Department did not properly implement the rules.

Tom Beauchamp provided an entirely different account of the standards applicable to judging the conduct of Cameron or other researchers in the 1950s. Professor Beauchamp differentiated between what he called transcendent standards identified by morally acute institutions and individuals and the standards adopted by the ethos of the research field. He agreed that morally acute individuals like Andrew Ivy, an author of the Nuremburg Code and a resolution adopted by the American Medical Association in 1946 that required consent of research subjects, demanded informed consent of human subjects in biomedical research around the time of World War II, but he denied that those principles had become part of the research ethos. As Professor Beauchamp put it in his deposition, "it was pretty much left up to the individual investigator to figure out in the context in which the individual investigator was doing the research the difference between it being decent or it not being decent, it being justifiable or not it being justifiable to do this."

Professor Beauchamp held that one could judge a person's conduct from either perspective, the morally acute individual or institution or the ethos of the time. But in the absence of binding regulations, he held that a person could be held professionally accountable in some relevant way (e.g., imposing a sanction) only by applying the standards derived from the ethos of the time. These in turn derive from "what the prevailing practices were, much in the same way in which a court might today consider whether or not it was a breach of practice by asking what the practices are." Put another way, judgments can be made based on what "the normal practitioner would know to be acceptable." And in his view, those prevailing practices among biomedical researchers did not require informed consent until well into the 1970s at the earliest. While allowing that we could choose to use a different standard of judgment, e.g., the morally acute individual, in general Professor Beauchamp did not believe it fair to do so. Rather, he claimed, judgments should be made only according to the then-current "ethos" in the medical research community.

Accordingly, in making judgments about Cameron (and government funding of Cameron) it was important to Beauchamp that "what [Cameron] was doing was fairly common in psychiatric research, not in terms of his particular methodology, but in the sense that there was a great need to come up with cures and therapies for desperate patients or very sick patients." He added, "And I would have thought that Dr. Cameron was a part of that ethos. And you might see it as a good ethos or as a bad ethos. Again I am not making any judgment on that. But he was a part of that."

Thus, Professor Beauchamp testified that in making a decision whether to hold Cameron or his funders accountable for the injuries to the nine plaintiffs because of lack of informed consent, the Nuremburg Code and its principles should not be applied because they were not part of the research ethos at the time. "Should Dr. Cameron have done better [in obtaining consent] … irrespective of what the prevailing practices were among psychiatrists at the time; should he really have transcended those standards and told his patients more than he did, that is the way I would interpret your question.… I think that would be asking too much of the man in my judgment."

I also asked Professor Beauchamp whether the government's decision to fund the research could be judged under the Secretary of Defense's memorandum for government-sponsored atomic, biological and chemical research with human subjects. He responded that since the Secretary's memoranda were largely ignored, they were not the basis for making a judgment. "I think that is the fallacy of thinking that something becomes a written rule, that it therefore becomes the prevailing practice." This exchange followed:

Q. But here, it was a government policy of the Secretary of Defense?

A. Yes. But you see, the critical question to me is not who passed the policy and is it on the books, but was it filtered down. Was there any mechanism for letting the people who had to then put it into practice put it into practice, other than say publishing it in a journal or something like that. To me, that is not an organ of dissemination that eventuates in any changes in practice.

III.

The court in the *Orlikow* case never resolved which standards of accountability should apply because the parties settled. There was a denouement, though. Harvey Weinstein, a psychiatrist and son of one of the plaintiffs, pressured the American Psychiatric Association for a condemnation of the conduct of its former President. Finally, in May, 1989, Paul Fink, President of the APA, issued a statement in which he said that "In my view, the research ... could not be conducted today, and should not have been conducted then, no matter how frustrated and desperate researchers were to find treatments for those with mental illness" (Weinstein).[16]

But what are we to make of the Rothman/Beauchamp disagreement? Does it illuminate any of our current controversies? At one level, the disagreement between them is simply about history: what was the status of the requirement of informed consent for human research subjects in the 1950s?

At another level, though, it sheds light on the nature of the disagreements that both led to and pervade this conference, disagreements that are based in part on the different concepts of accountability relied upon by Professors Rothman and Beauchamp. Professor Beauchamp essentially adopted a malpractice-type standard for accountability in research with human subjects, that is, whether the conduct at issue is consistent with the practices of colleagues in the research field. By contrast, Professor Rothman, like Beecher and the Tuskegee panel, viewed the obligation of practitioners far more broadly. In his view, standards of accountability do not and cannot derive exclusively from within the medical profession, but from the larger society. Moreover, researchers are expected to be aware of the principles of research ethics, including informed consent, that have been articulated through codes, journals, and other forms of discussion. Finally, in Rothman's view the fact that others in the field do not adhere to those principles in their research practices is no defense in a proceeding to hold a particular researcher accountable.

Even in a world now dominated by regulation of informed consent in research, the tension between the two views persists. The apparatus of regulation, including Institutional Review Boards and written consent forms, would appear to reflect the triumph of the view that the larger society, not the research community, controls the ethics of biomedical research. But the regulatory system operates through mechanisms that enable, even encourage, filling in the large interstices in the ethics of consent—the nature of the disclosures, the process of consent, the overall relationship to the subject—from within the research community. The values and approaches of that community thus can continue to exert a large influence over the approach to accountability for consent in research. Moreover, the very structure of the regulatory scheme tends to insulate the public from a meaningful role in the review process. The central vehicle for supervision, the IRB, is often dominated by institutional insiders and those who engage in research themselves. The public agency, the Office of Protection from Research Risks (OPRR), eschews oversight in the absence of a complaint. In such a system, the perpetuation of standards of accountability that emerge from within the research community itself should not be a surprise.

How great are the differences in the two worlds? Surely practice-based standards today are far different from what they were in 1950, or even 1970, and far closer to the autonomy-based approaches to informed consent generally embraced by the larger public. In this paper, I can only sketch some differences between public and research-community-based standards, but I do suggest that as evidenced in writings, in practices and at this conference, research-community standards sometimes differ from those that stress the autonomy of the research subject.

A brief paper written by William Wirshing, Director of the Movement Disorders Laboratory at UCLA, no doubt written in the wake of the controversy there, illustrates this point. Wirshing interprets his obligations as a researcher not to make informed consent and patient protection absolute requirements, but to engage in the task of "balancing rights and benefits" of research. He argues that "most clinical experimentation takes place in a setting that is safer, better funded, and far more enlightened and informed than nonresearch medical treatment" and complains that research is hampered by "a process that has become so legalistically elaborate that it is difficult if not impossible to 'properly' inform a psychiatrically ill patient." The result is for investigators to "favor to a fault the safety side of the risk equation" leading to the possibility, in his view, of "a decade of research marked by profound insecurity and constipation."[17] In short, Wirshing seeks to balance the rights of research subjects against the value of the research, an approach significantly different from one that asserts autonomy above all. The views expressed at the conference illustrate the point as well. Some of the researchers present responded to concerns about consent with a defense of the legitimacy of the research itself, as though the good ends of the research could temper the need for thorough disclosure and consent. They questioned both the wisdom of external standards of accountability and the content of such standards.

It is difficult to assess how representative these views are, but I believe we will not make progress until differences in the approaches to accountability are openly acknowledged and confronted. Ultimately, field-based standards of accountability in research that persist among researchers cannot be sustained because they cannot meet the standards the public demands—and has every right to demand. Researchers have no special expertise in research ethics and the public has too great a stake in the protection of human subjects and in the manner in which subjects are treated in the research process. Moreover, the researcher cannot possess special ethical expertise because of the enduring conflict between the researcher's fundamental interest in the advancement of knowledge and the subject's interest in disclosure and good treatment. As Katz has pointed out so often, this inherent conflict of interest underlies many of the dilemmas of the researcher-subject relationship.

In the absence of a recognition of differences, we can anticipate further conflict and calls for ever more detailed requirements, including specific changes in the OPRR regulations designed to improve the consent process (Goldner, 1993).[18] More formal regulation may be the only way to bring to an end the difference between socially imposed and research-community-based standards, but I believe it would be better for the research community to embrace standards from the larger public, and join with it in designing consent processes and in evaluating compliance with those processes that recognize the legitimacy of a public standard. It is time to abandon research community-based approaches to accountability for subject consent in research.

NOTES

1. United States v. Brandt (The Medical Case), II Trials of War Criminals Before the Nuremburg Military Tribunal Under Control Council Law No. 10, at 181–182 (1949).
2. Katz, J. Human Experimentation and Human Rights. *St. Louis University Law Journal* 1993; 38:7–54.
3. *New York Times*, March 10, 1994.
4. *New York Times*, March 14, 1994.
5. Appelbaum, P.S. and Roth, L.H. The Structure of Informed Consent in Psychiatric Research. *Behavioral Sciences and the Law* 1983; 1:9–19.
6. Beecher, H. Ethics and Clinical Research, *New England Journal of Medicine* 1966; 274:1354–1360.
7. Orlikow v. United States, 682 F. Supp. 77 (D.D.C. 1988).
   Although the case has received little attention in the United States, it was a cause celebre in Canada. Three books were published about the case in Canada, one of which was also republished here. The two exclusively Canadian books are Ann Collins, *In the Sleep Room, The Story of the CIA Brainwashing Experiments in Canada*, Toronto: Lester and Orpen Dennys, 1988; Don Gillmor, *I Swear by Apollo, Dr. Ewen Cameron and the CIA-Brainwashing Experiments*. Montreal: Eden Press, 1987. The third book, by a psychiatrist who is the son of one of the victims, is Harvey Weinstein, *Psychiatry and the CIA: Victims of Mind Control*. Washington: American Psychiatric Press, 1990.
8. Rothman, D. *Strangers at the Bedside*, New York: Basic Books, 1991.
9. Faden, R.R. and Beauchamp, T.L., *A History and Theory of Informed Consent*, New York: Oxford University Press, 1986.
10. Marks, J. *The Search for the Manchurian Candidate: The CIA and Mind Control*, New York: McGraw-Hill 1980.
11. Rauh, J. and Turner, J. Anatomy of a Public Interest Case Against the CIA. *Hamline Journal of Public Law and Policy* 1990; 11: 307–363.
12. Rothman, D. Ethics and Human Experimentation: Henry Beecher Revisited. *New England Journal of Medicine* 1987; 317:1195–1199.
13. The affidavit is quoted at Rauh and Turner at 339.
14. Rothman (1991) at 16–17.
15. Jones, J. *Bad Blood*, New York: Free Press 1981. p. 210.
16. Quoted in Weinstein at 278. The Canadian Psychiatric Association refused to go this far, stating only that Cameron's research could not have been conducted today.
17. William C. Wirshing, In a Perfect World None of This Would Concern Us, *The Journal of the California Alliance of the Mentally Ill*, 1994; 5:30.
18. See, for example, Jesse Goldner, An Overview of Legal Controls on Human Experimentation and the Regulatory Implications of Taking Jay Katz Seriously, *St. Louis University Law Journal* 1993; 38: 63–134.

Human Rights in Reference to Persons with Mental Illness

Adil E. Shamoo

Department of Biological Chemistry, University of Maryland, School of Medicine, 108 N. Greene Street, Baltimore, MD 21201

What is remarkable about our civilization at the end of the twentieth century is the slow pace by which we recognize fundamental human rights for all our citizens. The idea of individual freedom has been the subject of discussion in many major civilizations throughout the world's history (Laqueur and Rubin, 1979). It was not until the late 1940s that our civilization, through the United Nations, announced the "Universal Declaration of Human Rights." In the first sentence of the first article it states: "All human beings are born free and equally in dignity and rights" (U.N., 1948). In article two, the declaration makes clear that these rights are blind to all distinctions such as race, sex, and national origin. It is also clearly inherent in the intent of the Declaration that these rights cannot be taken away from the individual if the individual is: ill or on the site of the illness in the body—i.e. liver vs. heart vs. blood vs. brain or disability. This is why I think it is even more remarkable how slow our civilization is in recognizing the rights of persons with mental illnesses.

In this article, we will review the UN declaration on the protection of persons with mental illness and how the original UN Declaration would have been modified to include a non-discriminatory status for illness and disability. The article then reviews the historical context of human rights declarations by notable figures such as Locke and Jefferson. Further, the article reviews how one can extend the non-discriminatory nature of the "equal protection" clause to all human beings regardless of illness or disability. Finally, the article will address the issue of use of a group vulnerable and often incompetent human beings and persons with mental illness—as research subjects without any benefits to themselves.

UN DECLARATIONS

It was not until 1991 that the U.N. declared the "Principles for the Protection of Persons with Mental Illness and for the Improvement of Mental Health Care." Rosenthal and Rubenstein (1993) were the first to applaud and point out the

This article will also appear in the journal: *Accountability in Research*, Volume 4, pp. 207–216 (1996).

strengths of the declaration in the area of the protection of patients from harm. It is of interest to note that the U.N. felt compelled to issue a special declaration (relevant sections will be discussed later) for the rights of persons with mental illness. This moral force of world opinion even though welcomed, does further codify and stigmatize persons with mental illness. Even though I appreciate the good intentions and the enormous potential positive moral effect of such a declaration, it is nevertheless a separation that could have been avoided if persons with mental illness or disability had been seriously considered to be included with all other human beings as part of the "Universal Declaration of Human Rights."

The only place where the Declaration could have been modified to include the rights of persons with mental illness is in article two. It should have articulated these rights by declaring: "Everyone is entitled to all rights and freedoms set forth in this Declaration, without distinction of any kind, such as race, colour, sex, *illness*, *disability*, language, religion, political or other opinion, national or social origin, property, birth or status." The only addition (and emphasis mine) I have made to this profound statement are the words "illness" and "disability." In other words, all human beings should enjoy the same rights regardless of whether they are ill or not and regardless of the site of the illness or disability. This declaration would have been so broad as to add a nondiscriminatory status to all illnesses, regardless of site or origin—including such diverse maladies as AIDS, mental illness, physical or mental disabilities, or cancer.

Despite my reservations, I strongly support the UN declaration on mental illness. This is because it represents a giant step in recognizing that the world has not paid attention to the serious plight of persons with mental illness. In the first principle, sections 2 & 3, the declaration summarizes the fundamental human rights for persons with mental illness:

2. All persons with mental illness or who are being treated as such persons, shall be treated with humanity and respect for the inherent dignity of the human person.
3. All persons with a mental illness, or who are being treated as such persons, have the right to protection from economic, sexual and other forms of exploitation, physical or other abuse and degrading treatment.

Rights for individuals have been the subject of discussion throughout history. In the past few centuries, the concept of human rights has been at the forefront of declared intentions of a great many thinkers as well as political leaders (Henkin 1990, p. IX). The philosopher and human rights advocate Henkin states that: "Human rights is the idea of our time, the only political-moral idea that has received universal acceptance."

HISTORICAL PERSPECTIVE ON HUMAN RIGHTS

Historically, John Locke—the British philosopher and theologian—was considered one of the early individuals to advocate the protection of individual rights from abusive political structures (Henkin, 1990, p. 1). It has been the rebellious spirit of individuals which refused to be conquered by political entities that has prevailed. Those political entities not only suppressed individual rights but attempted to subjugate them into a lower class or even sub-class of human beings.

It is that individual spirit that screamed louder and louder to protect life, dignity, and the freedom of the individual in pursuit of their well being. It was Immanuel Kant in 1792 (cited in Laqueur and Rubin, 1979, editor, p. 84) who codified that the individual's inherent rights could not be taken away or even "voluntarily" given away regardless of the circumstances when he said:

And no-one can voluntarily renounce his rights by a contract or legal transaction to the effect that he has no rights but only duties, for such a contract would deprive him of the right to make a contract, and would thus invalidate the one he had already made.

The U.S. Declaration of Independence was one of the early political documents that codified these rights and their relation to the equality of all persons: "that all men are created equal, that they are endowed by their Creator with certain inalienable Rights, that among these are Life, Liberty, and the pursuit of Happiness."

The influence of Thomas Jefferson was clearly apparent in the Declaration of Independence. Jefferson stated that "[all human beings] are endowed by their Creator with certain inalienable rights." What is remarkable in all of these statements is the unwritten implication that "all human beings" should have these rights. There was no intent to discriminate against a group of people regardless of their specific characteristics—except, unfortunately, race. To my knowledge, there was no mention of excluding those persons with mental illness. I am defining mental illness here in the broadest sense from severe illnesses such as schizophrenia to less devastating maladies such as mild depression. Allen Gerwith (in Paul *et al.*, 1984, p. 3) in explaining the inherent human rights of each individual said: "The having or existence of human rights consists in the first instance not in the having of certain physical or mental attributes, but rather in certain justified moral requirements."

It is interesting to note that Alan Gerwith's term "mental attributes" could encompass the great variation in mental status among all of us in these attributes.

HUMAN RIGHTS AND COMPETENCE

Unfortunately, myths, lack of knowledge about persons with mental illness, and confusion about competence have influenced even few writers on human rights. For example Diana T. Myers (1985, p. 128) in her book on *Inalienable Rights: A Defense* equates "incompetent human" with "nonhuman animals" when she states:

In particular, nonhuman animals, children, and morally incompetent humans are not capable of moral agency but cannot be mistreated with moral impunity. We are not free to starve our pets wantonly, to beat our children brutally, or to use incompetent humans for experimental purposes. Yet, none of these kinds of individual seems to be entitled to the protection of inalienable rights.

This unfortunate inclusion of human beings with nonhumans reflects a lack of appreciation of our most fundamental value system which should include the equal treatment of all human people. The use of the term "incompetent" is on a slippery slope in which society could discriminate not just between incompetency from competency, but also between human beings and non-human beings. Furthermore, incompetency can be either intermittent (Shamoo and Irving, 1993) or a one time occurrence. Even if one follows Ms. Myers' logic of recognizing the rights of children as future moral agents with inalienable rights she does not recognize human beings who, as it is with children, are future moral agents when in remission, treatment or numerous other reasons. In addition, these "incompetent" persons can often become capable moral agents to differing degrees and at different times of their life. The vast number of human beings' contributions to humanity and civilization cannot be captured by an artificial and ill-conceived definition regarding "mental competence." As a matter of fact there are data to associate mental illness with some great historical leadership and creativity in all walks of life such as the arts, politics, religion, and the military (Goodwin and Jamison, 1990, p. 332). More recently, the contributions of a person with mental illness to our civilization was starkly demonstrated recently by John Nash, Jr., a mathematician, who was awarded the 1994 Nobel Memorial Prize in Economics (Nasar, 1994, p. 8, C3). Dr. Nash has been suffering from schizophrenia for over twenty years. Dr. Nash is again active in research in mathematics.

HUMAN RIGHTS AND DISABILITY

Destro (1986a, p. 79) argues against the slippery slope of the "quality-of-life" ethics in favor of the "sanctity-of-life" ethics. Stanley Herr (1992, p. 146), a Professor of Law at the University of Maryland School of Law, argued that each individual, regardless of illness or disability, has full human rights when he said: "There is now a working consensus that every man and woman, between birth and death, counts, and has a claim to an irreducible core of integrity and dignity." Some would even argue that the statement should have said: "between conception and death."

The fight for human dignity and justice for persons with mental illness is no less of a fight than the enormous struggles in our American history that helped shape our great nation. After all, some of our forefathers were willing to fight for the Bill of Rights, voting rights for women, and civil rights. The fight for the full rights of persons with mental illness *must* continue in intensity and in a similar fashion to those fights that gained us the Bill of Rights and voting rights, because, as Professor Herr (1992, p. 142) recognizes:

Persons with mental disabilities have historically had difficulties in convincing their fellow citizens of their personhood, let alone of their citizenship and possession of enforceable rights. As a result, the law has often sanctioned their inferior status and exclusion from society.

EQUAL PROTECTION

I am not a legal scholar but my thoughts on the subject are that the rights for persons with mental illness are inherent in our Constitution. This is consistent with the interpretation of the constitutional scholar Charles Black (1991) of Columbia University. He argues that unnamed human rights are protected. Although Black was not discussing persons with mental illness, I am. Moreover, human rights for persons with mental illnesses are protected under the Fourteenth Amendment's "equal protection" clause. Since persons with mental illness are not excluded by name they are, as any other unmentioned group, covered by the "equal protection" clause. Robert Destro (1986b, p. 55) aptly describes the role of government towards its citizens regardless of their individual characteristics with the following statement:

Equal Protection Clause is a limit on the power of government to impose disabilities or to grant benefits on the basis of characteristics that must be deemed irrelevant to the relationship between government and the individual. By its terms, it requires the government to be a protector that does not discriminate among individuals on the basis of those characteristics.

In the same article Destro emphasizes that the laudable social goals of helping minorities and the disabled is a legitimate function of the government outside the scope of the Fourteenth Amendment's equal protection clause. There is very little argument about whether physically disabled individuals should have full human rights. This is because of several factors. Among them is the fact that we can identify much more easily with physically disabled individuals. Therefore, we can accept the concept "Do unto others as you would have them do unto you." For some of us, we might think of this as an insurance policy just in case something like this happens to us or our family. However, this kind of reasoning falls apart when it comes to persons with mental illness not because of logical reasoning, but rather due to stigma, myths, and denials. No one thinks mental illness would strike them or their families until.it is too late. Therefore, declaring, as well as acknowledging in practice, the full human rights of persons with mental illness would contribute to alleviation of stigma, myths, and denials over and above its own inherent value of articulating these rights.

Some would ask what benefits would the declaration of human rights to persons with mental illness bring to this group of people? Some would argue none, because in practice persons with mental illness need to have additional protections of their rights. Furthermore, a declaration of human rights would not be enforceable on a day to day level in order to help them, moreover, in order to help them we need to have "healthy or good discrimination." All of these objections are reasonable.

However, insistence that persons with mental illness enjoy full human rights is of great benefit to this group, for it sets a moral as well as a legal tone for all subsequent discussions. This is extremely important in combating discrimination. Insurance companies use current laws to discriminate against persons with mental illness. Moreover, even the 1991 Americans with Disabilities Act (ADA) has been interpreted "as allowing" insurance companies to discriminate towards persons with mental illness. This, I believe, is in sharp violation of the constitutional clause of "equal protection." To be consistent with the constitution, the ADA should be interpreted as banning discrimination towards person with mental illness by insurance companies or anyone else.

USE OF HUMAN SUBJECTS IN RESEARCH

The use of human subjects in research has received attention in the past twenty-five years (for reviews see Shamoo and Irving, 1993, and Katz, 1994). Our focus here is on the use of persons with mental illness in research. Despite our focus, we cannot escape the strong relationships—historically, philosophically, and ethically—of the use of persons with mental illness in research with the use of persons in research in general. Our personal feelings as well as our ethical values have certainly been greatly influenced by the atrocities committed in Nazi Germany as illustrated in recent three books about that area (Muller-Hill, 1988; Proctor, 1988; Caplan, 1992) towards human beings in general as well as those with handicaps such as persons with mental illness. One cannot help but be influenced by the Nazi experience as well as other serious violations of human rights, such as those committed within the U.S. towards patients with syphilis in the Tusgekee Syphilis Study (Katz, 1972) which occurred between 1940 and 1960, and the radiation exposure of human subjects which took place in the 1940s and 1950s (Lee, 1993).

The use of human subjects in research brings with it a host of conflicting values and ethical norms. In an academic setting the freedom of inquiry is the bedrock of academic freedoms. However, the use of a human being for experimental purpose for the sake of acquiring new scientific knowledge is a serious undertaking which requires careful consideration and compelling justification in order to proceed. Over twenty years ago Katz (cited in Katz, 1993, p. 20) so eloquently stated. "When human beings become the subjects of experimentation ... tensions arise between two values basic to Western society: freedom of scientific inquiry and protection of individual inviolability."

However, Pellegrino (1993) gives a great deal of latitude to the individual investigator to resolve these conflicts. Resting the burden of resolving the awesome tensions between these possibly conflicting ethical values on the individual investigator, almost to the exclusion of the institutional policies, eventually results in lowered protections of human subjects in modern day research enterprises (Irving and Shamoo, 1994).

The use of vulnerable groups in research, such as those inflicted with mental illness, accentuates and magnifies all of the issues concerning the use of human subjects in research. Moreover, the use of persons with mental illness in research adds a few unique and important problems. The most prominent among these is the fact that persons with mental illness may have their cognition/comprehension impaired. This impairment, therefore, may present an enormous challenge to those involved in research who are trying to preserve the integrity of the patient while, at the same time, claiming that the patient has given informed consent to participate as a research subject. However, a more fundamental question which needs to be addressed concerns whether the patient should participate in research in the first place, and to whose benefit are the resulting data? Some would argue that acquisition of new knowledge is a sufficient justification.

Louis Henkin (1990) in response said:

If someone will have to be sacrificed it will be someone else; that others submit to the same risk for one's own welfare, and that the selection will be by lot or chance or at least according to some rational, neutral principle.

TO WHOSE BENEFIT?

There are a host of general questions that arise in the use of human subjects in research including:

1. Should there be direct benefits to the participant in research?

In addressing the first question, if there is no foreseen benefit to the participant then why should the person be "forced" to participate? As Jonas (1969, p. 222), when discussing voluntary consent, said:

"No one has the right to choose martyrs for science" was a statement repeatedly quoted in the November, 1967, Daedalus conference. But no one, not even society, has the shred of a right to expect and ask these things. They come to the rest of us as a grata gratis data.

In order to ensure voluntary participation, informed consent was introduced in order to avoid tainting the voluntary nature of participation. In a very lengthy and scholarly paper entitled "Human Experimentation and Human Rights," Jay Katz (1993) expounded this subject with a detailed historical account of how human experimentation has come close to violating basic human rights. Katz (1993, p. 10) in addressing the issue of the wide use of informed consent as the ultimate defense to perform research on human subjects states: "Clearly, consent is a necessary, but not sufficient, justification for using human beings as subjects for research."

2. Should the participants be subjected to control (placebo or double blind) experiment? This is a troubling issue since those on placebo will receive no benefit at all.

Jonas (1969, p. 240) in addressing the "double blind" control group (placebo) said: "Whatever may be said about its ethics in regard to normal subjects, especially volunteers, it is an outright betrayal of trust in regard to the patient who believes that he is receiving treatment."

3. Should the physician treating the patient be the same person as the researcher who is using the patient as a subject? No—this dual role of the physician corrodes and infects the physician–patient relationship. To quote Jonas again:

In the course of treatment, the physician is obligated to the patient and to no one else. He is not the agent of society, nor of the interests of medical science, the patient's family, the patient's co-sufferers, or future sufferers from the same disease (Jonas, 1969, p. 238).

Some would argue that those restrictions would slow science. Again quoting Jonas (1969, p. 237): "This price—a possibly slower rate of progress—may have to be paid for the preservation of the most precious capital of higher communal life."

Edmund Pellegrino (cited in Katz, p. 30) recognizes the problem of using beneficence towards the patient as the overriding utilitarian calculus for the use of human subjects in research when he states: "The possible loss of knowledge cannot outweigh the possibility of harm to the subject even if the utilitarian calculus indicates great benefit to many and harm to only a few."

In a recent commentary under the heading "Sounding Board: The continuing unethical use of placebo controls" in the *New England Journal of Medicine*, Rothman and Michels (1994, p. 397) stated:

All these parties should adhere to the precept that patients ought not to face unnecessary pain or disease on account of a medical experiment, and they should question the ethical legitimacy of using placebos in any experiment.

The huge public benefits that result from the utilitarianism of science drives the desire for quick progress in science. Deriving benefits to society does not constitute the sole compelling reason to pursue research. As David Walsh, chairman of the Department of Politics at Catholic University declared in a different context in an op-ed (1994, p. A23)—"Benefits Don't Make It Ethical." This strong drive for the acquisition of new knowledge in science for the public good has taken supremacy in some quarters at the expense of the "respect for persons" and autonomy as illustrated by Dr. Anthony Fauci. Dr. Fauci, who is the director of the National Institute of Allergy and Infectious Disease, states: "It's not to deliver therapy. It's to answer a scientific question so that the drug can be available for everyone once you've established safety and efficacy (cited in Hellman and Hellman, 1991, p. 1585)." Fauci appears to be willing to sacrifice the patients' interest for the sake of science (Hellman and Hellman, 1991). Hellman and Hellman (1991, p. 1587) go on to say:

We argue that such rights cannot be waived or abrogated. They are inalienable.... This rights based on the concept of dignity, cannot be waived. What of altruism, then? Is it not the patient's right to make a sacrifice for the general good?

The person's right to waive his own rights become more unacceptable when the person is vulnerable and his or her cognition may be impaired. Not only should such a person not waive those rights in such circumstances but society should not permit institutions to practice issuing such waivers. By asking such a vulnerable population to waive their rights we have committed a grievance action.

Destro (1994, p. 52) questions the validity of the whole concept of informed consent for patients with mental illness when he states: "the law of informed consent makes it clear that consents, by patients whose very capacity to make judgments is in question, cannot be trusted."

In reference to the autonomy of persons with serious mental illness, Eugene Brody (1985, p. 61), a psychiatrist with over thirty years of experience in psychiatry, former chairman of the Department of Psychiatry, University of Maryland School of Medicine, and the editor-in-chief of *The Journal of Nervous and Mental Disease* said:

Arguing for the autonomy of mentally incompetent persons without support in hostile or uncaring environments is increasingly considered a legal as well as an ethical error ... the patient suffers from a severe mental disorder, lacks capacity to make a reasoned decision concerning treatment, is treatable, and is likely to harm himself or others.

FINAL COMMENT

We are all and I mean *all* of us on this earth to be treated equally and fairly. In closing I would like to quote a short phrase from Nelson Mandela's (1994, p. C7) speech to a joint session of U.S. Congress. Even though Mr. Mandela was applying his remarks in the context of the human race in general, we can include persons with mental illness when he unified all peoples through the phrase: The Oneness of the Human Race.

Mr. Mandela's philosophy and declarations of treating all human beings equally and fairly embodies the highest ethical human values of respect to person. It is in this spirit that we all should rededicate ourselves to demand from our scientific and political leadership to adhere to these principles and put it into practice towards persons with mental illness.

ACKNOWLEDGMENT

The author appreciates comments of Stanley Herr of the University of Maryland Law School on an early version of this manuscript.

REFERENCES

Black, Jr., C.L. (1991) One nation indivisible: unnamed human rights in the states. *St. Johns Law Review* 66:17–57.

Brody, E.B. (1985) Patients' rights: a cultural challenge to Western psychiatry. *American Journal of Psychiatry* 142:58–62.

Brody, E.B. (1989) New horizons for liaison psychiatry: biomedical technologies and human rights. *American Journal of Psychiatry* 146:293–294.

Caplan, A.L. (ed.) (1992) *When Medicine Went Mad*. Totowa, New Jersey: Humana Press, pp. 1–359.

Destro, R.A. (1986a) Quality of life ethics and constitutional jurisprudence: the demise of natural rights and equal protection for the disabled and incompetent. *The Journal of Contemporary Health Law and Policy* 2:71–130.

Destro, R.A. (1986b) Equality, social welfare and equal protection. *Harvard Journal of Law and Public Policy* 9:51–61.

Destro, R.A. (1994) Law professionalism, and bad attitude. *The Journal* 5:50–53.

Goldner, J.A. (1993) An overview of legal controls on human experimentation and the regulation implications of taking Professor Katz seriously. *Saint Louis University Law Journal* 38:63–134.

Goodwin, F.K. and Jamison, K.R. (1990) *Manic-Depressive Illness*. New York: Oxford University Press.

Hellman, S. and Hellman, D.S. (1991) Sounding board: of mice but not men. *New England Journal of Medicine* 324:1585–1589.

Henkin, L. (1990) *The Age of Rights*. New York: Columbia University Press.

Herr, S.S. (1992) Human rights and mental disability: perspectives on Israel. *Israel Law Review* 26: 142–169.

Irving, D.N. and Shamoo, A.E. (1993) Which ethics for science and public policy. *Accountability in Research* 3:77–100.

Jonas, H. (1969) Philosophical reflections on experimenting with human subjects. *Daedalus* 98:219–247.

Katz, J. (1993) Human experimentation and human rights. *Saint Louis University Law Journal* 38:7–54.

Laqueur, W. and Rubin, B. (eds.) (1979) *The Human Rights Reader*. Temple University Press.

Lee, G. (1993). U.S. should pay victims, O'Leary says. *The Washington Post*, A1–A12.

Mandela, N. (1994) The oneness of the human race. *The Washington Post*, Oct. 9, C7.

Meyers, D.T. (1985) *Inalienable Rights: A Defense*. New York: Columbia University Press.

Miller-Hill, B. (1988) *Murderous Science*. New York: Oxford University Press. Translated by G. Fraser.

Nasar, S. (1994) The lost year of a Nobel Laureate. *The Washington Post*, Nov. 13, C3, p. 8.

Paul, E.F., Paul, J., and Miller, Jr., F.D. (eds.) (1984) *Human Rights*. Basil Blackwell.

Pellegrino, E.D. (1993) Autonomy, beneficence, and the experimental subjects consent: a response to Jay Katz. *Saint Louis University Law Journal* 38:55–62.

Proctor, R. (1988) *Racial Hygiene*. Cambridge, MA: Harvard University Press.

Rosenthal, E. and Rubenstein, L.S. (1993) International human rights advocacy under the "Principles for the Protection of Persons with Mental Illness." *International Journal of Law and Psychiatry* 16:257–300.

Rothman, K.J. and Michels, K.B. (1994) The continuing unethical use of placebo controls. *The New England Journal of Medicine*, 331:394–398.

Shamoo, A.E. (1994) Our responsibilities toward persons with mental illness as human subjects in research. 5:14–16.

Shamoo, A.E. and Irving, D.N. (1993) The PSDA and the depressed elderly: "intermittent competency" revisited. *The Journal of Clinical Ethics* 4:74–80.

Shamoo, A.E. and Irving, D.N. (1993) Accountability in research using persons with mental illness. *Accountability in Research* 3:1–17.

Shamoo, A.E. (1994) Our responsibilities toward persons with mental illness as a human subjects in research. *The Journal* 5:14–16.

United Nations (1947) Universal Declaration of Human Rights.

Walsh, D. (1994) Benefits don't make it ethical. *The Washington Post*, Oct. 27, A23.

Approximating Ethical Research Consent

Timothy J. Keay

School of Medicine, University of Maryland at Baltimore, Department of Family Medicine, 405 West Redwood Street, Baltimore, Maryland 21201

We are gathered primarily because of the perception that we need to change our understanding of what it means to give or obtain consent in medical research. The history of consent to participate as a subject in medical research is a history of learning from errors.

Let us start, somewhat arbitrarily, with the 1767 English decision *Slater v Baker and Stapleton*.[1] This case is well described by Faden and Beauchamp, who I quote as follows:

In *Slater*, the plaintiff hired Drs. Baker and Stapleton to remove the bandages from a partially healed leg fracture. Instead, the defendants, over the plaintiff's protests, refractured the leg and placed it in an apparently experimental apparatus to stretch and straighten it during rehealing. Slater claimed that the defendants had, in essence, breached the contract they had made with him by "ignorantly and unskillfully" breaking his leg and injuring him. [p. 116]

Slater won damages. The court stated, "… it is reasonable that a patient should be told what is about to be done to him, that he may take courage and put himself in such a situation as to enable him to undergo the operation." [quote from Faden and Beauchamp, p.117.]

One might say that the process could have been better approximated if warning had been given.

Closer to our own time and place is the case of *Schloendorff v Society of New York Hospitals*. In this 1914 case, a woman was found by her physician to have a mass in her lower abdomen. The physician was concerned that it might be a cancer, and urged surgery. The woman refused any operation. However, she did want to know what the mass was. The physician convinced her that if she submitted to an examination under anesthesia, he would be able to determine the mass's nature. She agreed to this examination. Once she was under anaesthesia the physician determined that the mass was probably cancerous and therefore required removal. The woman awoke to find that she had been operated upon and the tumor (which turned out to be benign) gone. What makes this case so memorable, above many like it, was that the Judge was the eminent jurist Benjamin Cardozo, who formulated the opinion "Every human being of adult years and sound mind has a right to determine what shall be done with his own body…"[2]

One might say that the process would have been better approximated if the physician had recognized the person's right to give consent.

With regard to research consent, there is the case of Walter Halushka, related by Dr. Veatch.[3] A student at the University of Saskatchewan, Mr. Halushka went looking for a student job, but was told that he could volunteer for a paid research experiment. He signed a consent form, agreeing to waive liability in return for $50 (Canadian). When he asked what harm might befall him, he was told that accidents, like falling down some stairs at home after the experiment, might happen. A catheter was floated in Mr. Halushka's veins to and through his heart and, after a time, his heart stopped. His chest was cut open by the physician/researchers and cardiac resuscitation was instituted to bring him back to life. After four days in coma and ten days in the hospital he was discharged with $50 (Canadian). This case among others helped to stimulate the formation of institutional review boards (IRBs).

One might say that the process of informed consent would have been better approximated if a committee had looked at the consent procedures and ensured their adequacy in giving detailed information to the potential subject.

We could spend a long time looking at case after case where a prevailing concept of informed consent in medical research was found to be in error. Willowbrook, Tuskegee, Canterbury, Salgo, Natanson, the XYY controversy at Harvard, the Jewish Chronic Disease Hospital Case, or a host of other key cases have shown us what informed consent in research should *not* be. Beecher,[4] in a landmark article documenting 22 studies with questionable ethical practices, and Katz, with the assistance of Capron and Glass, in the encyclopedic *Experimentation with Human Beings*,[5] chronicle the terrible price we have paid to learn lessons about what not to do when ethically including human subjects in medical research. At the recent Pittsburgh meeting of four ethics societies, I was struck by the fact that as much, or more funding for any ethics project ever—in the range of $10–$20 millions US—is going to study the ethics of research with radiation on unwitting human subjects performed by the United States government during the Cold War.

But what is most striking about each and every one of these cases is that they are all formulated in the negative. All are refutations of practices believed to be ethical by those who perpetrated them. It is as if we understand what ethical consent in human research is only by understanding what it is not, and approximate what ethical research consent is by avoiding past errors.

Which brings us to the focus of our conference, to the dead bodies of Tony Lamadrid and Susan Endersby, and probably many more unnamed dead or injured. A survey of published articles on relapse studies with schizophrenic patients raises serious questions about the ethics of some recent psychiatric research.[6] Should schizophrenic patients sign the same consent forms as adults without any mental or medical problems, without giving those schizophrenic patients any protections to determine if they are capable of giving consent?

Dr. Shamoo and I find that the general attitude of surveyed researchers towards obtaining consent might be inferred from the statements in the article by Barbee et al.:[7] The subjects were "twenty-eight acutely psychotic patients with schizophrenia" yet the authors go on to state that "all of the patients in this study were capable of informed consent and entered voluntarily." How acutely psychotic patients, with mean initial Brief Psychiatric Rating Scale scores of 35 and 31, could be considered capable of giving informed consent is not specified. It is deeply disturbing, especially in outpatient relapse studies, that surveyed authors did not report aspects of the expected carefully documented informed consent, explanation of risks and benefits, and assessment of subjects' capacity to give consent. They did not report a careful disclosure of the difference between experimental aspects of the study as against the patient's need for therapy, nor the attendant conflicts of interest. Instead the articles seemed to be continually conflating therapeutic and research goals, as if these were identical.

The research plan in many relapse studies was simple. The schizophrenic patient's medications were stopped and the patient was observed to determine if he or she went psychotic again. The rationale behind these studies was that the medicines used to treat schizophrenia have many serious side effects, and a minority of patients will do well off of any medicines. No one knows how to predict who those few patients are, however, so the studies were designed to collect data about who did well off of medicine and who did not.

The harms that this exposed subjects to is now becoming apparent. Besides those patients who killed themselves, besides those who attacked others in their psychotic confusion, besides those who are just not accounted for in the written reports, there are the harms of psychosis itself. Many subjects were hospitalized after relapse. As Branchey *et al.*[8] state about their relapsed subjects, "... 1 refused all food, 1 eloped, 1 started masturbating publicly, 1 became catatonic, 1 became mute, 1 described delusional experiences never mentioned before, and 4 became extremely aggressive." These severe symptoms associated with relapse are typical of most of the studies reviewed.

In essence then, what has happened is that many schizophrenic patients have been recruited, with questionable research consent practices, to participate in risky research, and many have been harmed.

One might say that the process of research consent would have been better approximated if the conflicts of interest between researcher/clinicians and subjects/patients had been better disclosed.

If we can only aspire to approximate proper research consent, there will be more cases like Slater, Schloendorff, Halushka, and Lamadrid. What we can do, however, is recognize that more problems will arise, that we should anticipate these problems, and that we should attempt to minimize them.

Three general methods for minimizing errors in ethical research consent come to mind. They are emphasizing informed consent, minimizing risk, and minimizing conflicts of interest. The first two methods are mentioned in a 1982 report of the President's Commission for the Study of Ethical Problems in Medicine and Biomedical and Behavioral Research.[9]

While the concept of informed consent is not subject to one simple definition, and is often defined differently by its application, there is general agreement that such a concept exists and that it should be sought from potential research subjects. Consent should not be presumed, nor should consenters be uninformed of the research to which they will be subject. Researchers have a duty to inform potential subjects and seek their informed consent, as difficult and time consuming as this process may be. Also, as Beecher stated, "In the publication of experimental results it must be made unmistakably clear that the proprieties have been observed."[10] Continued efforts to clearly define what these proprieties are is warranted.

Second, we should continue as a general policy to minimize the risks to which research subjects are exposed. IRBs are an effective way of reducing risk: they have been generally effective in improving the quality of research proposals while minimizing risks to patients. They should continue to disallow "no benefit/greater than minimal risk" studies except in extraordinary circumstances.

Third, a new concept of minimizing conflict of interest needs to be adopted. There are many potential areas of conflict of interest, each of which needs attention to determine if they can be practically eliminated, or if not eliminated, then disclosed. One of the first areas of conflict of interest, pointed to by Professor Katz, is the inherent conflict of interest of one person being both clinician and researcher. Many of the relapse studies show a tendency for the investigators to confuse their therapeutic and research roles, often conflating the two. If it is not financially feasible to separate these roles in academic settings, then serious consideration of explicit disclosure to all interested parties should be made. For patients who have a potential or real cognitive problem with giving consent, this should include routine notification of relatives or other proxy named by the patient, of the inherent conflict of interest of the clinician/investigator. Where there is a potential conflict of interest between the two roles of researcher and judge of a potential subject's competence to give consent, routine outside psychiatric evaluation of competence should be recommended. Finally, conflicts of interest in the IRB's composition should be minimized by inclusion of parties with a stake in the welfare of the potential subject group—or if this is not possible, with the clearly written disclosure to each potential subject (and proxies for potentially impaired subjects) of the IRB's conflicts of interest.

We should continue to search for a reasonable method of monitoring research for the harms which will inevitably occur. Many errors inherent in approximating ethical research consent have been shown to be preventable. Advocates of total quality management have sometimes referred to errors as "gems"—that is, a treasure to be carefully examined and valued.[11] Errors can allow us to analyze our current practices and improve our performance in obtaining ethical research consent. It makes more sense to mine for these gems, rather than stub our toes tripping over them when they come to the surface on their own.

And last, we as a society need to balance the costs of continued research with the costs of not doing the research. Progress has a price beyond research and development funding. The inexactitude of our formulation of the concept of ethical research consent necessarily leads to continued sacrifices. Yet these are the precious sacrifices that point us towards progress.

NOTES

1. 95 Eng. Rep. 860, 2 Wils K.B. 359 (1767), reported in Faden, R.R., and Beauchamp, T.L. *A History and Theory of Informed Consent*, New York:, Oxford University Press, 1986. pp 116–7.
2. 211 N.Y. 128, 105 N.E. 93.
3. Veatch, R.M., *Case Studies in Medical Ethics*, Cambridge, MA: Harvard University Press, 1977. pp. 291–295.
4. Beecher, H.K. Ethics and clinical research. *New England Journal of Medicine* 1966;274:1354–60.

5. Katz, J., Capron, A., and Glass, E.S. *Experimentation with Human Beings,* New York: Russell Sage Foundation, 1972.

6. Shamoo, A.E., and Keay, T.J. Ethical Concerns about Relapse Studies. *Cambridge Quarterly of Healthcare Ethics* (in press).

7. Barbee, J.G., Mancuso, D.M., Freed, C.R., Todorov, A.A., Alprazolam as a neuroleptic adjunct in the emergency treatment of schizophrenia, *American Journal of Psychiatry* 1992;149:506–510.

8. Branchey, M.H., Branchey, L.B., and Richardson, M.A., Effects of neuroleptic adjustment on clinical condition and tardive dyskinesia in schizophrenic patients, *American Journal of Psychiatry* 1981:138: 608–612.

9. President's Commission for the Study of Ethical Problems in Medicine and Biomedical and Behavioral Research. *Compensating for Research Injuries: A Report on the Ethical and Legal Implications of Programs to Redress Injuries Caused by Biomedical and Behavioral Research. Vol. 1: Report,* Washington, D.C.: U.S. Government Printing Office, 1982. pp. 25–39.

10. Beecher, *op cit.,* p. 1360.

11. Leape, L.L. Error in medicine. *Journal of the American Medical Association* 1994;272:1851–7.

An Institutional Response
to Patient/Family Complaints

Robert Aller and Gregory Aller

Patient Rights Network, 1301 Brockton Avenue, #9, Los Angeles, California 90025

This paper presents the viewpoints of a father and his son.

PREFACE

In 1979, Principal Investigator/Psychologist Keith Nuechterlein, at the University of California, Los Angeles (UCLA), Neuropsychiatric Institute, began conducting the "Developmental Processes in Schizophrenic Disorders" research project. This research began under a larger ongoing UCLA grant from the National Institute of Mental Health (NIMH) to Principal Investigator/Psychiatrist Robert Liberman. In the early 1980s Dr. Nuechterlein's project began receiving direct NIMH support to follow the course of schizophrenia with over 100 recent-onset patient-subjects, studying various developmental measures of the illness. "The objective of the research is to determine the independent and joint relationship of several psychological, cognitive, psychophysiological, and social environmental variables to relapse and outcome in schizophrenia."[1]

The study had three phases: 1) The first protocol involved experimentation on recent-onset schizophrenic subjects with a fixed dose of an injectable antipsychotic medication (12.5 mg of fluphenazine decanoate every two weeks) for a period of approximately one year; 2) For those subjects who completed the first protocol and were eligible for the second protocol, they entered a double-blind placebo-controlled crossover condition, randomly alternated from 12 weeks of placebo to 12 weeks of medication, then to "a neuroleptic withdrawal phase, in which fluphenazine decanoate is discontinued openly with one and one half years of follow through, for patients who are clinically stable throughout the crossover phase ... *Study Termination*: The study will continue through both conditions of the crossover study and the open withdrawal phase for patients unless 1) the subject withdraws permission for the study or 2) clinical relapse or psychotic exacerbation occurs."[2] 3) During the third phase of the research, patients were monitored, but no protocol existed.

This paper reports on the UCLA research from the perspective of some of the participating families. It may be that the family point of view is rarely presented in response to the reporting of data from high risk research projects where adverse outcomes occurred.

The UCLA researchers have published numerous papers about results from this project. Now, however, there are four families from the research who are speaking out because the grant applications, protocols, published papers and other documents have provided substantive evidence that the suffering endured by subject-patients and their families was a direct result of breaches of applicable ethical and legal requirements for federally funded research.

Following the release of an Office of Protection from Research Risks (OPRR) investigative file[3] (May 11, 1994) about the UCLA research, UCLA's Donald Rockwell, M.D., medical director of the Neuropsychiatric Institute, testified on May 23, 1994, before the House Subcommittee on Regulation, Business Opportunity, and Technology. Dr. Rockwell's 14-page statement was an institutional response to the OPRR report. This paper refers to a problematic assertion made by Dr. Rockwell in his Congressional testimony.[4]

We will examine problems in the UCLA research in three primary areas: 1) lack of proper informed consent; 2) failure to meet patient needs and; 3) failure of the research enterprise to respond properly to serious complaints.

But underlying these issues is a matter of deeper significance. Did the subject-patients suffer exacerbations and relapses as part of the research design?

Jay Katz, M.D., J.D., Yale Law School, addressed this specific issue in his paper "Human Experimentation and Human Rights," published in the *St. Louis University Law Journal* of Fall, 1993.[5]

Katz focussed on the overall design of the second protocol in the research—the crossover and medication withdrawal phase of the research: "In the UCLA study, on the other hand, all patients were withdrawn from medication, indeed required to do so, for research purposes until the needs of the study, and not those of the individual patients had been satisfied. The expectation of relapse was an integral of the research design; it was not an unfortunate consequence of the treatment but one which the investigators deliberately induced. This is particularly problematic because of the continuing controversy in psychiatric circles as to whether relapse leads to, at times, irreversible, injury."

After reviewing some of the researcher's goals (undisclosed to patients and their families in the written informed consents), one can speculate why the informed consent documents might not have candidly educated patient-subjects and their families about the risks and the alternatives to participating in this study.[6]

While we believe that human subject research is essential to improve the future treatment of those with neurobiological illnesses, we also believe that it is crucial for investigators to *adequately disclose* the purposes and the parameters of the research when obtaining written informed consent. Documents from this research project demonstrate that the proper medical needs of the patient-subjects may have been abandoned for research goals, without the knowledge of the patient-subjects and their families.

We hope that this paper will stimulate a more informed discussion among researchers, subjects, families, bioethicists, research administrators and others concerned with the integrity of high risk human subject research.

We begin with a chronology of events.

ALLER FAMILY'S INTRODUCTION TO THE UCLA RESEARCH

My wife, Gloria, and I are graduates of UCLA. As alumni, we've always had the utmost respect and allegiance for our university. This has not changed. Our difficulties have been with a small number of UCLA schizophrenia researchers and administrators.

In October, 1987, our 23-year-old son, Gregory, was diagnosed with "paranoid schizophrenia" by Robert Liberman, M.D., Director of UCLA's Mental Health Clinical Research Center for the Study of Schizophrenia. Dr. Liberman told my son and I that Greg could receive excellent free treatment for his illness at the Aftercare Clinic in UCLA's Neuropsychiatric Institute. Dr. Liberman handed us a brochure and suggested that we read the brochure.[7] The brochure stated that the Aftercare Clinic was sponsored by the National Institute of Mental Health. The purpose of the program was described as follows:

The goal of the Program is to assist persons in making a successful adaptation to life inthe community and to improve their daily living and social skills. An equally important goal is to facilitate the family's coping skills for dealing with mental illness. Where appropriate, consultation with other community agencies and social support networks is provided.

A range of outpatient services are offered through the Aftercare Clinic at the UCLA Neuropsychiatric Institute. Following discharge, patients and their families are provided: "Group Therapy: in small groups, patients learn problem-solving skills and interpersonal effectiveness."

"Family Education: counseling is aimed to upgrade the entire family's coping skills and understanding of the illness, and to facilitate use of resources both within the family and the community."

"Medication is administered at the lowest optimal dose to maximize coping with symptoms and stressors and to minimize side effects." (We would later learn that this claim was not true.)

"Fully integrated with these services is a research project aimed at increasing understanding and knowledge of the factors that are related to relapse and remission."

The brochure indicated that patients would receive excellent treatment.

We felt especially fortunate that our son could receive free care for this devastating illness at a premier treatment center, one of a small number of such NIMH-funded centers in the United States. While we understood that the program would be gathering data to better understand the nature of the illness, family interactions and treatment, we believed that the data gathering was benign and that Gregory would receive the best medical treatment possible, just a short distance from home. Gregory began treatment in this program in March, 1988.

GREGORY ALLER'S STATEMENT—STANDARD DOSE PROTOCOL

In the first phase of the experiment I was lucky. Though some patients had a poor response to the standard dose, the standardized dose practically eliminated all of my delusions and made it possible for me to think relatively clearly.

Since I had been diagnosed with paranoid schizophrenia, relief from the symptoms greatly improved by life. My family and I were ecstatic. I enrolled at Santa Monica College and during the next year I earned a 3.8 GPA, got on the Dean's list and worked 15 hours a week. I was proud of what I was accomplishing. It was certainly the best year of my adult life.

How were we treated? First of all, many of those in this program received financial public assistance. Patient-subjects had practically no extra spending money and were barely getting by. The Aftercare Clinic provided us free lunches with pizzas and sodas, a strong reinforcer so that we came in to the program for medication, groups, testing, and other special events. In addition, we were paid a nominal amount for taking tests.

The Aftercare staff would compliment us on our appearance or what we had recently accomplished. The staff planned activities such as board games that helped us to develop social skills. We had annual picnics with free food.

Jennifer Jones Simon, wife of the late industrialist Norton Simon, once led us on a tour of the Norton Simon Museum of Art. We had a gourmet outdoor lunch and we were each given gifts from the museum.

I appreciated this kind of treatment and I believe that others also felt that the Aftercare staff had our best interests at heart.

The Aftercare Clinic staff became like a second family for many of us. We had a social network and a place where we felt we belonged.

For over a year I had done well on the standardized medication dose of 12.5 mg of Prolixin decanoate. But entering the crossover and withdrawal phase of the research gave patients a a higher status because it was considered a sign of greater success. We were encouraged to believe that we were stronger if we didn't have to take medication.

Those who were in denial about having schizophrenia may have found the staff indirectly supporting their belief that they weren't really ill, because they were encouraged to go off medication.

Instead of providing us with full disclosure and facts about the crossover/ withdrawal phase, the staff used the group sessions to let patients talk about crossover/withdrawal. A patient who was in crossover/withdrawal and hadn't yet had an adverse reaction said that he wasn't having any problems. No one who had suffered an exacerbation or relapse in crossover/withdrawal ever spoke to my group. In retrospect, I believe that the staff withheld critical information about the seriousness of exacerbation and relapse and the after effects. None of the Aftercare doctors was every present to tell us about crossover and withdrawal. A few of the patients disappeared. The staff wouldn't talk about the missing patients. In fact, we were never informed about any adverse reactions that occurred to any patients in the crossover/withdrawal phase.

I had been convinced that this was the best thing to do[8] yet I had not been informed about the alternatives and serious risks.

GREGORY ALLER'S STATEMENT—MEDICATION CROSSOVER/WITHDRAWAL

In mid-1989, the researchers asked me to go off medication. The reason they gave was to prevent the onset of the potentially irreversible movement disorder called tardive dyskinesia (TD). Dr. Nuechterlein, the principal investigator, met with my parents and me, and in response to my mother's concern about potential relapse, he promised my parents that if I had a return of symptoms the medication would be reinstated. Dr. Nuechterlein said that since my parents knew me best, when they reported symptoms, I would get my medication back.

My mother asked Dr. Nuechterlein about this matter several times, stating that our family did not want to see a return of symptoms, especially since I was doing so well and was close to realizing my dream of transferring as a student to UCLA. We trusted the doctors and I agreed to go off the medication. I also made an agreement with my parents that whenever the doctors told me to go back on medication, I would do just that. But that never happened.

During the second phase of the experiment, my symptoms gradually returned. However, this time my symptoms were much more severe than the symptoms at the onset of my schizophrenia. What ensued was a nightmare.

My ability to concentrate fell apart. I was unable to do school work and I would later have to withdraw from all of my classes. I became manic and hyperactive. Some days I would hardly sleep at all. One night I woke up screaming, actually believing that I was sprouting another leg. Often, I wouldn't shave or shower or pay any attention to self-grooming. When riding on the bus I would growl like a dog. I became violent with my father and threatened to kill him.

At one point during my psychotic relapse, I thought that the Devil had taken possession of my mother by entering her body. I thought this was horrible for her, but I believed I could scare or cut the Devil out of her. I grabbed a butcher knife in the kitchen and called to my mother. Before I could reach her, she fled into the bedroom and barricaded the door. My father was able to talk to me so that I put down the knife. I was willing because I thought I had scared the Devil out of her. Looking back, it is horrifying to consider what might have happened if my father had not been present.

I started to believe that there were hidden cameras on the wall of my parents' living room. After a return of paranoia, I failed to self-report my true signs and symptoms to my case worker. I started to have paranoid delusions about government agents chasing me. While at school I remember typing a letter to the FBI while the other students were busy typing their term papers. Later, my sister intercepted my letter to the FBI.[9] (The letter had been returned because it hadn't been notarized.) Here are some excerpts:

My name is Gregory Benjamin Aller. My date of birth is September 13, 1963. Enclosed is a copy of my California identification. I need the following information.

A copy of the report from April 5, 1990 to date, of myself, for which I am under investigation.

Names, addresses, and phone numbers of informants.

Names and regional offices of F.B.I. agents who are investigating myself, in which I am talking on the phone.

Transcripts of phone conversations and present conversations with individuals who are maintaining surveillance on myself.

Budget allocated from all sources known to investigate Aller.

Persons in the Bush administration who are directly involved in the investigation of myself.

Photocopies of Presidential directives concerning the investigation into myself.

Much later, still in a highly delusional state, I started hitchhiking to Washington, plotting to assassinate President Bush with poison gas.

Though I had slipped back into paranoid symptoms, the co-principal investigator, psychiatrist Michael Gitlin asked me, matter-of-factly, if I felt I needed medication. He never told me to take medication, he only asked me what I wanted to do.

Since I was paranoid, I said I was fine, I didn't need medication. Dr. Gitlin never really delved into what was really going on in my life. He saw me for less than five minutes every two weeks. In my opinion, the self-report design of the Brief Psychiatric Rating Scale (BPRS), questions administered by my case worker, were not adequate measures for patients suffering from a return of paranoid symptoms. There were a number of patients with paranoia.[10]

In late 1989 my father called the project's staff, telling them that he thought I needed medication. My caseworker suggested that my father write to the researchers, listing my symptoms. My parents wrote to the researchers on January 12, 1990:[11]

Gloria and I strongly believe that Greg should have his medication reinstated at this time. We have observed a sharp escalation of symptoms in the last month. These symptoms are identical to the symptoms before his first full-blown psychotic episode: hyperactivity, agitation, red and wild-looking eyes, lack of empathy for others, frequent angry outbursts, grandiose schemes, lack of self-grooming, etc. In short, we have witnessed a major personality change. In addition, Greg's sister has been on break from Berkeley since December 20th, and Stephanie has also observed these severe personality changes when we were not present.

Shortly after my father delivered the letter to the Clinic, my parents met with Dr. Nuechterlein and the caseworker, Joseph Tietz. Dr. Nuechterlein said that my symptoms might be fluctuating and temporary and the staff would watch me carefully. Dr. Nuechterlein persuaded my parents to wait. Throughout my deterioration over the coming months my father persisted in his phone calls, but the researchers said they did not agree with my parents' assessment. My medical records show that I was telling the Clinic staff that everything was fine and they were believing me. They did not make the effort to verify the things that I was telling them, like how well I was doing in school. (In actuality, I wasn't able to complete my assignments at school.) The staff also rejected my parent's reports.

Schizophrenia is a complicated and baffling illness when you first experience the illness. My parents still respected the doctors and they deferred to the Clinic's opinion. Consequently, I suffered from months of psychotic experiences. When my parents asked me to take medication, I told them that I would whenever the doctors told me to take the medication. My parents could not cope with my psychotic behavior and my dangerousness and I had to leave home. Finally, I became indigent and homeless.

On May 15, 1990, after spending a night in a motel where I lapped out of the toilet like a dog, and after having a frightening delusion that my father had died of a heart attack, I went home and because I also needed money to survive I agreed with my parents to go in for medication. That day I received an injection of antipsychotic medication. The dose of medication that I had received before relapse was no longer adequate and had to be increased from 12.5 mg of Prolixin to 17.5 mg. Later, the medical records showed that I suffered from suicide ideation.

After I was back on medication Dr. Gitlin wrote a letter dated July 9, 1990,[12] so I could obtain Los Angeles County relief: "Gregory is unable to work at the present time due to an exacerbation of his illness including difficulty concentrating, severe anxiety, disordered thinking, and disabling negative symptoms. Since Mr. Aller has no place to live and is without any income, he is in immediate need of financial assistance to obtain food and shelter."

By this time, my family had been through an ordeal and it had been especially painful for my mother. But my parents stuck by. During the next six months I had many difficulties and still couldn't concentrate and think nearly as clearly as I could before my relapse. My parents waited approximately six months until they felt it was necessary to speak to a responsible party at the university with hopes that a review of the research project by UCLA might benefit other patients and families.

PRESENTING OUR CONCERNS TO UCLA—ROBERT ALLER

Following the advice of a bioethicist, on February 20, 1991, Gloria and I met with the Vice Chancellor of Research Programs, Albert Barber, Ph.D., the chairperson of the Human Subjects Protection Committee, Fred Montz, M.D., and the administrator of the Human Subjects Protection Committee, Brigitta Walton. After presenting our concerns, we told the administrators that based on our experience, we were also concerned that a murder or suicide could occur in this project. They assured us that they would look into the issues that we raised.

But just five weeks later, on March 28, 1991,[13] another aftercare patient, Tony Lamadrid, walked from his apartment in Westwood onto the campus and up to the roof of the engineering building, Boelter Hall. At approximately 8:35 in the morning, as students were on their way to class, Tony threw himself off the roof and committed suicide. Tony's medication had been withdrawn prior to his suicide. Tony had an appointment to see his caseworker that very day.

We wrote to UCLA asking for information about the research but we found that Vice Chancellor Barber and the university would fail to respond to our inquiries for lengthy periods of time, and would even issue false information.[14]

We received some of Greg's medical and research records where white-outs, obliterations and alterations had been used extensively. We viewed the original records and verified the white-outs, obliterations and alterations with UCLA staff. UCLA then blocked access to the balance of Greg's research records.[15] After consulting with administrators at other hospital record departments, we learned that the use of white-outs, alterations and obliterations constituted highly improper conduct. The act of altering Gregory's records triggered our family's decision to try to find out what really happened in this research.[16]

Subsequently, on May 2, 1991, we filed a formal complaint[17] with the National Institute of Health's Office for Protection from Research Risks (OPRR) that led to a three-year investigation focussing primarily on the informed consent documents.

INFORMED CONSENT FOR MEDICATION CROSSOVER/WITHDRAWAL PROTOCOL—COMMENTS OF ROBERT ALLER

First, the consent agreement[18] did not indicate any alternative treatment. But at that time we didn't know that alternative treatments were a requirement of informed consent. Nor did we know that there were any alternatives to the treatment Greg was receiving at the Aftercare Clinic.

The oral explanation for taking patients off medication, presented to us by Keith

Nuechterlein, was to prevent the onset of a movement disorder, TD. But this specific fact did not appear in the written consent. In retrospect, if the purpose of the research had genuinely been to minimize the risk for TD, a standard alternative clinical treatment would have been to titrate down the 12.5 mg of Prolixin to the lowest effective dose that would control symptoms. However, the consent omitted any alternative treatment. We simply believed that medication crossover and withdrawal was the best thing to do. We trusted the doctors because Greg had done extraordinarily well in the first protocol by following the advice of the staff.

What were the actual risks indicated in the consent form?

The consent form described only one risk in detail: "I understand that during blood drawing, I may experience pain from the needle prick, a small amount of bleeding, infection or black and blue marks at the site of the needle mark which will disappear in about 10 days." Our view of the consent at that time was that since the researchers had paid dutiful attention to such an inconsequential risk, they certainly had the patients' welfare at heart.

The consent form concluded with a summary statement that placed equal emphasis on three possible outcomes. "I understand that my condition may improve, worsen or remain unchanged from participation in this study." It is noteworthy that the researchers had withheld critical information that had already appeared in the 1988 grant application—88% of patients taken off medication had suffered an exacerbation or relapse at a mean time of 33 weeks (or eight months).[19] According to Jay Katz, M.D.:

In light of the high relapse rate it was misleading to aver "that my condition may improve, worsen or remain unchanged." The odds favoring relapse were far too great; few subjects would "improve" or "remain the same." ... Moreover, the patient-subjects were not presented with any information about the merits of not joining the research project. They were deprived of considering the alternative.

A review of relapse literature reveals that the researchers omitted other well-established facts about schizophrenia relapse.

The consent form failed to convey that a relapse off medication is usually more serious than a relapse on medication:

High risk behaviors almost certainly increase in those schizophrenic patients withdrawn from antipsychotic medications. Patients whose antipsychotic medications were withdrawn had seven times as many incidents of antisocial behavior, such as assault, criminal and property damage, arson, and stealing, as patients who were maintained on medications. The same patients had 2½ times as many incidents of self-injury, including wrist-cutting, poisoning, and attempted hanging. (Johnson et al., 1983)[20]

"Neuroleptics and the Natural Course of Schizophrenia," by Richard Wyatt, M.D., provides a useful literature review of the effects of relapse on patients (Wyatt, 1992).[21] The Wyatt paper leads the reader to consider the following:

The consent form failed to share with the patient-subjects the fact that scientific literature had already shown that relapses can trigger a step-like loss of intellectual and social functioning. The consent form failed to indicate that some published papers had already contended, though tentatively, that brain damage may result from relapse.

In summary, the consent form detailed the prick of a needle from drawing blood and yet it failed to describe the "panic, horror, and hallucinations" that Subotnik and Nuechterlein had aptly described at the 1986 American Psychological Association's Annual Convention.

A full disclosure of risks had been required by both the California Human Experimentation Act[22] and the Code of Federal Regulations for informed consent, 45 CFR 46.[23]

The California Human Experimentation Act of 1978 stated that a written informed consent include a description of any attendant discomfort and risks to the subject reasonably to be expected. The state law also required a description of alternative treatments, as well as benefits. But back in 1974, four years before the California Human Experimentation Act had become law, the Code of Federal Regulations had already implemented these same fundamental requirements; written consent, risks, alternative treatments, and benefits.

Finally, OPRR's May, 1994, determinations regarding the consents used in this experiment did result in UCLA's compliance with with the Code of Federal Regulations, 45 CFR 46. But UCLA complied reluctantly.[24]

In summary, if UCLA had provided full written disclosure, as required by state and federal law, Greg would not have participated in the crossover and withdrawal protocol. We share Greg's view that earning a 3.8 GPA in college, working, having a social life, was far more preferable than the almost certain probability of an adverse outcome. The risk of TD is a risk worth taking when compared to life in a psychotic state.

STANDARDIZED, FIXED-DOSE MEDICATION—COMMENTS OF ROBERT ALLER

Consider the experience of patient-subject Gabby Guardino. Her case example concerns the first protocol that used a standardized, fixed-dose of 12.5 mg of Prolixin decanoate. While the standardized dose was effective for Gregory, it was not effective for Gabby.[25]

The following statement is from Gabby's mother, Sydelle Guardino. Were the researchers meeting the medical needs of the patient-subject?

STATEMENT OF SYDELLE GUARDINO—MOTHER OF GABBY GUARDINO

After Gabby got sick, we convinced her to go into the UCLA Neuropsychiatric Institute as an inpatient. That was 1982. She was eighteen years old.

While an inpatient, the doctor suggested that we go into the UCLA Aftercare program for outpatient care. They took us right in.

They gave Gabby the standard dose of 12.5 mg of Prolixin. The prolixin made Gabby nauseous and she started vomiting frequently. She suffered from that medication for eight months while her behavior deteriorated.

I couldn't cope with Gabby's manic and psychotic behavior. I was smoking two packs of cigarettes daily, was unable to work, lost 25 pounds and had devastating headaches.

The Aftercare staff knew all this. It all came out in our therapy sessions that were videotaped in Dr. Goldstein's laboratory. Gabby begged them to put her on another medication. They wouldn't. She would bite her tongue and her mouth. The medication made her feel like her skin was crawling. Her psychiatrist, Dr. Gitlin, wouldn't change her medication or her medication dosage.[26]

Without adequate knowledge, I was calling the staff out of desperation ... asking for help. They saw what was happening. They told me I was wrong for getting so upset. Talking about this makes me want to cry ... In the medical records it states: "We employed paradoxical directives in our treatment, encouraging her [Gabby] to maintain this position as the scapegoat and to continue to upset her family."

The effects of the medication got worse over time. One day, after eight months, Gabby said, "I can't take it, it's going to kill me." The effects of the medication had gotten so bad she was shaking, biting, her tongue, rocking. We had a counseling session with David Miklowitz in a very small office. He was a Ph.D. candidate under Dr. Goldstein. He said, "Look, if you don't want to take this medication you're out of the program." I said, "Dr. Miklowitz, tell me what you can do, where can we go? She needs help." He said, "I don't know. If she doesn't want to take the medication, she's out." The meeting was very short and very cold. I'll never forget it.

When we were abandoned I had no one to turn to. I locked myself in the bathroom because I was afraid I might harm her. I remember screaming ... a primal scream. Gabby was the one who had enough sense to call the police, thinking I was going to do myself in. When they came a policeman said: "Mrs. Guardino, if I were you I would tell her to leave because she's not listening to you." I threw her out that day. I made my daughter homeless. God, I remember that!

The first week when she was gone, I didn't care. Then I started to worry and I worried all the time. Sometimes she'd call me. Once I went out and met her on the streets and I learned that a man had tried to rape her and she ran naked down the street. The police found her, wrapped her up, and gave her clothes. When I saw her she had been barefoot for about two weeks. I bought her some shoes.

Gabby was homeless for three years. As a mentally ill person she spent three months in jail. Gabby's life has been devastated ever since those eight months she spent in the UCLA program.[27]

SUICIDE OF TONY LAMADRID—COMMENTS OF GREGORY ALLER

I was in the Aftercare Clinic with Tony Lamadrid. We had both graduated from Santa Monica High School. Tony's mother had passed away when he was two-years-old and his father had died of a heart attack at Tony's fifteenth birthday party. Tony excelled in high school, receiving straight A's. After entering UCLA as a freshman, in the Fall of 1985, Tony had his onset of schizophrenia. Tony participated in the Aftercare Clinic for over five years.

Shortly before he committed suicide the staff had asked Tony to leave the program and seek another program. I learned that Tony was devastated by the loss of his social and emotional network at Aftercare. They also took away his medication. At the same time, Tony was losing his roommate—an added stressor in his life. Tony told a number of people that he was going to commit suicide. His case worker knew what was happening in Tony's life.

The *Los Angeles Times Magazine* report of September 11, 1994, stated:

A coroner's report included this notation: "This 23-year-old male with a history of depression and schizophrenia was being treated for same at UCLA Medical Center Psychiatric Department ... [his] social worker ... was trying to place him into a new psychiatric program and was trying to convince decedent on 3-22-91 to commit himself to UCLA because she felt he was suicidal ..."

Shortly after the suicide, police gave Tony Lamadrid's brothers a copy of an answering machine tape they found in his apartment near campus. It included a message from Debbie Gioia-Hasick, his social worker at UCLA's Aftercare Clinic. "Hi, Tony. It's Debbie," she began, "Even later today, [if] you're just feeling really bad, go over to UCLA and just check yourself into the hospital. Tell them you're suicidal and that you think you're really going to hurt yourself. Check yourself in, and then we'll go from there."[28]

I believe that the Clinic staff should have notified Tony's family about the seriousness of Tony's suicide ideation. No phone call was ever placed to any of the family members. Two of Tony's brothers and an aunt lived within 15 minutes of UCLA. With a simple call to any family member, they would have been there to help Tony. The researchers should have explained that Tony was not receiving antipsychotic medication and that he had acute suicide ideation. In any event, under the circumstances, the Aftercare Clinic staff should have hospitalized Tony.

I have never fully understood just why the UCLA researchers didn't notify the Lamadrid family about the seriousness of Tony's condition prior to his suicide. I have wondered, "Why didn't they hospitalize Tony?" I believe the researchers abandoned Tony.

Abandonment is a common theme in the treatment of the four families who are speaking out about the UCLA research conduct. I believe that Tony committed suicide after Dr. Gitlin and the Aftercare Clinic staff abandoned Tony. In my view, they may well have triggered his psychotic state by withdrawing medication and cutting him off from treatment. They enabled Tony to commit suicide by not intervening or warning Tony's family after they knew he planned to commit suicide.[29]

CASE EXAMPLE PROVIDED BY UCLA—COMMENTS FROM ROBERT ALLER

The *Los Angeles Times Magazine* article of September 11, 1994, "For the Sake of Science," included a case example provided by the UCLA researchers to the *L.A. Times* writer, Joy Horowitz.

One parent, whom the researchers asked to speak to me and who requested anonymity, told me how much she respected Gitlin, explaining that he would "break the blind" (intervene during the double-blind phase) at any time. The problem was she lacked authority over her adult son, who was relapsing on the placebo but "fooling everyone"—hearing voices yet denying it because he wanted to stay off medication. She began crying, "I'll tell you the truth," she said of the study during which her son was hospitalized. "I don't resent the fact that my son went through it. But I wish he hadn't. It caused him a lot of pain ..." At one point during the recuperation, she said, "he tried to jump out of his car on the freeway."

In this case example, provided by the UCLA researchers, the mother stated: "I don't resent the fact that my son went through it. But I wish he hadn't ..."

GREGORY ALLER'S CONCLUSIONS

One day my father was trying to lighten me up since things weren't going well. He asked what I might like to be in the next life. I replied: "I don't want to come back."

I was responding to the day-to-day experiences that I have to contend with. After my relapse, the most personally distressing outcome has been the persistent difficulties I experience in trying to concentrate. Before my relapse, I earned excellent grades, but afterwards school has been painful and frustrating. I wish to hold a steady job and have not been able. My condition after my relapse has made me doubt my own competence. My goal had been to become a lawyer. In high school I had served on a Los Angeles County Commission, I had been president of a statewide high school Latin language association and I had been active in student body politics. I was not short on ambition.

For many of us with schizophrenia, we frequently ask ourselves "What can we do with our lives?" This remains a difficult question.

To this day, the researchers at UCLA deny they have done anything wrong. In his Congressional testimony, Dr. Rockwell claimed that patients received the "best treatment possible." From what I have seen, these assertions are simply not truthful. The consequences of relapse are far more severe than the UCLA researchers acknowledge. For example, I think that Tony Lamadrid's escape from the relapse he suffered was suicide.

I recently visited with a former patient from the Aftercare Clinic. Elizabeth DeBalogh, at the board and care facility where she lives. Prior to her relapse, Elizabeth had been living successfully at home. But in 1985 she had an extremely severe relapse. I quote from a complaint filed with OPRR by Frank DeBalogh, Ph.D., Elizabeth's older brother.[30]

One Saturday morning in May, 1985, I received an urgent call from my mother, who requested my help, saying that she could not handle Elizabeth. I found Elizabeth not speaking, not eating, not sleeping, pacing about, pushing her mother away, turning lights on and off, opening and shutting doors and windows. In short, she was psychotic.

When I called the UCLA phone number for help, all we got was a message that they would be available during business hours Monday through Friday and they could not be reached on the weekend.

Dr. DeBalogh states that UCLA had not provided emergency instructions or plans in the event of a relapse. Since immediate medical care was required, Dr. DeBalogh checked Elizabeth into a Los Angeles County Hospital and Elizabeth has remained institutionalized since that episode in 1985.

Currently, Elizabeth is unwilling to visit any of her family members' homes because she has a fear of cars and freeways. She is afraid to see her cousin, whom she thinks is a witch. She has suspicions about her food being poisoned. She doesn't watch television in her room because she claims that she watches television in her mind. In short, she remains gravely disabled, institutionalized and under conservatorship.

I was lucky to survive as well as I did, but I am left with a feeling of concern for the many others who suffered much more.

Yet, I often feel valueless myself...and I know that I may not be able to accomplish nearly as much as I would like to in my life. Though it is difficult for me, I do try to speak out. I hope that my efforts will benefit others.

ROBERT ALLER'S CONCLUSIONS

OPRR determined that UCLA did not obtain proper informed consents for this research. The omissions made by UCLA in the consents were not trivial—they were material. The failure to describe alternative treatments, and the failure to describe the almost certain probability of an adverse outcome in the withdrawal phase of the research made it impossible for patient-subjects and their families to make informed decisions about participating in this project.

The medical treatment needs of patient-subjects were seemingly abandoned for the research goals, but patients and families were not informed about this practice.

UCLA had failed to respond properly in 1985 to a complaint filed by Ms. Sydelle Guardino and addressed to the researchers, administrators, and the Vice Chancellor of Research, Albert Barber, Ph.D. That complaint had focussed on some of the same issues that our family encountered in 1989 and 1990.

The university's disregard for the welfare of experimental subjects permitted the problems to continue.

The university has continued to deny any wrongdoing in the face of considerable evidence provided in grant applications, protocols, published papers, deficient informed consents, and NIMH summary reports. The university now even claims that a published paper does not mean what it says.[31]

Enrique Lamadrid, Ph.D., the eldest brother of Tony Lamadrid, wrote to OPRR on June 9, 1991,[32] requesting an investigation of Tony's suicide. He closed his letter with these thoughts:

I feel that our trust in the mental health professionals involved in this case was betrayed. I understand now the reasons for their lack of interest in treatment and their lack of concern for their subjects' families. I have the horrible feeling that my brother has simply become another statistic in their research findings.

We share Enrique's feelings.

We have been guided in our evaluation of this UCLA research project by the insights and comments of clinical and research psychiatrists and psychologists, bioethicists, law professors, members of IRBs at other universities, Congressional Subcommittee staffers, concerned family members and friends. We deeply appreciate the efforts of all those who have helped to stimulate a needed public discussion of these issues.

NOTES

1. Keith Nuechterlein, Developmental Processes in Schizophrenic Disorders Project, Protocol: Developmental Processes in Schizophrenia, Human Subject Projection Committee (HSPC) 86-07-315, Protocol for first phase of research (on file with authors).
2. Keith Nuechterlein and Michael Gitlin, Research Protocol for Developmental Processes in Schizophrenic Disorders Project: Protocol; Double Blind Crossover and Withdrawal of Neuroleptics in Remitted, Recent-Onset Schizophrenia, HSPC #86-07-336 (on file with authors).
3. Evaluation of Human Subject Protections in Schizophrenia Research Conducted by the University of California Los Angeles, Office for Protection from Research Risks, Division of Human Subject Protections, May 11, 1994, (available under National Institutes of Health Freedom of Information office (on file with authors).

"Summary (1) The "Developmental Processes" study maintains schizophrenic subjects on fixed dosages of fluphenazine decanoate. OPRR determined that the Informed Consent documents (a) failed to disclose the "fixed dose" nature of the study, (b) failed to discuss either the risks of such fixed doses or the availability of clinically individualized doses as an alternative treatment, and (c) failed to address the rationale for use of fluphenazine decanoate and the availability of alternative medications or treatments. (2) The "Double Blind Crossover and Withdrawal" study involves schizophrenic subjects in medication-free conditions over lengthy period of time. Annual reports to the IRB indicate a high rate of relapse or exacerbation (75–80%) had occurred in the study. OPRR determined that the Informed Consent documents failed to discuss thoroughly either the risks associated with lengthy withdrawal of medication or the alternatives to such withdrawal. (3) OPRR determined that the investigators had published clinical data obtained outside the parameters of IRB approved research protocols. This OPRR document was prepared on 01/11/93, (on file with authors).

4. Donald Rockwell's Congressional testimony of May 23, 1994, before the House Subcommittee on Regulation, Business Opportunities, and Technology, available from the Subcommittee (on file with authors).

5. Katz, J., Human experimentation and human rights. *St. Louis University Law J.*, Fall 1993; 38:7–54.

6. After we filed a complaint about this research with NIH's Office of Protection from Research Risks, my wife, Gloria, found a paper that elaborated upon the research design and objectives, specifying just how achieving severe or extremely severe relapse symptoms was an integral part of the research design (Subotnik and Nuechterlein, 1988).

The paper, "Prodromal Signs and Symptoms of Schizophrenic Relapse," published in the *Journal of Abnormal Psychology*, claimed that the scientific data for this paper was gathered prospectively, comparing this research to previous retrospective studies of the relapse process.

"Retrospective studies of the relapse process may involve a biased view of past events, as the negative outcome is known when data is collected.

"The present study is a prospective examination of prodromal signs and symptoms of schizophrenic relapse, using a systematic and carefully controlled research design. One important improvement over the previous studies is that relapse is defined as the elevation of psychiatric symptoms to the severe or extremely severe level. Thus, minor fluctuations that might often be inconsequential were not considered relapses. In contrast to the studies that defined the period of observation by the necessity to increase medication to avoid a possible relapse, we can be certain that any prodromal changes that we isolated actually did precede a clear relapse. Another methodological improvement is that the use of reliable systematic BPRS ratings every 2 weeks allowed us to follow the course of the prodrome from early signs all the way to relapse. This ensured that we were able to examine subtly changes in symptomatology that may portend the relapse.

"Psychotic relapse was defined as an elevation to severe or extremely severe [6 or 7] from nonpathological levels to one or more of the BPRS items: Hallucinations, Unusual Thought Content, and Conceptual Disorganization.

"One strength of the present approach is that we are certain that the symptoms we examined actually were prodromal to a psychotic relapse, not merely judged to be signs of an impending relapse that presumably would have occurred if not aborted through pharmacologic intervention."

Thus, the UCLA researchers were not constrained by the community standard of attempting to prevent relapse. Quite the opposite, these researchers planned for exacerbation/relapses and only intervened after the patient-subjects reached the severe or extremely severe relapse measurements on one or more of the BPRS items; unusual thought content, hallucinations, or conceptual disorganization.

7. The brochure was used for recruiting patient-subjects for this project. This brochure is available from NIH's Freedom of Information Office (on file with authors).

8. At the 1986 American Psychological Association's Annual Convention, researchers Kenneth Subotnik and Keith Nuechterlein presented "Prodromal Signs and Symptoms of Schizophrenic Relapse," 97. *Journal of Abnormal Psychology*, 1988, 97:405–412.

In the paper presented, the researchers described just what happens to patients during the continuum of decompensation, from exacerbation to extremely severe relapse.

They cited Docherty, J.D., Van Kammen, D.P., Siris, S.G., and Marder, S.R. (1978), Stages of Onset of Schizophrenia Psychosis. *American Journal of Psychiatry*, 135:420–426.

The paper describes five stages of decompensation. This information was never presented to patient-subjects and families.

"A person in stage one, overextension begins to feel overwhelmed. This stage is characterized by overstimulation, persisting anxiety, irritability, decreasing performance efficiency, and distractibility.

"Stage two, restricted consciousness, encompasses mental phenomena that limit the person's range of thought. Symptomatology includes boredom, apathy, social withdrawal, obsessional and phobic symptoms, somatization, hopelessness, dissatisfaction, and dependency.

"The third stage, disinhibition, is a period of increased impulse expression. Risk taking, sexual promiscuity, rage attacks, and impulsive dissociative phenomena and elevation of mood.

"In stage four...as the external world breaks down there is increasing perceptual and cognitive disorganization, speech production and comprehension becomes difficult...primitive sexual and aggressive images intrude freely into consciousness, and severe anxiety, panic, horror, and hallucinations are present.

"The final stage is psychotic resolution. The panic and horror of psychotic disorganization decreases as delusional organization occurs."

This detailed description of the onset of schizophrenia psychosis characterizes some of the experiences I would suffer during relapse.

9. Letter from Gregory Aller to the FBI (on file with authors).
10. NIMH Summary Report for Developmental Processes in Schizophrenic Disorders, 1985. Dr. Nuechterlein reported: "Out of 57 patients diagnosed by RDC, (Research Diagnostic Criteria), 48 were paranoid schizophrenics ...," [over 84%] (on file with authors).
11. Letter of January 12, 1991, from Robert and Gloria Aller to Joseph Tietz, Ph.D. candidate, Caseworker at Aftercare Clinic (on file with authors).
12. Letter of July 9, 1990, from Michael Gitlin, M.D., To Whom It May Concern (on file with authors).
13. Los Angeles County Coroner's Report on the death of Tony Lamadrid (on file with authors).
14. On October 25, 1991, I wrote to UCLA asking precisely when Gregory was taken off medication. (Because of the double-blind placebo/medication crossover, we never knew, with certainty, when Gregory was first taken off medication.) UCLA replied to my letter four months later, on March 2, 1992: "The dates in question are: on placebo—June 29, 1989 through December 27, 1989; on active medication—September 27, 1989 through December 27, 1989." This letter, claiming that Gregory was on placebo and medication simultaneously, had been written on behalf of Drs. Nuechterlein and Gitlin, and Vice Chancellor Albert Barber, by Campus Counsel Patricia Jasper. UCLA had claimed that Greg was on placebo and medication simultaneously.
15. UCLA violated California law and withheld most of Gregory's patient "research" records for 2½ years. The State of California and County of Los Angeles Health Facilities Licensing Division issued a formal deficiency to the UCLA Neuropsychiatric Hospital on December 15, 1993, for blocking access to patient "research" records. The issuance of this deficiency constituted a state administrative ruling that the Neuropsychiatric Institute had committed a "misdemeanor" (on file with authors).
16. The OPRR Report on the UCLA research contains UCLA's admission to improper record alterations. OPRR Report available from NIH's Freedom of Information Office.
17. Letter of May 2, 1991, from Robert and Gregory Aller to Melody Lin, Ph.D., OPRR, concerning failure of the UCLA research project to conform to ethical and legal federal requirements for human subject research under The Belmont Report, under UCLA's Assurance with DHHS, under 46.108 IRB functions and operations, under 45 CFR 46 related to informed consent (on file with authors).
18. Nuechterlein, Keith, Informed Consent Agreement for Patients: Double-Blind Drug Crossover and Withdrawal, 1982–1992, available from NIH's Freedom of Information Office (on file with authors).
19. Keith H. Nuechterlein, Grant Application: Developmental Processes in Schizophrenic Disorders, RD 1 MH 37705-07, 1, 84 Nov. 15, 1988 (on file with authors).
20. Johnson, D.A.W., Pasterski, G., Ludlow, J.M., Strett K.,, and Taylor, R.D.W. The discontinuance of maintenance neuroleptic therapy in chronic schizophrenic patients: drug and social consequences, *Acta Psychiatrica Scandinavica* 1983; 67:339–352.
21. Wyatt, R.J. Neuroleptics and the natural course of schizophrenia. *Schizophrenia Bulletin* 1992; 1.
22. California Human Experimentation Act (on file with authors).
23. Code of Federal Regulations, Title 45, Part 46—Protection of Human Subjects, Department of

Health and Human Services, National Institutes of Health, Revised as of March 8, 1983 (available from OPRR, on file with authors).

24. Throughout the OPRR investigation, UCLA denied any wrongdoing, claiming that their consents had always been appropriate. For example, after OPRR required UCLA to evaluate their consents, Albert Barber, the Vice Chancellor of Research wrote to OPRR on May 23, 1991, stating the following: "Based on my review of the records in this matter and my conversation with both the staff of the IRB and the principal investigator, I believe that this study continues to meet all human subjects protection criteria, including sections 103 (b) (4) (iv) and 113 of 45 CFR 46."

On October 15, 1991, Dr. Barber again wrote to OPRR denying any wrongdoing: "Based on my review and lengthy discussions with the two investigators, HSPC staff and the Associate Dean of the School of Medicine, I find UCLA in full compliance with 45CFR46. It is understandable that a family member of one of the subjects might question, on hindsight, certain aspects of a complex and sensitive research protocol but these questions cannot be misconstrued as allegations of non-compliance with 45 CFR 46." Approximately a year later, on September 17, 1992, another UCLA administrator, the Senior Vice Chancellor of Academic affairs, wrote to OPRR again claiming that the consents were appropriate, (all letters on file with authors).

25. The protocol for the standardized dose phase of the study The Developmental Processes in Schizophrenic Disorders Protocol: (Developmental Processes in Outcome), states: "6. The dosage of 12.5 mg of Prolixin Decanoate every two weeks is a standard outpatient maintenance dosage that is judged to provide the optimal level of protection against psychotic relapse."

This claim, made in the protocol approved by the IRB and NIMH, is contradicted by other members of the UCLA research team. Asarnow, R.F., Marder, S.R., Mintz, J., VanPutten, T., and Zimmeran, K.E. Differential effect of low and conventional doses of fluphenazine on schizophrenic outpatients with good or poor information-processing abilities. *Archives of General Psychiatry*, 1988; 45:822–826. "Decisions about the best drug or the best drug dose for an individual patient are made on a trial-and-error basis, since there are no methods currently available that are clinically useful in predicting the best drug and optimal drug dose for a particular patient ..."

The FDA approved product insert sheet for Prolixin Decanoate, distributed by Princeton Pharmaceutical Products, states: "Appropriate dosage of Prolixin Decanoate (Fluphenazine Decanoate injection) should be individualized for each patient and responses carefully monitored.... The optimal amount of the drug and the frequency of administration must be determined for each patient, since dosage requirements have been found to vary with clinical circumstances as well as individual response to the drug."(Princeton Pharmaceutical Products, Product Insert Sheet, Prolixin Decanoate, Fluphenazine Decanoate Injection. AHFS 28:16.08, J-4-150A. Revised May, 1987 (on file with authors).

The NIMH Summary statement (page 9 of 1 RO1 MH 37705-01) states: "In study 1 [standardized dose of 12.5 mg] subjects will be kept on a constant injected dose regardless of their size or how they metabolize the drug," (on file with authors).

26. Michael Dawson, Ph.D., who had been a co-principal investigator, reported on the design of this research. (Dawson, M.E. Psychophysiology at the interface of clinical science, cognitive science, and neuroscience. *Psychophysiology*, 1990; 27:243–255.

Dawson reported on the possible significance of electrodermal abnormalities in schizophrenic patients. Dawson stated: "Both tonic and phasic electrodermal activity increased when schizophrenic patients entered into a psychotic episode, and preliminary evidence suggests that the increase may serve as early prodromal signs of an impending relapse."

"The patients were seen every two weeks in an outpatient clinic. Here they received a standardized low-dose of neuroleptic medication (12.5 mg of Prolixin decanoate) and their symptoms were assessed with the Brief Psychiatric Rating Scale. Thus, it was possible to document states of clinical remission and relapse based on the biweekly Brief Psychiatric Rating Scale and to do repeated electrodermal testing with medication held constant in the two states." Dawson stated that "the most recent analyses of the electrodermal data from this ongoing study involved 22 patients tested in both clinical remission and in the active psychotic state.

27. In his Congressional testimony, Dr. Rockwell stated: "A crucial point that has often been overlooked is the 'clinical override' aspect of the study. At any point in the study, the research could (and in fact did) override the study criteria to provide medication or other treatment if needed by the patient. The patient's best interest always came first." Dr. Rockwell's claim is not consistent with Sydelle Guardino's experience with her daughter in this research.

In 1985, after joining the Alliance for the Mentally Ill, Sydelle Guardino learned more about schizophrenia from other family members. On October 20, 1985, she wrote to Keith Nuechterlein, explaining what happened to Gabby. Ms. Guardino sent copies of her letter to Michael Gitlin, M.D., Michael Goldstein, Ph.D., the Vice Chancellor of Research, Albert Barber, and others at UCLA. The institution never wrote back to Sydelle Guardino, responding in writing to the issues she had raised.

On October 27, 1994, Ms. Guardino wrote to Dr. Rockwell, asking for the complete patient records for Gabby, including the videotapes of each of the videotaped family sessions. The patient records were sent to Ms. Guardino and she states that the records corroborate Gabby's deterioration in the program while she was on the standardized dose.

But in violation of applicable California state law regarding patient records, UCLA did not provide all of the videotapes from the family sessions. Currently, the Los Angeles County Health Facilities Division and OPRR are investigating UCLA's failure to provide all of the missing video-tapes (all letters on file with authors).

28. I would like to correct a misleading statement about Tony Lamadrid that appeared in a letter of Dec. 29, 1994, from David Kupfer, M.D., President of the American College of Neuropsychopharmacology. This letter was sent to the members of the Baltimore Conference Organizing Committee on Ethics in Neurobiological Research with Human Subjects. Dr. Kupfer stated the following about the suicide of Tony Lamadrid: "The patient was not in a research protocol at the time of his suicide and had no contact with the research project for over a year after completing his part in it." I can only presume Dr. Kupfer believed he had obtained accurate information that he was transmitting in his letter.

For the record, Dr. Kupfer's comment that Tony Lamadrid "had no contact with the research project for over a year after completing his part in it," is contradicted by other evidence. The American Psychiatric Association's *Psychiatric News* of March 17, 1995, reported: "However, OPRR determined that Lamadrid was a research subject at the time of his death because clinical monitoring data obtained from the AfterCare Clinic participants was used for research purposes." The coroner's report also indicates that Tony was being seen the researchers. While it may be true that Tony was not in a research protocol, Dr. Kupfer omits OPRR's finding that Tony and a significant number of other patient-subjects, including myself, participated in a phase of this research experiment where no protocol existed, even though the research had been ongoing for over a decade. (Kupfer letter of December 29, 1994, on file with authors).

29. On February 3rd, 1995, Tony's psychiatrist, Michael Gitlin, gave a lecture about suicide issues at UCLA's Neuropsychiatric Institute. The lecture was part of a Continuing Medical Education (CME) series presented at the institute. This lecture presented Dr. Gitlin's attitudes toward a patient who indicated that he was planning to commit suicide. (The following text was transcribed verbatim from a videotape recording of Dr. Gitlin's presentation made by the UCLA Neuropsychiatric Institute media staff.)

Dr. Gitlin: "This was a fellow who I had seen for numbers of years, again in psychotherapy and medication. Surprisingly, he did not have a mood disorder. He had the worst chronic generalized anxiety disorder I have ever seen in my life, and his anxiety became progressively crippling, uuhhh. And I everything I knew how to do. For a guy with GAD, I actually had him get ECT, to show you how desperate I was. He had been on every medication, I had two consultations with anxiety mavin colleagues. I had done everything. I had taken one gun away from him, he had ordered one and I kind of blackmailed him, telling him that I would fire him as a patient if he didn't give me the gun. He gave me the gun and then he just got worse and worse and more and more desperate and his psychological pain clearly became unbearable. For real. He bought a second gun, and I knew, and he told me about it, and I realized he was out of insurance. I couldn't get him in anywhere other than Harbor, and I knew that there was nothing Harbor could offer him. I mean, I had done everything that I knew and asked everybody and had him see everybody. And basically I was out of, I was out of therapeutic techniques, and I was also out of psychological gas.

"This guy called me continually. I mean he didn't call me in a borderliney way, but he was desperate and I was worn down, and one weekend I got, I mean I was becoming convinced that he was going to kill himself, soon (soon was emphasized in presentation). I got a call from a friend that said you know I called your patient and he said he didn't want to make a plan to see me next week 'cause he wasn't sure he wanted to stay alive and he'd call me Monday if he decided to live. And this friend called me on a Saturday. And I thought about it and thought about it and I elected not to do anything because I had nothing more to offer this fellow, neither internally in terms of my own ummfff, nor externally in terms of treatment. And I realized at that point what I decided as best I could. And as I look back, and I have many times, I can't think of what else to do, I could have done, is that this person at that point needed to decide to live or not to live. It was his. I could no longer rescue him. I had nothing with which to rescue him. And sure enough he did kill himself on Monday with the second gun he had bought after I had taken away the first one ..."

30. Frank DeBalogh, Ph.D., complaint filed with OPRR regarding the treatment of Elizabeth DeBalogh (on file with authors).

31. In his Congressional testimony, Dr. Rockwell commented about the paper, "Prodromal Signs and Symptoms of Schizophrenic Relapse," *Journal of Abnormal Psychology*, 1988; 97:405–412.

    While the paper emphasized that the data was collected under a "prospective" research design, UCLA claimed that the paper has been misinterpreted.

    Dr. Rockwell: "Some of the critics of this study have confused the design of this study with a "secondary retrospective analysis" published by Subotnik and Nuechterlein in the *Journal of Abnormal Psychology*, (1988) that used different definitions of relapse." Dr. Rockwell claims, six years after the paper was published, that the paper was a "secondary retrospective analysis." This paper, however, makes no such claims.

    In the *Los Angeles Times Magazine*, report on this research, "For the Sake of Science," September 11, 1994, the reporter states: "Later, Nuechterlein would say that the choice of words in the article had been bad, and that it didn't accurately reflect the criteria for relapse and remedication in the main study."

    Was the choice of the word "prospective" a poor choice? Consider that the use of the word "prospective" or the use of the word "retrospective" are not words with slight variations in meaning. These two words are not likely to be chosen as a result of a momentary or careless oversight by the authors. These two words have opposite meanings.

    In addition, the paper stated: "The authors wish to thank Michael J. Goldstein for his valuable comments on an earlier version of this manuscript." Drs. David Fogelson and Michael Gitlin were thanked for "their supervision of the medication of project patients." The paper was received on March 2, 1987, peer reviewed, and accepted on April 26, 1988, by the *Journal of Abnormal Psychology*. Did all of the reviewers fail to notice that the paper was a "retrospective" rather than "prospective" study?

    The paper unequivocally states that the underlying research was a "prospective design." Was Dr. Nuechterlein's initial claim of "prospective design" correct? Or, was Dr. Rockwell's claim made six years later of a "secondary retrospective" analysis correct? Both claims cannot be correct.

32. Formal letter of complaint from Enrique Lamadrid, Ph.D., to OPRR's Melody Lin, Ph.D., June 9, 1991. (Letter available from NIH's Freedom of Information Office, on file with authors.)

Statement of the UCLA Clinical Research Center

UCLA Clinical Research Center

We at the Clinical Research Center welcome the opportunity to comment upon allegations that have been publicized in the media, misrepresenting the facts of treatment provided to research subjects in studies at our Clinical Research Center. For example,

- It has been publicly reported that we purposely and deliberately withheld antipsychotic medication from subjects whose clinical status required such treatment. This is absolutely false.
- It has been publicly reported that a subject in our studies committed suicide while participating in a medication protocol. This is absolutely false.
- It has been publicly reported that we failed to adhere to accepted clinical, scientific and ethical standards in our treatment studies. This is absolutely false.

It is well known that young people with schizophrenia often discontinue their medication, or refuse medication initially or over the course of their illness—even against medical advice. It is also well known that such young people frequently abuse street drugs. This poses a challenge to clinicians. Family members and significant others must recognize that, at times, we may not be able to tell them when subjects refuse medication or abuse drugs because the subjects refuse to give us permission to disclose this information.

The official report by the NIH Office of Protection from Research Risks found our study of treatment of young persons in early stages of schizophrenia to be "scientifically and ethically justifiable." False and unfounded allegations have been made by a few individuals who are in litigation with the University of California. These individuals have chosen to hide behind a shield of confidentiality, prohibiting us from discussing the details of the treatment that is the focus of their complaints. This has, unfortunately, lead some to engage in baseless speculation about the course of illness and treatment of certain subjects. Significantly, OPRR also reviewed the clinical treatment of the two subjects whose families had complained to OPRR and found "no evidence that the clinical treatment" of these two subjects "failed to adhere to currently accepted clinical standards."

We are striving to forge a successful alliance among those afflicted with mental illness, their clinicians, and their families. We meet regularly and frequently with families, research subjects and patients at our research clinics and hospitals. For a decade, we have sponsored an annual Schizophrenia Research Update at UCLA for AMI members, research subjects, clinicians and others. In fact, we are currently developing a new module entitled, "Involving Families in Services for the Seriously Mentally Ill." This program is aimed at enhancing the partnership between clinicians and families in treatment planning and implementation to the extent that mutual agreements will allow. In addition, on our own initiative, our CRC has formed a Bioethics Advisory Panel to advise our research groups on ways to further enhance the informed consent process. We are confident that we will all benefit from and contribute to this effort.

We are determined to learn from our research findings and clinical experiences and to share our knowledge with colleagues, research subjects and their families. We will continue to welcome input from our research colleagues, bioethics advisors, patients, research subjects, families, advocates for the mentally ill, OPRR and others.

UCLA has released information about the OPRR review of one of our projects. Copies are available for those who are interested. Once again, thank you for the chance to make these comments. We look forward to the Conference proceedings and to engaging in the emerging dialogue on these important matters.

Independent Family Advocates Challenge the Fraternity of Silence

Vera Hassner Sharav

Quality of Treatment Chair, Alliance for the Mentally Ill—NYS, 142 West End Avenue, Apt. #28P, New York, New York 10023

The question which has been evaded for too long, and which must be addressed is: **If there is no therapeutic benefit, no reward or compensation that would induce a well person to participate in experimental research, is it ethical to exploit the mental incapacity of persons who cannot make an informed, self-serving decision?** This does not mean that we oppose the inclusion of persons suffering from brain diseases in research. It does mean that we are against exploiting or abusing these human beings in and out of the research arena.

I speak on behalf of the Alliance for the Mentally Ill of New York State, and represent thousands of families who share the pain and agony of severe mental illness. Until recently, our organization had given its complete, unqualified support for all neurobiological research. Neither we nor our national organization (NAMI) bothered to ask whether (and how) mentally impaired patients are protected from harm.

Our trust in the integrity of those who pursue research was unequivocal. But an ever-increasing number of disturbing reports (in the press and media) have brought to light psychiatric studies which exploited the vulnerability of mentally impaired persons, recruited them into experiments which cruelly exacerbated the painful symptoms of their illness, and which significantly increased the risk of serious harm. These experiments offered them no therapeutic benefit whatsoever.

These, and other disturbing revelations have given us cause for grave concern, but also resolve to continue to raise troubling questions and never again to remain silent. We are determined to open the psychiatric research arena to independent oversight—which will include patient representatives—and to establish meaningful implementation and enforcement procedures that will protect persons with psychiatric illnesses from harm and exploitation. Although we still hope and do believe that the majority of neurobiological researchers do not violate ethical principles, and do not cause their patients unnecessary pain and suffering, we also believe that there are psychiatrists who have violated a sacred trust—with their patients, with families and with the public.

As clinicians and families know, it takes a great deal of effort to win the trust of patients with severe psychiatric disorders to accept treatment. For a physician, therefore, to violate that trust by aggravating symptoms instead of alleviating them, is a betrayal that has long-lasting adverse clinical consequences.

When the risk for the patient-subject is high and the level of pain and discomfort is severe, is it ethically possible for the researcher, who has a professional and financial interest in the project, to calculate a risk/benefit ratio? Some physicians have conducted experimental studies which intentionally induced schizophrenia relapse in stable veterans (another vulnerable group which is inadequately protected). One such study was conducted at the Bronx VA Medical Center by Mount Sinai research investigators, and was published in *The American Journal of Psychiatry.*

In the article, the investigators inform their peers: "This study attempted to predict time to relapse in 28 schizophrenic patients withdrawn from neuroleptics and challenged with L-dopa for 7 days, then followed until relapse."

Unfortunately for the patients, the consent form (which we obtained under the Freedom of Information Act) failed to inform them (or their families) about the purpose of the study, about the consequences of drug withdrawal, or about the known and potential risks and side-effects of L-dopa which, psychophamacologists tell me, is countertherapeutic for schizophrenia—L-dopa induces severe psychotic symptoms. The *Physicians Desk Reference,* which provides physicians instant access to information about all FDA-approved drugs—including FDA-required disclosure of warnings by the manufacturer—forewarns physicians abut severe psychiatric risks with L-dopa.

The most common risks are:

… involuntary movements … mental changes including paranoid ideation and psychotic episodes, depression with or without development of suicidal tendencies … confusion, insomnia, nightmares, hallucinations, delusions, agitation, anxiety, euphoria …

This information was withheld from patients who were in a state of remission prior to the experiment. The consent form they signed misled them even further by stating: *"We think that by giving you this drug and evaluating your response to it, we may be able to tell if your regular medication is safe for you."* Predictably, all 28 patients relapsed as a result of the experiment.

Omissions of this magnitude have led some of us to wonder whether such signed consent forms do not make a mockery of the requirement for "informed consent." In light of the American College of Neuropsychopharmacology leadership's aborted attempt to sabotage this conference at the last minute (with a memo exhorting its members and other participants, "It is inappropriate for the ACNP to support, or to participate in, the Baltimore Conference on Research Ethics …") we wonder whether, in some circles, psychiatric researchers believe that their professional standing and affiliations entitle them to immunity from public discussion and review for publicly funded experiments involving disabled human subjects. Do they believe that mentally impaired persons deserve to be used experimentally without any public limits or restraints, "all for the sake of science"? Surely, most psychiatrists do not condone experiments which induce schizophrenia relapse. Why then, has no one from the psychiatric community come forward to protest these experiments? *Is this a fraternity of silence?*

In the absence of a clearly articulated position that defines in unambiguous language "what's ethical, what isn't," by the American Psychiatric Association (or any other psychiatric organization), the Board of Directors of the Alliance for the Mentally Ill of New York State unanimously adopted a resolution a year ago qualifying our support for research. It states: "Our support for research is contingent upon it being conducted in accordance with ethical and professional standards of treatment."

We hope that this conference will begin to set forth and to define those standards. We are encouraged that one eminent leader in the psychiatric community, Dr. Herb Pardes, chairman of Psychiatry and dean of the College of Physicians and Surgeons of Columbia University, (responding to our question at the NYSOMH Psychiatric Research Conference in Albany, on Nov. 6, 1994) stated emphatically that he fully subscribes to our philosophical position and the need to implement increased controls to police the few "rotten eggs" in the field who (he said) are operating outside the pale. We are heartened that Dr. Pardes recognizes the value of an alliance between citizens and professionals to work toward the same goal. We concur fully with his statement: "We want research to advance, but we want it done right."

Institutionalized psychiatric patients are the most "administratively available" group of human guinea pigs. Though they have committed no crime, they are currently far less protected, and, therefore, more vulnerable and disadvantaged than prisoners—they are utterly powerless and trapped. Like prisoners, they are restrained captives on locked wards; they are totally dependent upon the institution and the state system for their care; they are economically disadvantaged; and they cannot, if they're dissatisfied with the facility, decide to transfer to another. Furthermore, unlike prisoners or pregnant women, institutionalized mentally ill patients have diminished powers of reasoning, judgment, and comprehension. These overwhelming deprivations should have earned them greater protections; instead, they have been disenfranchised and cheated of their rights.

Whereas it is illegal to use prisoners (and other vulnerable groups) in publicly funded, nontherapeutic, invasive experiments, mentally impaired persons who lacked independent advocates, were left unprotected by federal regulations from such experimentation. This, we have come to realize, was in large part due to the lobbying efforts of the psychiatric research establishment, most prominently, Dr. Frederick Goodwin, at that time, Director of the National Institute of Mental Health. In his testimony before the National Commission for the Protection of Human Subjects of Biomedical and Behavioral Research, Dr. Goodwin warned that protections for human subjects might dilute the physician-patient relationship and he expressed his concerns about singling out persons for special protection, just because they have a certain psychiatric diagnosis. This, I believe, is but a specious argument which implies that providing protections to prevent abuse will increase stigma.

Those who suffer from major mental illness were thus left with the "privilege" of enjoying an altruistic satisfaction (as some researchers have tried to explain), that of serving as guinea pigs to advance knowledge. We have not, however, noticed a stampede by noninstitutionalized citizens (or aliens) competing for this privilege. It was, no doubt, his quest to expand the window of opportunity for mentally impaired human beings (to serve humanity) that motivated Dr. Goodwin to recommend to the Commission that: "Research unrelated to conditions of mental illness should not be categorically prohibited for institutionalized persons ..." This philosophical position by the Director of NIMH helps to explain why such a large proportion of "civic minded" citizens who are "eager" to give their consent to uncomfortable, often disorienting biomedical experiments—for the sake of science—are congregated in mental institutions. (I believe Shakespeare was wrong to single out lawyers ...)

New York State is taking full advantage of institutionalized psychiatric patients through its State Office of Mental Health Research Regulations. Those regulations permit nontherapeutic, invasive, high-risk research—which is not federally funded—to be performed on incapable psychiatric patients—including children—at state facilities without their consent or the consent of their immediate family. (A lawsuit filed by New York Lawyers for the Public Interest is pending, TD v NYSOMH)

The tragedy is that the mentally ill have been betrayed and exploited by the very professionals, institutions, and organizations who claim to be "advocating" on their behalf. As a result, patients are sacrificed for dubious pseudomedical experimentation which passes off as science.

An experimental study at the University of Cincinnati College of Medicine conducted "psychophysiological pain rating procedures." Ninety-three random electric shocks were applied to strapped psychiatric subjects of research who had been washed-out for seven days. The protocol states:

... a random assignment with crossover of 2 mg. naloxone or equal volume of saline intravenously will be given in double blind fashion, followed 5 minutes later by the psychophysiological pain rating procedure. The procedure involves single shocks administered to the left forearm by a computer controlled constant current stimulator with a Tursky concentric electrode. Subjects will receive 3 shocks at each milliamperage increment from one to 31 milliamps for a total of 93 shocks in a random sequence at 2½ second intervals. Subjects will rate each shock in one of four categories: noticeable, distinct, unpleasant or very unpleasant.

These pseudomedical experiments are not from the Nazi era: these experiments were conducted by contemporary American psychiatrists at accredited institutions. All currently hold responsible positions in academic institutions which receive public funds. These research protocols and procedures passed IRB review, were awarded financial grants by the National Institute of Mental Health, and were published in respected, peer reviewed professional and academic journals, including the American Psychiatric Association's *American Journal of Psychiatry*.

If most psychiatric researchers disapprove of unethical experimentation, why have they remained silent, and why have they failed to come forward in support of improved safeguards for patients? How many more individuals will be hurt before the scientific community disavows these practices? To our knowledge, no professional organization of psychiatrists has publicly criticized or even questioned the ethics of these experiments.

Until independent family members began to raise questions about experiments that induced schizophrenia relapse, some through drug withdrawal, others by giving patients compounds known to induce psychotic symptoms (e.g., L-dopa, amphetamines, and apomorphine), experimenters duly charted the total collapse of these desperate, uncomprehending human beings. **For whose benefit were these patients put through hell?**

The steadfast silence of the psychiatric community, and the attempt by some to shoot down family members and patients who question the ethical justification for certain experiments, has left us with a deeply felt sense of betrayal. Does their silence indicate support for colleagues who have violated the very first and ancient ethical principle of every physician, namely; "First, do no harm"? Does their silence also indicate support for colleagues who have used tactics of intimidation, whispering innuendos, insinuating they have confidential information about family members who speak out against unethical human experimentation? **What else, we wonder, is being covered-up by this fraternity of silence?**

To date, no physician/investigator nor institution has been held accountable (by the current regulatory process) neither for inflicting pain and suffering, nor for putting uninformed, vulnerable psychiatric patient-subjects of research at risk of suicide. Even when the federal Office of Protection from Research Risks (OPRR) found gross violations in the informed consent process, no penalty has so far been imposed, no professional reprimand has even been issued: in fact, violators are openly unrepentant (see the *Los Angeles Times* and *Time*).

It is clear to us, that the welfare of psychiatrically disabled persons who participate in research cannot be entrusted solely to those who conduct research. Without a mechanism to ensure independent oversight and enforcement, vulnerable human beings are at risk of exploitation by those more powerful than they. We hope that leaders in the psychiatric community like Dr. Pardes will join our efforts to secure patient safeguards.

THE WELFARE OF MENTALLY DISABLED PATIENTS WHO BECOME SUBJECTS OF RESEARCH IS UNDERMINED BY THREE MAJOR UNACKNOWLEDGED CONFLICT OF INTEREST FACTORS

First: biomedical research is strongly driven by financial interests—pharmaceutical companies in collaboration with federal government agencies—those interests often conflict with the interests of vulnerable patients who need individualized care and protections.

Second: the physician/patient relationship has (in many cases) been compromised by an unacknowledged conflict of interest against which the individual patient has no way of protecting him/herself: the primary focus of the physician/researcher is conducting the study, gathering and evaluating the data, and publishing the findings. Careers and reputations are advanced as a result of publishing research findings—unfortunately for the patients, no one has (until now) questioned the means they employ to obtain the data.

Third: The exclusion of independent patient advocates from government and institutional policy, decision-making and oversight bodies—including institutional review boards—has left the mentally ill without any dependable safeguards to ensure that they are not harmed or exploited. Instead, vaguely defined (and therefore easily circumvented) safeguards which are poorly enforced accommodate the research enterprise, not the needs of patients. Multiple forms, a minutia of details, and conflicting federal requirements add confusion, but do not focus on the areas of abuse. In too many cases, a cursory, uninformative—even misleading—

signed consent form is accepted by an IRB in lieu of the process of obtaining and maintaining the person's informed consent.

NIMH is now entering a new era and we hope that our endeavors to improve the safeguards and protections for vulnerable human subjects of research who are disabled by severe mental illness will be fully supported by the agency. Following are suggestions which we hope the entire psychiatric community will support:

1. Inclusion of family and recipient advocates on institutional review boards.
2. Mandating a position for an independent research intermediary (family member or other independent member of the community) whose responsibility is to monitor vulnerable subjects of research to ensure their well-being and their continued consent. such as the initiative at the University of Texas Health Science Center, Houston.
3. Federal regulations expanding the safeguards for mentally impaired persons, including setting limitations on nontherapeutic experimentation which carries more than minimal risk. Safeguards should be at least as protective as those mandated for other vulnerable populations.
4. Defining accountability of physicians and institutions for patient-subject welfare during and following experimental research, and establishing enforcement guidelines and procedures.
5. Requiring financial disclosure on consent forms.

Finally, we offer the following 12 questions which we recommend should be asked by every prospective patient-subject of psychiatric research (and/or their families) prior to signing a consent form:

1. Is there an independent monitor (who is not connected with the research team) assigned to ensure that patient safeguards are followed? What authority does that individual have? How will this person ensure the patient-subject's continued consent?
2. Who is authorized and available during off hours (i.e., weekends, evenings, holidays) for patients to contact in the event they wish to withdraw from a protocol?
3. During participation in a research protocol, what services and activities are available to the patient? Are these services part of the institution's normal rehabilitation and treatment program or are they available only to research subjects?
4. Are drug wash-outs required? For how long? Is this a placebo control study? If symptoms return and I am uncomfortable, will I be given medications immediately? Who is authorized to prescribe therapeutic medications upon my (or my family's) request ?
5. Are non-FDA approved, investigational medications given in this protocol? Is this study being funded the pharmaceutical company? If yes, ask to meet with amember of the IRB to discuss the known and potential risks involved.
6. If non investigational drugs are involved, are these drugs approved (by the FDA) for my condition? Ask for a photocopy of the PDR reference discussing the risks and side effects of this medication, and ask whether you will be given the standard recommended dose?

7. If the drugs used in the study prove to be beneficial to me, am I assured that I will continue to be treated with the medications after completing the study?
8. Are there procedures that involve radiation in this study? If yes, ask to speak to a member of the IRB to discuss the limitation of risks.
9. Will this experiment involve any procedures that are painful or uncomfortable (e.g., injections or iv studies, spinal taps, long periods of immobilization, sleep deprivation, dietary restrictions, bed/unit restrictions).
10. Does the consent form cover more than one study? Is this an "umbrella" consent that will allow several procedures to be performed concomitantly or successively?
11. If there are any therapeutic benefits learned from the protocol, will I and my family be informed of this in writing? Will my patient record indicate these beneficial findings for the purpose of follow-up care?
12. What follow-up care is offered to patients who complete a study? To those who drop out?

Expanding on *A Mother's Testimony*

Janice Becker

I was asked to write *A Mother's Testimony* about my family's experiences, when my younger daughter Laura was a research subject at the Maryland Psychiatric Research Center (MPRC) for over four years. My article appeared in *The Journal* (Vol. 5, No. 1), published by the Alliance for the Mentally Ill in California. Today, I will elaborate on my testimony.

I want to begin by telling you about the responses to my article. I was surprised. Some challenged my experience and suggested that my perceptions must be wrong. Others thanked me for having the "guts. " It was not "guts," but anger and frustration that prompted writing this testimony.

Initially, my impression of the MPRC was positive. My husband and I had a tour of the inpatient facility. It was spacious and attractive with carpeting, soft furniture, and personalized bedrooms.

We met with staff several times to learn about the center's program. During our interviews, staff emphasized that the patient's quality of fife was as important as the research. We were also told individual families would be kept informed of protocols of their relative and the families would meet as a group with staff each month.

We had looked forward to hearing our daughter's progress as well as meeting and talking with other families. However, my husband and I were very disappointed that we were not correctly informed of Laura's status and for over two years there were no family meetings just rare invitations to parties or cook-outs. When the MPRC's staff finally held meetings they were on an irregular schedule. This raised the questions about our daughter's quality of life, the quality of her medical care, the research methods, and her safety.

During our weekly visits, my husband and I saw that activities for patients were inadequate and sometimes inappropriate. There was no trained activity therapist; instead nursing staff was responsible for providing activities. Many of the activities were not geared to research subject's capabilities. At one point, families were asked to contribute items for a kitchen that was being installed for patient activities. We contributed, but almost as soon as the kitchen was completed, the MPRC's administration put another program in the same wing and declared the kitchen off limits to patients and staff.

As disappointing as the lack of activity was for patients, we were more deeply concerned about the quality of Laura's medical care. For years, psychiatrists had medicated Laura with numbers and combinations of drugs. Despite their treatment, she did not improve. We had hoped that the center's research would result in improvements for Laura so that she could have a better life than being warehoused in the back ward of a psychiatric hospital.

We were told, before beginning research, "washouts" were necessary. Laura was taken off all medication. My husband and I watched as our daughter became very psychotic. It was heartbreaking for us to see this as her condition worsened. When we asked when research would begin and what protocols were planned, the answers varied as follows:

1. The research drugs were in, but the protocol papers were not.
2. The reverse—the protocol papers were in, but the drugs were not.
3. Some protocols were not appropriate for her.
4. It takes longer for drugs to leave some peoples' systems.
5. Meanwhile, videotaping of her tardive dyskinesia was being done.

We were not told that what Laura was experiencing was a "drug-free interval." We did not know what a drug-free interval was nor did MPRC staff tell us what this implied. We found this out much later. Laura was kept in a drug-free interval for over a year. During this time, she was very psychotic. Laura suffered and we suffered with her. What we could not understand was: "If our daughter is actively psychotic in a drug-free interval and kept in a drug-free interval, the MPRC administration must have a purpose. What was the purpose and why weren't we told?"

Laura did not suffer alone. We knew of another patient whom we were told that the center staff kept in this locked unit in a drug-free interval, which we discovered, lasted for nine years.

During Laura's hospitalization the Inpatient Director resigned. Staff protested by a sick-out. Orders were given that all subjects not on a drug protocol be given Haldol even though it was known that some had a dystonic reaction to the drug. When families became aware, 15 of us met and composed a letter of our concerns to the chief of the Inpatient Program. Another program director was eventually hired, then after a while replaced. These events had a disturbing effect as did other incidents.

Laura told me of being given wet sheet wraps and cold baths. Twice, when I visited her, I found her tied to a chair with sheets and towels. These kinds of restraints were prohibited in state hospitals, yet, apparently, these regulations did not apply to the research program. I have seen Laura held down by six staff people until she was quiet. One incident of wrestling her down resulted in a cut on her face that required sutures.

You must wonder why I did not take my daughter from the MPRC. My trust in the center, that the quality of her life was important, was gone. Staff had told us that research subjects could leave the program if they wanted. How could someone actively psychotic leave? How can anyone, who is so vulnerably ill, be expected to comprehend choices, articulate their feelings, or be able to leave. When I questioned the social worker about discharge plans, she had one offering—an adult foster care home in Baltimore City. This offer, I believe, was more threat than solution. My daughter could not survive in that setting, even in her preresearch condition.

Bringing her home was not a solution. At this point, Laura was off medications, the most ill I had ever seen her and, at times, dangerous. When it appeared she was going to be kept in this drug-free interval, even with her florid symptoms, I wanted her out of there. However, I was afraid of what might happen to Laura. What are a family's choices? or Laura's? What would staff do? Would they find a medication to reduce Laura's symptoms at least to her admission state? If so, would they try to find an appropriate placement? And when? Would she be rehospitalized? And where?

These were just some of the issues that angered me enough to speak out. After my *Journal* article was written, members of three other AMI families talked to me about their research experiences. Our family members were not all in the program at the same time, yet we share many of the same concerns and feelings. Their stories are for them to tell.

We shared feelings of anger and frustration. We felt lied to and used. We were sorry. We had not expected miracles. We did hope our sick ones would benefit as well as research. We did not expect harmful incidents.

There is something that still puzzles me. What scientifically justifies keeping a research subject on a locked unit—off medication for nine years? I wonder how he feels about those nine years of his life now that he is in the community and doing well.

I want to know, who benefited from Laura's pain and our anguish? What did her suffering and the suffering of others, in the name of research, accomplish? Since the inception of the inpatient research program, what has it accomplished for the mentally ill? I have asked about the findings of this research. My questions have never been answered.

Consumer Representation on IRBs— How It Should Be Done

Peggy Straw

Past President, Alliance for the Mentally Ill of New Hampshire; 2 Ridgeview Drive, Hookset, NH 01306

In 1989, representing the Alliance for the Mentally Ill of New Hampshire, I was asked to be on the Institutional Review Board (IRB) of the Division of Mental Health. There were representatives from the law, the state hospital, the research division of Dartmouth College, the Division of Mental Health and, in addition, a consumer and a family member. The Board was chaired by the psychiatrist medical director of the Division. There was talk about including an ethicist or representative from pastoral counselors in an attempt to be all-inclusive. Our duties, as they were spelled out to us, were not to comment on the merit of the research but on the effect, if any, on human subjects and the legality of the effort. We were, however, allowed comments if we felt that the research was poorly constructed, not properly documented or with a specious hypothesis, but the research proposal had already been approved by the educational institution to which it was submitted before it got to us, either Dartmouth College or the University of New Hampshire and our opinions would be just that, opinions, but forwarded to the parties involved.

Dartmouth Medical School had already formed an IRB Committee and would be doing most of the research since they had a contract with the state of New Hampshire to provide psychiatric services to the State Hospital in Concord where the preponderance of research would be taking place. We were given the Dartmouth Medical School "Assurance of Compliance with HHS Regulations for Protection of Human Research Subjects." This treatise, among other assurances, outlined the basic elements of informed consent that included:

1. A statement that the study involves research, an explanation of the purposes of the research and the expected duration of the subject's participation, a description of the procedures to be followed, and identification of any procedures which are experimental.
2. A description of any reasonably foreseeable risks or discomforts to the subject.
3. A description of any benefits to the subject or to others which may reasonably be expected from the research.
4. A disclosure of appropriate alternative procedures or courses of treatment, if any, that might be advantageous to the subject.

5. A statement describing the extent, if any, to which confidentiality of records identifying the subject will be maintained.
6. For research involving more than minimal risk, an explanation as to whether any compensation would be forthcoming and an explanation as to whether any medical treatments are available if injury occurs and, if so, what they may consist of, or where further information may be obtained.
7. An explanation of whom to contact for answers to pertinent questions about the research and research subjects' rights, and whom to contact in the event of a research-related injury to the subject.
8. A statement that participation is voluntary, refusal to participate will involve no penalty or loss of benefits to which the subject is otherwise entitled.

In addition, the research investigator was mandated to include statements about danger or risks involving pregnancy, circumstances under which the subject's participation may be terminated, any additional costs to the subject, the consequences of a subject's decision to withdraw from the research and procedures for the withdrawal, the number of subjects involved and a statement that new findings may influence the willingness of subjects to continue the research.

The informed consent document had to be in understandable language, give the subject enough time to decide whether or not to participate, and not waive any of the subject's rights. The Committee reviewing the research was to ascertain that risks to subjects were minimized, selection of subjects was equitable, and the research plan makes adequate provision for monitoring the data to insure the safety of subjects.

I saw my job as being sure that the informed consent document was in language that a layman could understand, that the person involved was not in such a psychotic state that they could not fully understand what they were getting into, that there was no coercion, that the risks were fully spelled out, that there were sufficient benefits or "what is in it for me," because I believe that research subjects should be given some sort of reward for being guinea pigs, and, of course, that there should be some follow-up if there were any negative after-effects and a therapist or staff member other than the researcher should be available to explain any confusion that the subject may have before, during, or after the research. When at all possible, families should be notified when the patient is considering being a participant in research, especially if there is any serious risk factor involved such as a double-blind study in which some would be given a placebo or medication would be withdrawn and, in that case, there should be follow-up to make sure that the person in the study has stabilized and is connected to a support system after the research is terminated. Also I was alert for any family-blaming sort of research that would have a negative effect on the relationship of the subject with their family, the only permanent support system they have.

We inherited one research project that had been started at New Hampshire Hospital (NHH) before our committee was convened. Its title was "The Impact of Systemic Family Processes and Structures of Mental Illness and Family Violence." and the primary and only investigator was a social worker at the hospital with a M.S.W. who believed that he could prove the relationship between violent families and mental illness. Another study, this time by an investigator from the Harvard

Department of Psychology, was entitled, "The Origins of Verbal Hallucinations" and had as its hypothesis that the contents of hallucinations would be similar to the content of statements made about patients by family members. There were leading questions about criticism and emotional overinvolvement on the parts of families that were taken from research into expressed emotion theories. Although I objected to the hypothesis on the basis of its merit, I agreed to the consent form when the word "family" was replaced with the word "person or persons with whom you live." I felt that the implications that the family was somehow to blame for the patient's destructive voices would be therapeutically counterproductive, and I was convinced that the hypothesis was way off base considering the experience I had had with families. As it turned out, the investigators later admitted that, to their surprise, there was no correlation between things families said or did and the content of the voices. It seemed to me that social workers and psychologists were still operating on the old theories and, although it was good to prove that those theories had no basis in fact when the research was completed, if patients at NHH were to be used for research, there ought to be more valid work being done by medical research into forms of treatment outcome rather than fixing blame for symptoms of the illness. One of the most valid research projects conducted was by a Dartmouth psychiatrist, Dr. Robert Drake, into the forms of treatment most effective for the dual-diagnosis patient. The consent form was very closely scrutinized by the IRB since the drawing of blood was a provision for participation. This study has proved very important nationally in developing treatment models for the homeless mentally ill dual-diagnosis patient as well as those who were not homeless.

In every case, consent forms were gone over very carefully to make sure that all the provisions were included and the language was understandable. Sometimes research scientists tend to write, perhaps inadvertently, in jargonese in their forms which tends to confuse subjects. In one protocol requiring a magnetic resonance imaging (MRI) examination, I requested a sentence be included that explained the possibility of claustrophobia for those who might be susceptible. In a protocol that required a double-blind study, I objected to the use of Haldol in the control group without a sentence that explained that if the subject had any negative reactions to the medication, they should not participate in the study. In one study, in which one control group would be taken off all medication, our whole AMI group had input into under which terms they would consider their loved ones being involved in such a protocol. Among other things, If anyone were put on an experimental medication, we felt that it should be continued without cost to the patient if it proved to be effective, and it should be available to those in the control groups as well for at least a year; however, that particular study was aborted because of adverse reactions in another pilot area. Some of the research required responses to long questionnaires which I felt was too demanding and time-consuming for those who had problems with concentration.

There were several areas in which I felt the subject might object to procedures and the consumer representative on the board disagreed. She said that most of the consumers of her acquaintance liked to be part of research, if for no other reason than that they got extra attention from professionals, which was a sad commentary. There is an article in the recent September/October 1994 issue of the National

Alliance's "Advocate" by a research subject who speaks of her experience positively. Margaret Brender is a volunteer subject at the National Institute of Mental Health, living for a year at their clinical center in Maryland. She has an affective disorder and has participated in everything from medicine trials to positron emission tomographic scans, blood and spinal fluid testing, MRIs, sleep-deprivation studies and magnesium infusion studies. She admits that she is not given information that might affect the research results, which are blinded in most cases, and does not seem to resent it; in fact, she says that most patients go through a great deal of discomfort for research that does not directly benefit them so that someday others may benefit.

The *Time* article in the November 7, 1994 issue entitled "Madness in Fine Print" wrote about tactics used to coerce participation in research by a Veteran's Affairs facility in Los Angeles that is now undergoing an ethics review by the American Psychological Association. Other researchers are reviewing their consent forms to comply with the new climate brought about by media attention to abuses of the past. It is a sad fact that some scientists regard "any means to an end" in their research efforts without considering the so-called subjects of that research as human beings with feelings. Unfortunately people with mental illness have always been considered somewhat less than human, objects of either ridicule, sick jokes, or unsubstantiated fears which, of course, has contributed to the horrendous stigma that these people have had to face heroically, on a daily basis, in addition to the debilitating pain of their illness.

Fortunately in New Hampshire, Dr. Robert Drake, Dr. Thomas Fox, Medical Director of the New Hampshire Division of Mental Health, and Dr. Robert Vidaver, medical director of the New Hampshire State Hospital have had personal experience with mental illness in their families which gives them a special sensitivity to the needs of research subjects. Families of the mentally ill in New Hampshire were pleased at the outset that the quality of treatment would improve with Dartmouth providing needed psychiatric services to the state hospital and pleased that research was being initiated, but were understandably worried about the potential for their loved ones being used as guinea pigs in research projects. Some of us were worried about the "ivory tower" academic approach which we deemed too "precious" for the severely mentally ill that we represented, with past psychiatric teaching leaning on the "blame family" approach and most psychiatrists graduating from medical schools more interested in dysfunctional families and Freudian concepts of behavior. Fortunately, IRBs were mandated and the Division of Mental Health under the leadership of Don Shumway had always worked closely with the Alliance for the Mentally Ill and insisted on its representation on both the Division IRB and the New Hampshire State Hospital Board. I am privileged to serve on both boards and feel that my input and the input of a primary consumer has given professionals and administrators on the boards a valuable perspective that they might not otherwise have had and highly recommend the same representation on all IRBs throughout the country.

A Consumer/Professional's View of Ethics in Research

Frederick J. Frese, III
Western Reserve Psychiatric Hospital, 283 Hartford Drive, Hudson, Ohio 44236

As a long-time clinician in a state hospital and one who is both trained at the doctoral level as a psychological researcher and who has spent the past three decades being diagnosed with and repeatedly hospitalized for schizophrenia, I hope that my comments will represent a merging of the views of consumers, researchers, and clinicians.

During the past few years, members of my family have been active in producing video materials aimed at education and lessening the stigma of mental illness. As a result I have become accustomed to animated teaching aids which are not readily available for a presentation such as the one today. So if you will have the kindness to bear with me I would like to use myself as such a device.

If you would all imagine me as turning into an object, the object I am turning into is a scale. I am not a fish scale or a bathroom scale but a scale of justice. If you would all stare at this silver spot in front of my heart as I speak, I will tell you that with such a scale, the balance shifts as conditions change. (The speaker's arms are outstretched and moving up or down to demonstrate as he speaks.) In the course of this talk and of this conference we will be considering the balance of factors impinging on the conducting of psychiatric research. As conditions change with time, rules and regulations need to be periodically reexamined. For example, if the *Diagnostic and Statistical Manual of Mental Disorders* has needed to be altered every seven years recently, and if our knowledge, definitions, and characterizations of mental disorders change, is it unreasonable that the rules by which we conduct investigations into these disorders also be reexamined?

It may be particularly opportune for us to engage in this exercise at this time because, as was said by a former Iowa governor at the Carter Center Mental Health gathering last year, divisiveness among constituents makes the jobs of our political friends much easier. When an issue comes to their attention, they typically call representatives of the interested groups together and ask how they can be of assistance. When the constituents disagree or otherwise begin to squabble among themselves, the politician simply needs to tell them to go and work out their differences and when they decide what it is they all want, they should then come back for his help.

Hopefully during this three-day conference we can begin to chip away at some

191

of the divisiveness we may have among the stakeholders affected by psychiatric research.

Currently we are in Baltimore, this is Saturday, January 7, 1995 and I am Fred Frese. I trust this establishes that I am oriented in three spheres. I am oriented and, unlike when I delivered the presentation at the American College of Neuropsy- chopharmacology in Puerto Rico last month, I am not fragmenting. I am not fragmenting, but this speech is fragmented into three parts.

In this introduction I trust I have established my credentials as both a consumer and a professional in the mental health field. In this regard, I would like to state that as a consumer, I am and have been available for any reasonable research protocol having to do with the study of serious mental illness. I have been saying that I am available for almost thirty years now, but with the exception of one study by John Strauss about three years ago, no one has yet taken me up on my offer to be a research subject. Once again I am ready to be a research subject in virtually any reasonable study, but I have a condition, and that condition is that I would like to be able to choose the members of my institutional review board, or at least three of the members of such a board.

The members that I want to serve on my institutional review board are as follows.

First I would like Carol North, M.D. Dr. North (1987) is the author of *Welcome Silence*, which has the best description of delusions and of the side effects of psychotropic medications that I have yet seen. She is a person has been repeatedly hospitalized for schizophrenia, as well as being a faculty member in the Depart- ment of Psychiatry at the Medical School of Washington University in St. Louis. She has the experience both as a consumer and provider of psychiatric services.

Second, I would like as a board member Daniel Fisher, Ph.D., M.D. Dan has his Ph.D. in biochemistry from a Big Ten university and became a researcher in schizophrenia about twenty-five years ago at the National Institute of Mental Health. He produced some twelve research papers on the biochemistry of schizo- phrenia and one day he came to the conclusion that indeed all people in the world were but biochemical robots, all but himself, that is. Shortly thereafter he found himself a patient of the Bethesda Naval Hospital, on the neuropsychiatric unit. After being there a while he decided he did not like the experience and decided to try to see it he could effect some changes. As a first step in this process he gained entry to medical school and continued there as a student until he had another breakdown and hospitalization. This time he imagined himself to be one of the electrons on one of the molecules that were the basis for schizophrenia. He was again rehospitalized. With some difficulty, after his discharge he was able to continue his medical studies, eventually graduating, completing a residency in psychiatry, and becoming the chief psychiatrist at one of the community mental health centers in the greater Boston area. Dr. Fisher is also very active in the national consumer/survivors' movement, serving part time as director of the National Empowerment Center during the past several years.

As a third member of my Institutional Review Board I would like to see John Forbes Nash, Jr., Ph.D. Dr. Nash had been called one of most brilliant mathemati- cians in the country in the late 1950s shortly after completing his dissertation at Princeton University at age 21. After holding faculty positions at M.I.T. and Prince- ton, he came down with schizophrenia and spent some forty years without work, spending much of his time wandering around the mathematics building, Fine Hall, at Princeton, scribbling incomprehensible messages on the blackboards in empty rooms. In December of 1994, however Dr. Nash was awarded the Nobel Memorial Prize in Economics on the basis of the work he had done prior to his illness. Dr. Nash may not be a mental health professional, but I feel his extensive personal experience with schizophrenia gives him more than sufficient qualification for me to trust him as a member of my board.

In the event that I could not manage to have these three individuals serve on the institutional review board for the research project for which I would be volunteering, I would find perfectly satisfactory the substitution of their mothers, or their brothers, sisters, spouses, children, or any other family member who had a close and caring relationship with these persons as they suffered through their illnesses. The point being, that I far more trust persons who have had personal experience with schizophrenia than I do persons lacking in such experience.

As I mentioned earlier, I come to this forum as a person who has been diagnosed with schizophrenia and who was repeatedly treated as an inpatient for this condition. While I still carry the diagnosis of schizophrenia and periodically have breakdowns I have not been an inpatient for this condition during the past twenty years. During most of this time I have functioned as the Director of Psychology at Western Reserve Psychiatric Hospital, where my duties have included both supervising research projects and serving on the hospital's research review committee, as well as having clinical supervision duties for the hospital's psychology staff.

As I have stated elsewhere (Frese, 1994), I feel very strongly that when someone is disabled with schizophrenia or some other form of serious mental illness, he or she is subject to losing the capacity to make a rational decision concerning matters concerning his or her own well being. When faced with the prospect of entering into a research project as a subject, it must be assured that persons with mental illness have given their informed consent for participation. For individuals with less disabling, episodic forms of mental illness, it is reasonable to assume that they may grant consent for such participation when they are in remission or they are in a state where their reasoning processes are not impaired in such a manner as to interfere with their ability to make a rational decision concerning the informed consent process.

For individuals who are more seriously impaired, to the extent that the disorder they have substantively interferes with their ability to rationally participate in the informed consent process, then the question arises, how is it possible for such a person to participate in clinical research as a subject? This gives rise to the further question as to whether such research can be ethically conducted at all.

Balancing the requirement that subjects must grant informed consent in order to participate in clinical research, however is the very real consideration that if we are to develop improved treatments for these devastating disorders, there is no question that we must conduct research in order to do so. If we are to consider the well being of seriously mentally ill persons as a whole, certainly it would be intolerable to put ourselves in a position whereby we cease research and abandon these persons due to an impasse concerning our ability to elicit appropriate informed consent from the persons who must be involved as subjects in order for the research to be conducted.

Since obviously we cannot in conscience abandon these research efforts, the question becomes how do we go about addressing the informed consent question? On the basis of my limited experience I suggest the following.

As alluded to above, whenever there is likelihood that a person may periodically return to a rational state, we must take advantage of these times to have the person state his or her desires in the event that they again relapse into an irrational psychosis. Such arranging for consent in advance has been addressed in detail elsewhere (Rosenson and Kasten, 1991).

For persons who do not return to periods of rationality, and these tend to be the more disabled patients, of course, provisions must be made so that they can be included in research efforts. I recommend that this can be approached by taking two steps.

The first step is to allow a trusted family member or "significant other" to be appointed as guardian for the purpose of making such decisions. Such a trusted person should be one who is capable of weighing the risks of possible harm to the patient with the benefits which could accrue for the patient and others suffering from similar disorders. If such a trusted guardian feels that the possible benefits outweigh the risks then the process should be initiated to include the patient in the research being considered.

A second step should be taken as well, however. After consent from such a guardian is obtained the informed consent procedure should be reviewed. I strongly suggest that as part of that review process persons should be involved who themselves have personally experienced serious mental illness. This is a needed step because no matter how caring and thoughtful a trusted guardian may be, unless that person has personally experienced psychosis and some of its various treatments, they cannot put themselves in the place of the psychotic patient. The recovered person, to a great degree, should be able to function in this capacity. A person who has "been there" has experience similar to that of the person in question. Unless this sort of personal experience is brought into the process, the endeavor loses the benefits of valuable insights that can make the process substantively more humane and valid.

REFERENCES

Frese, F.J. (1994). Informed consent and the right to refuse to participate. *The Journal of the California Alliance for the Mentally Ill.* 5(1). 56–57.

North C.S. (1987). *Welcome silence: My triumph over schizophrenia.* New York: Simon and Schuster.

Rosenson, M.K., & Kasten, A.M. (1991). Another view of autonomy: Arranging for consent in advance. *Schizophrenia Bulletin, 17,* 6–14.

Ethical Considerations in Medication-Free Research on the Mentally Ill*

Adil E. Shamoo

Center for Biomedical Ethics, and the Department of Biochemistry, School of Medicine, University of Maryland at Baltimore, 108 N. Greene St., Baltimore, Maryland 21201

Thank you for inviting me to share my views with you on this important subject. I fully realize and appreciate the anxieties some of you have in simply discussing this issue. I also appreciate the fact that all of you are certain that your research is serving a noble cause. Moreover, you entered the field of medicine in general and of research in particular to do good—to serve the public good, to serve humanity—to advance knowledge which will serve future generations. For all of these noble reasons, you therefore rightfully demand the freedom of inquiry.

This is the formal text prepared and delivered on Dec. 13, 1994 at the Plenary Session on "Science, Clinical Care, and Ethics of Medication-Free Research in Mental Disorders" Co-Chaired by Donald Klein, M.D., and William T. Carpenter, Jr., M.D., during the annual meeting of the American College of Neuropsychopharmacology held in San Juan, Puerto Rico, December 12–16, 1994.

*I am compelled to publish the text of my talk at ACNP meeting with these conference proceedings because several speakers and a manuscript quoted this talk. Unfortunately, I have been either misquoted or quoted out of context. Therefore, the text will provide the reader a more accurate source for deciphering the facts on the ongoing discussion.

During my presentation, I departed in one place from the text and I so stated during my presentation. I skipped the section entitled "Brief History" until the paragraph beginning with the word "Recent allegations.... " I skipped this part because my paper (Shamoo and Irving, 1993) reviews this well and another speaker before me at this meeting, Dr. Richard J. Bonnie covered it during the session. I then told the audience about the recent case of Susan Endersby of Minneapolis. I was consultant to the program and have copies of all of the original documents (Lorraine Roe, Producer, 1994, Channel 5 KSTP TV News, Minneapolis, MN. Investigative Report titled: "Dangerous Experiments", October 26, 1994). I stated that I will be able to report to you only what the TV segment discloses.

Ms. Endersby is a 40-year-old patient diagnosed with schizophrenia for over 15 years. Ms. Endersby was stable on medication with severe bouts of depression. All her records indicated that she is suicidal and her record in numerous places recite why she is suicidal by describing how she will commit suicide by jumping off a specific bridge in downtown Minneapolis. She was enrolled in a study of a new drug called Sertindole (where patients were washed out of their medication and placed either on Sertindole or a placebo). The exclusion criteria clearly states that no patient who is suicidal or with suicidal ideation should be enrolled. Furthermore, the research protocol states that if a patient is enrolled, the patient will remain for 26 days as an in-patient. Ms. Endersby was enrolled into the research program. On day 11, Ms. Endersby signed a "no-harm" contract and was given a pass out of the hospital. Ms. Endersby left the hospital and committed suicide by the method she had described for years.

Freedom of inquiry is the rock bed of academic life for most of you. Because of this, the question naturally arises—why in the world do we have to listen to criticism by the media and the general public concerning our research protocols? These research practices, which to the scientist seem benign and morally correct, get blown up to horrific proportions, leaving the scientist with a public perception that approaches Nazism. Even if errors are made, our motivation is to serve the patient and the public.

I fully agree with all of your contentions on this subject. However, the possession of a noble cause and the desire to work for the public good does not justify every act. History has taught us on numerous occasions that a noble cause and the desire to serve the public good does not safeguard our society from harm whether it be from the actions of governments or of individuals. To our dismay, these abuses have occurred both abroad and in our own country. For example, in our own country we have had the unfortunate experience of the Tuskegee Syphilis episode and, more recently, the radiation exposure experiments among others. It is a necessary component of our democratic system that in order to change a policy a public discussion must ensue. Public discussions carry with them the price of being misunderstood, and exaggerated; sometimes, these discussions even serve to perpetuate falsehoods. Nevertheless, it is the only method in a democracy we have for effectuating changes in policy.

I personally am not interested in a scorched earth policy towards this issue. More importantly, I am interested in changing policies for a better future. I would like to read to you what I think is a very pertinent quote.

The experience of seeing him off all medication is not one that I ever care to see again; certainly not until that day comes when he will no longer need it—if that day ever comes.

This mourning is also stranger in that as painful as it is, it coexists with a lingering hope that one day Gary will be returned to his former self, a hope strengthened by the improvement on clozapine and the knowledge that new atypical antipsychotics are not far behind. (Martin S. Willick, 1994, p. 10)

Willick, a psychiatrist, is speaking about the clinical treatment of his son. This statement, however, may as well have been made if his son was a research subject. This statement captures the ethical dilemma we face whenever we induce pain and suffering on an individual. The issue then becomes whether there are:

1. Direct benefits now or in the very near future for the individual.
2. Future benefits for the individual.
3. No foreseeable benefits for the individual and potential benefits only for others (e.g., for "others of the same class").
4. Benefits only for others (e.g., not "in the same class") in the future.

In a nutshell, these are the ethical dilemmas we face when using psychiatric patients in research as human subjects. Whenever we proceed in a manner which results in using an individual in research, several safeguards are invoked either to justify the act as ethical or to ameliorate our ethical concerns. The safeguards usually invoked are in compliance with standards that include informed (valid) consent and the resolution of conflict of interest.

I will get back to these topics at a later time in my presentation. But for the benefit of some, I will discuss briefly the history of the use of human subjects in research as well as the U.N. resolutions and our recent findings. Finally, I will make a few personal recommendations.

BRIEF HISTORY

An explosion in research involving the use of human subjects occurred after the second World War. However, the world was shocked when the Nuremberg trials revealed the use of human subjects by Nazi physicians/scientists in the most abhorrent manner that included torture, degradation of human dignity, and death. It was the result of the tribunal that the Nuremberg Code evolved, with its primary objective being the protection of human subjects (Shamoo and Irving, 1993).

The most important and enduring principle stated in the code reads as follows: *The voluntary consent of the human subject is absolutely essential.*

After three years of deliberation, the U.S. National Commission issued a report known as the Belmont Report, which identified vulnerable groups—namely children, pregnant women, prisoners, and persons with mental illness.

In the late seventies, The President's Commission also recommended that special vulnerable status be given for persons with mental illness.

These federal regulations which the National Intitutes of Health (NIH) promulgated into law declared vulnerable groups to include: children, pregnant women, and prisoners—but *not* persons with mental illness. The regulations established that each research institution would be required to establish an institutional review board (IRB) to review all protocols involving the aforementioned human subjects and pass judgment as to their suitability for each particular project.

However, the regulations left the clear impression that those vulnerable groups not covered by regulation (i.e., persons with mental illness) should be covered as follows:

If an IRB regularly reviews research that involves a vulnerable category of subjects, including but not limited to subjects covered by other subparts of this part, the IRB shall include one or more individuals who are primarily concerned with the welfare of these subjects. (Code of Federal Regulations 45 CFR 46, 1981, 1983, 1989, p. 7)

In the 1991 revisions the clause "individuals who are primarily concerned with the welfare of these subjects" was changed to "persons knowledgeable about the patients' illness."

The best description of what happened to the issue of the use of persons with mental illness in research is summarized in a section on "special populations" in a book by Paul Appelbaum, M.D., Charles W. Lidz, Ph.D., and Alan Meisel, J.D., published in 1987.

None of those recommendations were implemented. This outcome was the result in large part of opposition from researchers on mental disorders, who claimed that the populations in question were no more vulnerable than most persons with severe medical disorders and that the suggested limitations would seriously restrict research on mental disorders. (Appelbaum, Lidz and Meisel, 1987, p. 228)

However, persons with mental illness have still never been declared as a vulnerable group in the Office for Protection from Research Risks (OPRR) regulations even as I speak to you today. It would appear that the issue of using persons with mental illness as human research subjects has been lost in the shuffle, due in part to the lobbying efforts of some researchers on mental disorders, and in part to the relatively small size of the constituency who have advocates for persons with mental illness when the above developments occurred during the mid-seventies to mid-eighties. Especially given their vulnerability and the reality of their use in on-going research, the situation of these persons with mental illness remains a concern. As one historian has recently and succinctly put it: "... it is the socially powerless and disadvantaged who are most likely to be subjected to unethical research" (Gillespie 1989, p. 13).

Recent allegations about a protocol conducted by UCLA Neuropsychiatric Institute, which were publicized by the mass media and congressional hearings, have forced us to focus attention on potential abuses of persons with neurobiological disorders when used as subjects in research. There is a legitimate need to conduct research on human subjects with neurobiological disorders. For example, there is a need to maximize the effectiveness and minimize the side effects of the existing medications, and to find better medications. However, important questions have been raised about experiments involving washout studies. In brief, stable and, at the time, functional patients were subjected to placebos or no drug treatment for periods ranging from a few days to a few years. It has been reported that a patient in the UCLA study was withdrawn from medication and subsequently committed suicide. On September 11, 1994, the *Los Angeles Times Magazine* section in a lengthy cover story described the circumstances surrounding the suicide. In addition, a few months ago an editorial in *The New York Times* called the informed consent currently practiced a "charade" and went on further to describe the main function of the institutional review boards was to act as a "rubber stamp."

We surveyed 41 U.S. studies performed on schizophrenic patients spanning the past three decades. We found that 2,471 subjects entered the program with 936 (approximately 38%) suffering relapse and 245 (10%) dropping out without any mention as to their whereabouts. The most troubling segment of these research protocols was the use of "challenge doses" of amphetamine or L-dopa to induce relapses in patients.

In 1991, the U.N. (in which the U.S. is clearly the leader) adopted a resolution entitled "Principles for the Protection of Persons with Mental Illness and in the Improvement of Mental Health Care." The relevant sections are:

Consent to treatment
(a)   [Information about] Alternative codes of treatment, including those less intrusive; and
(d)   Possible pain or discomfort, risks and side-effects of the proposed treatment.

These are serious public policy issues that require careful attention in order that all members of our society openly discuss these ethical dilemmas, and hopefully come to a constructive resolution. Our efforts must succeed because the alternative is to risk draconian rules in response to public outcry which will not serve the best interests of patients or society.

Let me quote the American philosopher Jans Jonas: "This price—a possibly slower rate of progress—may have to be paid for the preservation of the most precious capital of higher communal life."

He also said:

In the course of treatment, the physician is obligated to the patient and to no one else. He is not the agent of society, nor of the interests of medical science, the patient's family, the patient's co-sufferers, or future sufferers from the same disease.

I will now return to the four issues concerning "benefits" I noted earlier.

Issue 1—If the patient can derive direct benefits now or in the very near future from the experimental drug, then the balance of the scale is in favor of enduring the pain and suffering by the individual for the presumably better medications. This is of course with the provision that the appropriate issues of valid consent and conflict of interests have been resolved satisfactorily.

Issue 2—If the benefit for the patient lies in the future then the balance of the scale is slightly in favor of conducting the experiment. However, the burden is on the investigator to illustrate and document the potential future benefits. Furthermore, more careful attention should be paid to the design, protocol, and patient selection to enhance the potential benefit to the participating individual. Also, the design should minimize pain and suffering as much as possible. In this connection, the placebo control should be designed so as to be administered for a minimum period of time, with early termination whenever there is pain and suffering. I do not subscribe to Dr. Fauci's notion as described by Hellman and Hellman in 1991 in the *New England Journal of Medicine*.

As Fauci has suggested, the randomized clinical trial routinely asks physicians to sacrifice the interests of their particular patients for the sake of the study and that of the information that it will make available for the benefit of society. This practice is ethically problematic.

Issue 3—If there is no foreseeable benefit for the individual but potential direct benefits to others, then the balance of the scale weighs against such experiments. Why should a patient with compromised cognition endure pain and suffering for someone else?

Issue 4—If the benefits are for others and in the future, then the balance of the scale weighs against such experiments. This is basically designing and exploiting an experimental protocol whose sole purpose serves only to gain knowledge in science, with dubious benefits in the future to other members of the society. Louis Henkin, a contemporary philosopher, said: "An individual's right can be sacrificed to another's right only when choice is inevitable, and only according to some principle of choice reflecting the comparative value of each right."

In conclusion, there is an urgent need for additional safeguards such as: The prohibition of the most abusive experiments, which include the washout experiments associated with relapses. In addition, we plan to recommend a set of recommendations for others to adopt to improve and maintain future ethical and successful research endeavors. Finally, I would like to support the continuation of research on human subjects. However, the investigator must shoulder the burden of convincing the public that there are unique and compelling reasons why such a research protocol should proceed.

National Institute of Mental Health Human Subject Activities

Lana Skirboll,[†] David Shore,[‡] Andrea Baruchin,[§] and Paul Sirovatka[§]

[†]*Now at Office of Science Policy, National Institutes of Health, Bethesda, Maryland*
[‡]*Office of Science Policy and Program Planning, National Institute of Mental Health, Rockville, Maryland*
[§]*Division of Clinical and Treatment Research, National Institute of Mental Health, Rockville, Maryland*

Research conducted to discover the causes of mental illnesses, to improve treatments, and, eventually, to prevent these illnesses, requires the ongoing participation of people who live with the daily burdens posed by these illnesses. Individuals who choose to participate in research contribute their time and effort as well as that of their family members; more critically, research subjects affirm their trust in the competence of the clinical investigator who is conducting a particular study and in the integrity of the larger research enterprise—that is, those responsible for supporting and overseeing biomedical research.

The National Institute of Mental Health (NIMH) attaches high priority to the need to protect the rights and safety of participants in clinical research protocols. All clinical research funded and conducted by the NIMH must comply with Federal human subject protection regulations contained in 45 CFR 46, and guidance issued by the NIH Office for Protection from Research Risks (OPRR).

Beyond this formal adherence to Federal regulation and policy, however, the NIMH is actively seeking means of further enhancing the protection of human subjects in psychiatric research and recently has mounted several initiatives toward that end. Much of the impetus for these initiatives stems from two meetings, sponsored in 1993 by the NIMH and the National Alliance for the Mentally Ill (NAMI), to discuss ethical issues in psychiatric research involving human subjects. Participants in these meetings included NIMH staff, along with representatives of mental health clinical investigators, institutional review boards (IRB), and family and consumer groups.

A principal objective of the meetings was to examine and to further improve informed consent processes in mental illness research. Informed consent may be viewed simply as a requirement of Federal regulation that is designed to ensure the rights of research subjects; but NIMH views it also as critical to the benefits and the productivity of clinical research on mental illnesses.

Individuals at the meetings agreed on the importance of targeting educational efforts to clinical investigators, IRBs, and consumers and families. The following activities have been implemented.

- Efforts to heighten the awareness of members of the clinical research community to their ethical responsibilities for research subjects;
- Efforts to apprise IRB members of contemporary considerations regarding human subjects in psychiatric research; and
- Efforts to advise research participants, and candidates for research participation, of their rights and prerogatives in the clinical research setting.

These initiatives, discussed in turn below, reflect an appreciation by NIMH and advocates for mentally ill persons that human subject protection must extend beyond the "letter of the law" and be imbued in the knowledge, attitudes, and behaviors of all individuals who are involved in biomedical and behavioral research.

A FOCUS ON CLINICAL INVESTIGATORS

Widespread recognition of the fundamental importance of ensuring the dignity and protection of human subjects in biomedical and behavioral research has prompted increasing academic attention to the topic in recent years. The National Institutes of Health requires that universities receiving NIH training funds devote formal curriculum time to issues relevant to integrity and misconduct in research, and young clinical investigators are familiar with these concerns by the time they enter research careers. In furtherance of the NIH requirement, the NIMH, in 1993, corresponded with all program directors of NIMH-funded institutional research training grants, stressing the importance of informed consent and other concerns relevant to the conduct of ethical research, including the need for research trainees to appreciate the role and information needs of patients and family advocates in clinical research.

All NIMH-supported researchers who work with human subjects must comply with Federal regulations now in place to protect human subjects. These regulations, first promulgated in the U.S. Code of Federal Regulations nearly a quarter century ago, described essential requirements for voluntary informed consent and delegated the primary review of human subject research to local review groups known as IRBs.

Although the regulations have been reissued several times, most recently in 1991, they continue to comprise a broad set of principles rather than a specific list of "dos and don'ts." While the principles have been consistent over the years, the standards by which research protocols and consent documents are judged have changed considerably in recent years; research protocols and explanations of risks and benefits that were considered appropriate five or ten years ago might well be found unacceptable now. That is, human subject provisions must be interpreted and applied by clinical investigators and by IRB members in a societal context that is characterized by an evolving commitment to informing research subjects about risks, benefits, and alternatives.

Primary responsibility for ensuring an appropriate informed consent process rests with the principal investigator (PI) who designs a particular protocol. A PI's local IRB evaluates informed consent documents and the process of their administration, to ensure that risks will be presented understandably and consistently with the Federal regulations. With IRB approval in hand, the PI may apply for Federal funding. At this point in the process, an NIMH initial review group (IRG), which constitutes an independent group of experts who are convened to judge the scientific merits of the research proposal, provides a second-level review of the human subject protection provisions. The NIMH is authorized to ensure that any human subject concerns identified in this review are satisfactorily addressed by the PI and his or her institution before a grant is funded.

Recognizing the importance to NIMH's own scientific and public health mission of ensuring that all clinical researchers it supports fully understand and appreciate the requirements of 45 CFR 46, the Institute has stepped up its efforts to reach out to the clinical research community. One approach has entailed "brokering" information between the OPRR and the field. Toward this end, for example, the NIMH recently mailed to more than 200 clinical investigators a packet of information describing recent OPRR findings and interpretations of the regulations regarding informed consent. This May, 1994 report was stimulated by an OPRR investigation of an NIMH-funded schizophrenia research grant.

Also, the NIMH now is developing, jointly with the American Psychiatric Association, an investigators' resource manual that will address a wide range of ethical issues in clinical research. The manual is envisioned as a "working tool" for clinical investigators; that is, it will not discuss principles of research ethics in the abstract, nor will it prescribe guidelines for clinical studies. Rather, the manual will explore the pros and cons of various approaches to responding to the concerns of all parties in the research process. Issues to be covered will extend from research design considerations, to methods of assessing the quality of care in research settings, to strategies for effective and ongoing informed consent procedures.

A FOCUS ON IRBS

A second target of NIMH outreach efforts is the approximately 2,000 IRBs responsible for reviewing and approving the human subject protections required in all research proposals involving human subjects. Under Federal regulations, the IRB system is purposely decentralized, with oversight responsibilities for human subject protection assigned to local (i.e., community) control. IRBs are required to include both scientists and non-scientists, experts in clinical, legal, and/or ethical issues, and at least one individual who is not affiliated with the institution carrying out the research. While there is no requirement for an IRB to include members with expertise in any particular illness, NIMH encourages advocates and others with specialized knowledge of mental illness to seek opportunities to work with local IRBs.

More recently, the NIMH has begun an innovative collaboration with the OPRR to highlight contemporary issues regarding the participation of mentally ill subjects in biomedical and behavioral research. As part of an ongoing series of OPRR-sponsored regional workshops for IRB members and researchers, NIMH has convened panels of clinical scientists expert in various facets of psychiatric research and human subject protection issues. Panels to date have addressed several mental illness research concerns including special ethical challenges that may be encountered in research involving children and adolescents, women, and minority group populations.

A FOCUS ON INTERNAL NIMH PROCESSES

As noted above, the study sections, or IRGs, that conduct peer review of NIMH grant applications also consider procedures proposed by a PI to protect subjects and to assure that informed consent is obtained. If an IRG notes a "human subject concern" during the review process, the grant cannot be funded until that concern is properly addressed. Moreover, even if a particular grant is "approved but not funded," a frequent occurrence in a period when all Federal grant support is highly competitive, the applicant's institution is notified of the IRG's finding of a human subject concern.

Given the responsibilities associated with the review and administration of research grants and contracts, it is imperative that NIMH staff be well versed in human subject protection issues. Accordingly, NIMH, in conjunction with OPRR, is conducting in-service training for Institute staff whose duties require such knowledge. These sessions provide continuing education about Federal regulations concerning human subject protection, specific informed consent requirements, and other relevant human subject issues.

Moreover, in the interest of maintaining a constructive dialogue with the research field broadly and with consumers and families, NIMH staff are encouraged to participate in scientific and consumer meetings and to write and submit for publication articles focused on a variety of ethical issues relevant to clinical neurobiological research—for example, confidentiality, ownership of research data, competency to consent, and appropriateness of substitute judgment for those not able to give consent.

A FOCUS ON CONSUMERS AND FAMILIES

Mental health service consumers, along with advocates in NAMI and other organizations, are understandably very interested in regulations and procedures aimed at protecting subjects in mental illness-related research, and clearly want a more active, participatory role in the overall research process. Increasing recognition that patient volunteers are not merely "subjects" of research, but are active participants in the scientific quest for new knowledge has underscored the need for patients and advocates to make decisions about research participation on the basis of accurate information about all aspects of research. Participants must be informed about the foreseeable risks, anticipated benefits, alternatives to, and voluntary nature of a given research protocol. Consumers and family members have also expressed substantial interest in participating in discussions of research design and planning of clinical studies before the IRB and peer review take place.

Public attention to ethical concerns in clinical research on mental illness recently has been stimulated by controversies surrounding particular types of research protocols—for example, the use of placebo controls in medication trials and drug discontinuation studies. In order to prevent misconceptions or misinterpretations of published research findings, it is important not only for NIMH and the larger research community to re-examine the scientific rationale for particular research designs and methodologies, exploring alternative strategies where these might be indicated, but also to promote improved public understanding of strategies that stand up to such scrutiny.

Accordingly, in conjunction with NAMI and an advisory task force comprised of consumer representatives, the NIMH is developing a "consumer's guide to clinical research." This guide will provide information to aid potential participants and their families in making decisions regarding research projects. The guide will include an overview of the clinical research process, a discussion of ethical and legal issues in research, questions patients should ask and consider before participating as a research subject, and information about obtaining results of studies in which individuals have participated.

In the interest of facilitating continued discussion of research design and human subject protection issues in psychiatric research, the NIMH sponsored a symposium at the 1995 NAMI annual meeting. The symposium featured a panel of NIMH-supported clinical researchers who presented and discussed with consumers and family members who have had research experiences, a variety of innovative "partnership" models for broadening and enriching collaborations among patient-subjects, advocates, and researchers in the design and conduct of clinical studies of mental illness.

A FOCUS ON THE FUTURE

Despite NIMH's ongoing efforts to encourage discussion, education, and where needed, improvements in the protection of human subjects in psychiatric and behavioral research, there are limits on the Institute's authority in this arena. Regulatory authority over IRBs rests with OPRR, the Federal entity established as "a process for the prompt and appropriate response to information provided to the Director of NIH respecting incidences of violations of the rights of human subjects of research." In addition, 45 CFR 46 lists requirements for informed consent, delegates the primary review of research utilizing human subjects to IRBs, and specifies the composition of IRBs.

On the other hand, NIMH, through its IRGs, does review and approve human subject issues prior to funding a grant. The Institute ensures that human subject concerns are satisfactorily addressed prior to a grant's being funded, and that an IRB has approved a clinical research project within one year prior to funding. In addition, the NIMH will continue to disseminate information about human subject protection issues and the informed consent process.

The role of the NIMH in the arena of human subject protection is not in the form of regulatory authority, but rather in the strength of relationships it maintains with all parties involved in NIMH-supported research, and in its leadership role in the mental health field. It is in this capacity that NIMH articulates positions on specific issues and encourages discussion throughout the field of concerns that have bearing on the well-being of patients with mental illness and on the vitality of the research enterprise.

The Use of Placebo Controls in Psychiatric Research

John M. Kane[†] and Michael Borenstein[‡]

[†]*Chairman, Department of Psychiatry, Hillside Hospital, Division of Long Island Jewish Medical Center, 75-59 263rd Street, Oaks, New York 11004; and Professor of Psychiatry, Albert Einstein College of Medicine, Bronx, New York*
[‡]*Director, Biostatistics, Hillside Hospital, Division of Long Island Jewish Medical Center, 75-59 263rd Street, Oaks, New York 11004*

TREATMENT RESPONSE TRIALS

Overview

The controlled clinical trial remains one of the most important strategies for advancing medical science. The history of treatment research is replete with examples where erroneous conclusions were based on anecdotal reports or uncontrolled treatment trials. The introduction of double-blind, random assignment methodology combined with sophisticated data analytic techniques is one of the major achievements in modern medicine.

The extent to which placebo controls are necessary continues to be the subject of considerable debate and varies enormously depending upon a number of factors including, but not limited to: the illness being treated, the particular patient population, the nature of the outcome measure(s) of interest, the availability of proven safe and effective treatment, the predictability of treatment outcome, the natural course of the disease, etc.

There are a number of cogent arguments that can be made against the routine use of placebo controls in clinical trials. Rothman and Michels (1994) argue that when an effective treatment exists for a particular disease, the use of a placebo control is inappropriate on scientific and ethical grounds. They suggest that the appropriate question of interest is whether or not a new treatment is better than the existing treatment(s), not its superiority over placebo.

Supported by the Mental Health Clinical Research Center at Hillside Hospital MH-41960.

FUNCTION OF PLACEBO CONTROLS

It is generally acknowledged (Rothman and Michels, 1994) that the use of placebo controls is appropriate in situations where an effective treatment has not been established. In this case the question of interest is whether or not the new treatment is better than placebo. The ethical concerns are different in that the patient is not being deprived of the potential benefit of an effective treatment.

At the same time, placebo controls may serve a number of different purposes. As Kierman (1986) pointed out "the placebo control group represents a control for a 'package' of effects which potentially could obscure and confound the demonstration of drug efficacy." In this group of effects are included the passage of time, particularly in many psychiatric conditions where there is waxing and waning of the illness over time or even spontaneous remissions; the increased support and attention given to a patient participating in a clinical trial; the patient's expectations and hope for improvement.

Placebo washouts and controls may take on additional importance in patients who have already been taking some type of psychoactive drug (either therapeutic or recreational). The carryover effects of previous treatment can be very difficult to ascertain and may be confounded with the putative effects of the experimental treatment (or active control). These effects could be therapeutic or adverse. In the case of some patients, improvement may be noted when medication is discontinued. This may be evident for example in patients suffering from chronic schizophrenia.

As Leber (1986) and others have argued, placebo provides an estimate of what would have been had no pharmacologic treatment been given. In combination with an active control, an estimate can be made of the intrinsic responsiveness of the particular patient sample in question (to a "proven" treatment) as well as estimating the "sensitivity" of the "assay" system (the trial design, the investigators, the assessment methodology, etc.).

The discussion so far has focused on only the issue of efficacy. It is also important to recognize that there are no comprehensive, universally accepted definitions of efficacy *and* safety. How much of an effect needs to be demonstrated before we would consider something effective? How safe does a treatment have to be? What if there are disparities or even contradictions when multiple effects are examined?

In the case of complex psychiatric illnesses like schizophrenia and partially effective and frequently toxic treatments like neuroleptic medication, the issue of placebo controls cannot be dismissed solely on the basis that proven therapies exist.

In terms of safety, available antipsychotic drugs are associated with an array of adverse reactions. The neuroleptic side effects are the most troublesome and include: acute dystonia; akathisia; parkinsonism and tardive dyskinesia or tardive dystonia. Approximately 50% of patients will develop some degree of neurologic side effects in the first several weeks of neuroleptic treatment (Ayd, 1983). These effects can be subjectively distressing and often lead to subsequent noncompliance.

Tardive dyskinesia, a syndrome of abnormal involuntary movements, occurs later in treatment. Incidence estimates in young adults suggest that approximately five percent of patients develop new signs of tardive dyskinesia with each additional year of drug treatment (i.e., the cumulative incidence after five years of treatment is in excess of 25%) (Kane *et al.*, 1992). Fortunately most of these cases are mild, but some can be severe and disabling. Even mild cases, however, can contribute to the odd appearance or behavior that many people associate with the severely and persistently mentally ill.

The fact that such adverse effects occur so commonly is a fact of life not only in the treatment of this disease, but also plays a major role in extending the challenges of treatment development research. A major goal of new drug development has been to identify compounds that might have little or no propensity to produce these adverse effects. Suppose, for example, that an experimental drug appeared to be less effective than an established treatment, but had no neurologic side effects. Could we make an informed judgement as to whether or not to utilize this treatment without knowing whether it was in fact superior to placebo (or no treatment).

The argument that active controls are adequate for establishing the efficacy of a new treatment is countered by the fact that even when a trial establishes that the two are, for all intents and purposes, equally "effective," it may tell us very little about efficacy, since in that particular sample both treatments may be ineffective.

There is a tenuous assumption in dismissing the need for placebo controls in many psychiatric conditions, that the proven treatment will always demonstrate efficacy. In fact, even in a disease as potentially severe and chronic as schizophrenia, there is enormous variability in placebo response as well as response to "proven" standard treatments.

We (Kane and Borenstein, unpublished data) recently compared the response rates (defined for these analyses as a 30% improvement in Brief Psychiatric Rating Scale ratings) in eight clinical trials. These trials were run by various pharmaceutical companies for the purpose of evaluating new drugs, but our analyses focus on the placebo groups and the active control groups (haloperidol) that were included in these trials.

Response rates ranged from three percent to 26% for placebo, and from 12% to 45% for haloperidol—of note, the response rates for placebo overlap with the response rates for haloperidol. In the six trials that included both a placebo and a haloperidol group the treatment effect (i.e., the difference in response rates between the two conditions) could be computed and ranged from eight percentage points to 19 percentage points.

Rothman (1995) suggested that if we had a clear understanding of the factors responsible for this diversity, it would be possible to dispense with the placebo control, even if (as in the earlier example) the comparison group of interest was the placebo group. The argument is that we could use an active control, coupled with our knowledge of treatment effects, to determine what the response rate would have been for placebo. The response rate for the new treatment could then be compared with this inferred rate. As a practical matter, however, such a solution would be difficult to implement because we still do not understand all of the factors contributing to the diversity of response (some of which were outlined earlier), nor are we likely to develop this understanding in the near future.

SAMPLING ERROR AND SAMPLE SIZE

Above, we have outlined some reasons why placebo controls continue to be important in clinical trials. To the extent that these issues can be addressed, and in specific trials they sometimes can be (e.g., when we are working with a population where the treatment effect is consistent), it will be possible to dispense with the placebo control. It is, therefore, helpful to review two arguments that are often made against the use of placebo controls and serve as a source of confusion.

Trials comparing an active treatment with a placebo sometimes fail to yield a significant effect despite the fact that the active treatment is superior. It follows (the argument goes) that a study which finds no difference between a new drug and an active control could be similarly meaningless.

In fact, the failure to find a difference can stem from two sources and it is important to distinguish between them. Studies may fail to find a statistically significant difference because of low power. The proponents of active controls would require not merely the absence of a significant difference, but rather a demonstration that the active treatment and the new treatment were, for all intents and purposes, equally effective. However, the argument is incorrect in that even a demonstration of equivalence may be misleading, as outlined above, if the samples are drawn from a treatment refractory population.

The recognition that the use of an active control requires a demonstration of equivalence leads logically to the next argument. Studies that attempt to show the superiority of the new treatment (using a placebo control) require fewer patients than studies which attempt to demonstrate equivalence (using an active control). The difference in sample size can be substantial.

This argument is correct in the sense that current statistical practice for a placebo control requires only a demonstration of statistical significance—i.e., that the drug effect is not nil. The required sample size is driven primarily by the size of the expected effect and in many cases is relatively small. By contrast, current statistical practice for establishing equivalence with an active control requires that the difference in response rates between the two groups be reported with a small margin of error, which requires a larger sample.

However, this distinction is arbitrary. Just as we require that equivalence be established with a small margin of error (which is an appropriate criterion), we should require that the size of an effect *vis à vis* placebo be established with a small margin of error (rather than being satisfied merely that the effect is not nil). If we were to adopt this approach, then the sample size required for a placebo control would more closely resemble the sample size required for an active control.

MAINTENANCE TREATMENT

The course and outcome of schizophrenia has been shown to be very variable. Some persons with this illness experience a chronic deteriorating course, some experience periodic remissions and exacerbations, while still others may achieve a relatively stable remission for long periods of time.

Once antipsychotic drugs were shown to be effective in alleviating some signs and symptoms associated with acute psychotic episodes, the question arose as to what value continuation and long-term maintenance (or prophylactic) drug treatment might have in sustaining the improvement and preventing the occurrence of subsequent episodes of exacerbation or "relapse."

Over a period of time a number of double-blind, placebo-controlled trials were conducted attempting to determine the efficacy of maintenance treatment (most such trials lasted one year or less). Overall the results of these clinical trials suggested that maintenance medication could significantly reduce rates of relapse and/or rehospitalization.

Gilbert *et al.* (1995) recently reviewed the literature on neuroleptic withdrawal in schizophrenic patients and found 66 studies involving 4,365 patients. The mean cumulative relapse rate was 53% in patients withdrawn from neuroleptics and 16% in those maintained on neuroleptics over a mean follow-up period of 9.7 months.

It is of particular interest to note that the rate of relapse in the 66 studies of drug withdrawal ranged from 0 to 100% underlining the variability of outcome in this context. It should also be emphasized that there is no universally accepted and applied definition of "relapse" employed in these clinical trials or in clinical practice. A relapse could mean for example a mild and transient increase in anxiety and irritability; a predetermined change on a rating scale; or a dramatic increase in delusional thinking and suicidal ideation requiring hospitalization. Lumping all of the potential outcomes under one label makes little sense except for attempting to conduct metaanalysis of outcomes of numerous clinical trials. A practicing clinician, a patient or a significant other would never equate these very different clinical situations.

At the same time that the potential impact of maintenance medication in reducing risk of symptomatic exacerbation became apparent, it also became clear that long-term treatment was associated with serious adverse effects (e.g., tardive dyskinesia) in what appeared at times to be an alarming rate.

Despite the established effectiveness of antipsychotics, numerous individuals within and outside the field called for a reevaluation of the use of long-term treatment and focused on the substantial proportion of patients who did not relapse within 6–12 months following drug discontinuation in many of the reported trials. In addition, it was also evident that despite data arguing for maintenance, a substantial proportion of patients were not complying with long-term treatment recommendations.

As a result of these factors, a number of studies were undertaken over the past 15 years to attempt to improve the overall benefit-to-risk ratio of maintenance treatment. The focus of many of these trials was to attempt to reduce cumulative drug exposure, to identify minimum effective dosage requirements, and/or to attempt to identify those patients who might not require continuous long-term drug treatment. The results of these trials have been reviewed in detail elsewhere (Davis *et al.*, 1989; Kane and Marder, 1993) and the relevance to the present discussion relates to the issue of the types of methodology and control or comparison groups that are necessary to address these issues.

If one accepts the need to attempt to prevent tardive dyskinesia as well as other adverse effects (and also possibly attribute some degree of noncompliance to adverse effects or subjective discomfort), how can this be achieved? What outcome measures are appropriate and what controls are necessary? Or should such research not be conducted at all?

Drawing meaningful conclusions regarding the relative merits of dosage reduction strategies (as was discussed regarding acute treatment studies), may require not only active controls, but also placebo controls in order to determine when and if an alternative (potentially less toxic treatment approach) is superior to no drug treatment at all. For example, suppose a new medication had no risk of causing tardive dyskinesia, but appeared to be somewhat less effective than standard treatments in preventing relapse. Would not it be critical to know to what extent this treatment was superior to placebo in order to determine whether or not it should be utilized?

As was discussed in the context of acute treatment trials, the problem of multiple effects becomes even more critical when addressing long-term treatment. Although the outcome measure "relapse" or "exacerbation" was used to contrast some aspects of drug treatment efficacy, measures of psychosocial and vocational adjustment, subjective well-being, compliance and adverse effects are essential in order to evaluate outcome in a meaningful way.

It is important to note that in many of the studies involving substantial neuroleptic dosage reduction (which was associated with significantly higher rate of "relapse"), there were few if any differences in terms of psychosocial and vocational adjustment (Schooler, 1991). In fact in one study (Kreisman et al., 1988), a patient cohort with a 56% cumulative rate of "relapse" (as defined by a specified change on a rating scale) were viewed by their families (who were blind to treatment) as somewhat better off in terms of psychosocial adjustment. This group were also rated as having less emotional withdrawal, blunted affect, motor retardation and tension as well as having fewer early signs of tardive dyskinesia at endpoint. This is an example of how a simple measure of relapse is insufficient to assess the overall impact of a particular treatment strategy. The fact that disparities or even contradictions may be evident when multiple effects are examined argues against simplistic notions of what types of research are appropriate and necessary.

CONCLUSION

One of the most difficult dilemmas in medical research is the evaluation and balancing of risk and benefit involved in the design and conduct of clinical trials as well as the informed consent process involved in requesting subject participation. Many would argue that the more important the question being addressed, the more likely there will be significant risk to the subjects involved.

This issue was highlighted by the initial controversy regarding controlled clinical trials in the treatment of AIDS. Many people afflicted with this lethal disease argued very strongly that any potential treatment should be made available to those affected outside of the context of controlled clinical trials. Given the nature of this illness, it is certainly understandable that patients would insist on having access to some form of treatment. Many of these same individuals ultimately came to bemoan the fact that inadequate data were available to assess the potential benefits of putative treatments because adequate controlled trials had not been done.

In the case of complex psychiatric illnesses such as schizophrenia which are treated with partially and inconsistently effective and frequently toxic treatments like neuroleptic medications, the issue of placebo controls cannot be dismissed solely on the basis that proven therapies exist.

REFERENCES

Ayd, F.J., Jr.: Early-onset neuroleptic-induced extrapyramidal reactions: A second survey, 1961–1981. In J.T. Coyle and S.J. Enna (eds.) *Neuroleptics: Neurochemical, behavioral and clinical prospectives*. New York: Raven Press, 1983, pp. 75–92.

Davis, J.M., Barter, J.T., and Kane, J.M.: Antipsychotic drugs. In H.I. Kaplan and B.J. Sadock (eds.) *Comprehensive Textbook of Psychiatry*. Baltimore: Williams and Wilkins, 1989. pp. 1591–1626.

Gilbert, P.L., Harris, M.J., McAdams, L.A., and Jeste, D.V.: Neuroleptic withdrawal in schizophrenic patients: A review of the literature. *Archives of General Psychiatry* 52:173–188, 1995.

Kane, J.M., Marder, S.R.: Psychopharmacologic treatment of schizophrenia. *Schizophrenia Bulletin* 19: 287–302, 1993.

Kane, J., Jeste, D.V., Barnes, T.R.E., Casey, D.E., Cole, J.O., Davis, J.M., Gualtieri, C.T., Schooler, N.R., Sprague, R.L., and Wettstein, R.M.: *American Psychiatric Association Task Force on Tardive Dyskinesia*. Washington, D.C.: American Psychiatric Press, 1992.

Klerman, G.L.: Scientific and ethical considerations in the use of placebo controls in clinical trials in psychopharmacology. *Psychopharmacology Bulletin* 22:25–29, 1986

Kreisman, D., Blumenthal, R., Borenstein, M., Woerner, M., Kane, J.M., Rifkin, A., Reardon, G.: Family attitudes and patient social adjustment in a longitudinal study of outpatient schizophrenics receiving low-dose neuroleptics: The family's view. *Psychiatry* 51:3–13, 1988.

Leber, P.: The placebo control in clinical trials (A view from the FDA). *Psychopharmacology Bulletin* 22: 30–32, 1986

Rothman, K.J.: Presentation at Conference on Ethics in Neurobiologic Research with Human Subjects, Baltimore, MD, January 7–9, 1995.

Rothman, K.J., and Michels, K.B.: The continuing unethical use of placebo controls. *The New England Journal of Medicine* 331:394–398, 1994

Schooler, N.R.: Maintenance medication for schizophrenia. *Schizophrenia Bulletin* 17:311–324, 1991.

Schizophrenia Research: A Challenge for Constructive Criticism

William T. Carpenter, Jr.

Professor of Psychiatry and Pharmacology, University of Maryland School of Medicine; Director, Maryland Psychiatric Research Center, P.O. Box 21247, Baltimore, MD 21228

Several individuals have directed harsh criticism toward clinical investigators responsible for schizophrenia research. Assertions of harm to subjects, failure to protect patients autonomy, and the unethical conduct of science are placed in public discussion without documentation. Of particular concern are medication withdrawal studies involving first-psychotic-episode patients, placebo-controlled clinical trials, and pharmacologic-challenge studies.

The clinical realities and empirical data pertinent to these studies fails to support assertions of harm to subjects. It is argued that critics misunderstand key issues, fail to address the substantial literature containing empirical data relevant to safety, and ignore rules of evidence concerning cause-and-effect relations.

The ethics community is challenged to base criticism on knowledge and understanding, and to require evidence when asserting harm or cause of harm. Criticism provided in this framework will prove invaluable in evolving the optimal approaches to human experimentation in schizophrenia research.

Keywords: Ethics, schizophrenia, research, drug withdrawal, human experimentation, criticism

SCHIZOPHRENIA RESEARCH: A CHALLENGE FOR CONSTRUCTIVE CRITICISM

Introduction

Schizophrenia research has recently received intense negative criticism in leading magazines, newspapers, network talk shows, Congressional hearings, and other forums. Along with this criticism, ill-conceived proposals that would substantially alter research procedures, hinder the development of new knowledge in schizophrenia, and substantially increase the cost of investigative work have been promulgated. A deplorable aspect of this debate is the *ad hominem* attack on individual investigators in a manner damaging to their personal lives and professional careers. In the absence of any evidence of improper conduct, leading spokespersons from the ethics community and self-appointed critics engage freely in pronouncements on issues about which their information is either incomplete or based on tenuous cause and effect relationships. Unfortunately, as a result of this set of circumstances, private and public institutions responsible for funding schizophrenia research have been barraged with demands that they stop funding specific investigators, institutions, or particular areas of research.

Despite the activity of critics, the extensive media attention and a widely circulated California Alliance for the Mentally Ill publication (Weisburd, ed., 1994), specific allegations of unethical research involve less than a handful of instances, and in these, formal investigation by authoritative bodies have found neither unethical science or unethical care, nor has any evidence of harm directly attributable to the research been forthcoming. In the most discussed cases, critics continue to allege unethical conduct of science and research causation of harm to individuals, including suicide, despite formal findings unequivocally contrary to these claims (Office for Protection from Research Risks, Division of Human Subject Protections, 1994).

With the development of a hostile environment in research occurring at a time when the potential for knowledge is so great, the clinical research community is alarmed with the potential for disruption in schizophrenia research and the consequent damage to individuals suffering from this illness. Our community agrees in principle in the regulatory and ethical considerations governing the conduct of human experimentation in medicine, and believes that the determination of optimal procedures should be an ever-evolving process. Many elements of society have a voice in establishing ethical standards for human experimentation, and criticism is essential to the evolution of optimal procedures. However, uninformed and misdirected censure derails the process. I contend that criticism should be based on an adequate knowledge and appreciation of schizophrenia, of the clinical realities surrounding the diagnosis and treatment of persons suffering from this illness, and a realistic view regarding the actual purposes, risks, benefits, and clinical management involved in clinical research. An understanding of terms such as study attrition and relapse, whose meaning depends on study-specific methodology, is essential. So, too, is the ability to evaluate empirical data contrasting the course of illness in research-based care with that observed in nonresearch clinical care. Such knowledge and appreciation enable critics to raise their level of discussion in a manner conducive to an active collaboration of all parties in determining which aspects of current practice are satisfactory, which need improvement, and how best to implement change. This is a difficult challenge to meet in view of the fact that the center stage of criticism is now occupied by those who not only misunderstand the purpose and methods of studies and the quality of the informed consent process, but also engage in tactics which discourage constructive interaction with the clinical research community.

Missing in the debate of the last two years is any indication from the ethics community as to quality standard expected in this discussion, the nature of documentation required to support accusations, and whether assertion of cause and effect relationships should be substantiated with evidence. Dispassionate and fair discussion of the issues necessarily require that the ethics community also conduct its affairs with scholarship, honesty, and intellectual and personal integrity.

CLINICAL CARE ISSUES

There are many types of schizophrenia research, but drug withdrawal protocols are under intense scrutiny. Critics of research in schizophrenia involving periods off medication have been particularly concerned with three situations: 1) first episode patients who experience a relapse; 2) drug withdrawal in stable out-patients; and, 3) pharmacologic-challenge studies. The central concern expressed in all of these cases is that patients are either allowed or made to have a relapse which otherwise would not occur. Concerns are heightened with assertions that such relapses invariably result in harm such as suicide, homelessness, permanent brain scarring, or diminished responsiveness to future treatment. These issues can be addressed from the perspective of ordinary (nonresearch) clinical care and from a substantial body of empirical data.

1. *Relapse and First Episode Patients.* From a clinical care perspective, consider the following:

A diagnosis of schizophrenia during the first psychotic episode is often tentative, and schizophrenia accounts for only a minority of first psychotic breaks. While the longitudinal pattern of illness is diagnostically clarifying, first episodes are often confounded with psychoactive drug use, stress, mood disturbance, and other diagnostic confounds. Diagnostic uncertainty during an initial episode necessarily complicates the clinician's long-term treatment plan. Most clinicians would agree that antipsychotic medication is crucial to acute treatment and early maintenance of clinical stability, but many young, first episode patients experience dysphoric reactions to medication and are significantly noncompliant over time. Even when schizophrenia is carefully defined, a minority (perhaps 30%, depending on diag-nostic criteria and cohort characteristics) will not have a recurrent psychotic illness (Lieberman, 1993; Watt *et al.*, 1983; Scottish Schizophrenia Research Group, 1992; Rabiner *et al.*, 1986). Once clinical stability is established, those patients without a recurrent form of the disease have an unacceptable risk/benefit profile with con-tinued medication. Substantial risks of continuing drug therapy include irrevers-ible brain changes and affective, cognitive, and interpersonal dysfunction. How-ever, the majority of patients display a relapsing form of the illness and, in these cases, the benefits of prophylactic antipsychotic medication are substantial. Unfor-tunately, the doctor does not have *a priori* knowledge as to which group an individual patient belongs, nor can he/she be confident of medication compliance if continuous antipsychotic drugs are recommended. Which treatment approach is wise, ethical, and scientifically established in this situation? Could any critic iden-tify in advance which patients should be on neuroleptic drugs for life, and which should not? Should the clinician be protected from the critics who only focus attention on cases of tardive dyskinesia in patients who turned out not to have a recurrent illness? Should they be protected from critics who only select cases of relapse off medication? Is there a cogent objection to gaining better knowledge through clinical science?

From a clinical research perspective, careful study aimed at discovering indica-tors of relapse potential, determining effectiveness of low dose drug treatment, ascertaining the effectiveness of rapid antipsychotic drug intervention as an alter-native to continuous prophylactic medication, and addressing other similar ques-tions are vital. There is no doubt that future first-episode patients and society will benefit from such knowledge. Assuming that proper procedures are undertaken to minimize risk, obtain informed and voluntary participation, and assure appropri-ate review of scientific and clinical care procedures, diagnostic and treatment studies involving medication withdrawal are not inherently unethical. Support for this view was found in the recent OPRR investigation of the UCLA study (Office for Protection from Research Risks, 1994). In spite of inadequacies noted in the signed consent form (which were correctable with modifications), the science and clinical care were determined to be ethical, and no harm to participating patients was found. Despite these findings, critics continue to assert, and the *New York Times* continues to publish, that this work was unethical (Hilts, 1994a, 1994b).

2. *Drug Withdrawal in Stable Outpatients.* Shamoo has been cited in the press (Hilts, 1994a) over the past several years as reporting a literature review involving 2500 patients in what he describes as intent to cause relapse studies. Although this review is not published in the professional literature, Shamoo (1994b), and Irving and Shamoo (1995) cite these studies, and Keay reported aspects of this material at the Baltimore Conference (1995). They review a series of placebo-controlled studies where the effectiveness of antipsychotic drug treatment was established, a next generation series of studies where prophylactic efficacy was established, and a more recent series of studies aimed at ascertaining minimal effective dose and evaluating the comparative effectiveness of continuous vs. discontinuous medication applied at the onset of symptom exacerbation. In all cases, the rationale for these studies was unequivocal. The investigators had no intent to cause relapse, but rather sought to establish and scientifically validate effective treatment for persons with schizophrenia. All the drugs used to treat psychosis are associated with high risk and poor patient compliance. Therefore, the more recent studies which evaluated minimum effective dosing strategies have provided patients and their clinicians with information relevant to benefit-to-risk considerations and given clinicians alternative strategies for patients unwilling to accept "standard" treatment. Clinician choice standard treatment has resulted in excessive medication for the majority of patients with schizophrenia. As a result, they have suffered more depression, more weight gain, more Parkinsonian symptoms, more sedation, more negative symptoms, more tardive dyskinesia, and more self-induced medication withdrawal. Depression is a risk factor for suicide, and medication noncompliance places patients at higher risk for relapse! The recent generation of studies aimed at determining minimum effective dose have demonstrated that these drug-related problems can be reduced while still gaining therapeutic benefit. It is vexing that Shamoo (1994b), Irving (1994), and Keay (this conference) misrepresent the purpose of these studies as an intent to cause relapse. Of greater concern is the assertion that these studies were unethical, that patients were killed [Keay's explicit assertion at the Baltimore Conference] (Keay, 1995), and that other serious harm had been suffered because of participation in these studies.

As a clinical investigator, I am troubled by three aspects of these assertions: 1) misrepresentation of investigators' motivation and purposes; 2) the absence of any documentation; and 3) the failure to consider the very extensive data directly addressing these issues. At the conference, Keay was only able to cite newspapers as his source for documentation of harm, and even here, he conceded that the newspaper reports did not involve any of the 2471 patients cited in the review (Irving and Shamoo, 1995). Concerning credible data, the following can be noted:

1. In a meta-analysis of 4365 patients in drug withdrawal studies, Gilbert *et al.* (1995) found no evidence for lasting harm. Patients responded to reintroduc-

tion of medication, usually returning to their pre-drug withdrawal clinical baseline within three weeks.

2. Even in studies with large differences in relapse rates between drug and placebo groups, no evidence was found for any adverse effect on long-term course of illness (Curson *et al.*, 1986).

3. Compared to continuous drug therapy, random assignment of patients to drug withdrawal followed by noncontinuous drug therapy is associated with a disadvantage in relapse rates, but not in longer-term symptomatic or social course. Advantages accrued to the noncontinuous medication groups with respect to negative symptoms and cumulative drug exposure (the established risk factor for tardive dyskinesia).

4. When research-based clinical care involving increased risk for symptom exacerbation during a drug withdrawal protocol is compared to the standard care of similar patients, no evidence for poorer long-term outcome for research patients has been forthcoming. In my own medication-free research, where direct comparisons have been possible, the results support this contention (Carpenter *et al.*, 1977, 1987, 1990).

5. Studies requiring a period off medication often have several design components which reduce risk and enhance benefits compared to standard clinical care. These include more intense clinical monitoring, concurrent implementation of psychoeducational and psychosocial therapeutic procedures, and a built-in capacity for rapid intervention if the patient's condition deteriorates. Furthermore, patients may participate in research-based clinical care before and after the off-medication period. In this circumstance, they can receive enriched clinical care plus antipsychotic drug therapy. The monitoring procedures may decrease adverse events such as severe relapse and suicide, may detect early signs of tardive dyskinesia, and may substantially increase medication compliance. In the research outpatient clinic at the Maryland Psychiatric Research Center (MPRC), close clinical monitoring with psychoeducational and psychosocial techniques, weekly pill counts and drug dispensing, close work with patients regarding side-effect issues, involvement of significant others, and outreach in response to missed appointments result in a medication compliance rate of 98%. As a consequence, patients receiving clinical care in our research clinic, despite participation in drug withdrawal and placebo controlled studies, spend more time protected with antipsychotic medication, receive closer supervision when off medication, are less likely to be rehospitalized (the rate is seven percent per year), and have a low rate of suicide (none during off-medication protocols since the clinic opened in 1977).

    3. *Pharmacologic Challenge Studies.* Harsh criticism is directed at pharmacologic challenge studies, which are characterized by critics as giving drugs to make symptoms worse (Hilts, 1994a; 1994b; Irving, 1994; Shamoo, 1994a; Sharav, 1995; Hassner, 1994). What are the rationales, purposes, and actual consequences of these studies? First consider three key biochemical hypotheses:

1. psychosis emerges from hyperdopaminergic limbic system circuitry;
2. antipsychotic drugs induce hyperdopaminergic motor circuitry resulting in tardive dyskinesia; and

3. negative symptoms emanate from hypodopaninergic dorsolateral prefrontal cortical circuitry.

Investigators using pharmacologic challenges do not wish to cause harm to patients by inducing severe symptom exacerbation. Rather, these challenges have been designed to test the above major hypotheses and to ascertain the diagnostic and therapeutic value of perturbation of dopamine systems. These studies involve brief exposure to drugs such as methylphenidate, amphetamine, apomorphine, and L-dopa, each of which can increase dopamine neural transmission.

The first of these pharmacologic challenge studies (Janowsky and Davis, 1976; Janowsky et al., 1973) was conducted when a diagnosis of schizophrenia was associated with lengthy hospitalization. As psychotic symptoms subsided, many patients were observed to have a more subtle disorganization of thought which might reflect a suppressed active psychotic process rather than a true remission. The question arose as to whether methylphenidate could produce a mild and transitory worsening in patients with an active, but suppressed, psychosis, and whether failure to worsen would identify patients in true remission who might be safely discharged. Results confirmed a brief and mild symptom exacerbation in some, but not all, patients, a worsening that was sometimes so subtle as to be missed on symptom assessments but detected with a psychological test. No adverse effects were discernible the next day.

In today's climate of extremely brief hospital care and high outpatient relapse rates, the clinician needs effective tools to distinguish highly stable from relapse vulnerable patients. To address these questions, these same pharmacologic tools are currently being used with notable success (Lieberman et al., 1984; van Kammen et al., 1981, 1982a, 1982b). Others have used these tools in an effort to detect hidden or covert dyskinesia (the neuroleptic drugs that cause tardive dyskinesia also suppress the abnormal movements until they become more severe and persistent) in the hope that early detection would be associated with greater reversibility, while other studies have attempted to downregulate dopamine receptors with therapeutic intent, or to determine if negative symptoms improve, or to study the basic biology of schizophrenia (van Kammen et al., 1977; Angrist et al., 1975, 1981; Davidson et al., 1987; Tamminga et al., 1978).

One critic of these studies (Shamoo, 1994a), asserting that subjects in these studies have been harmed, admits that no evidence supports this accusation. Nonetheless, this critic states that he will continue to condemn these studies because investigators intend harm and are conducting mechanistic studies with no potential benefit to the subject. I offer three points to counter this argument:

1. With substantial data supporting safety, and no evidence of pain or lasting harm, this type of study would be ethical even if only conducted to study etiology and pathophysiology without diagnostic or therapeutic intent in a given case.
2. Stating that the purpose of these studies is to cause relapse represents an idiosyncratic interpretation of the unambiguous diagnostic and clinical care purposes that have been put forward by the investigators (Janowsky and Davis, 1976; Janowsky et al., 1973; Lieberman et al., 1984; van Kammen et al., 1977, 1981, 1982a, 1982b; Angrist et al., 1975, 1981; Davidson et al., 1987; Tam-

minga *et al.*, 1978). These studies have been conducted to test the aforementioned dopamine hypotheses, and diagnostic implications include assessing vulnerability to relapse and detection of covert tardive dyskinesia. The therapeutic implications of these studies relate to clinical care decisions based on a relapse and dyskinesia vulnerability assessment and to therapeutic hypotheses which suggest that dopamine receptor down-regulation might enhance antipsychotic drug efficacy, be directly therapeutic for tardive dyskinesia, and provide a treatment approach to both primary and secondary negative symptoms. Not only might patients in such studies benefit from participation, but also clinically applicable findings could eventually impact favorably on their future care. Keep in mind that patient subjects are usually expected to live with their disease for 30–60 years after protocol participation and, therefore, have a vested interest in improved clinical care based on new knowledge. Consider the years of future benefit to the patients who participate in the early drug/placebo neuroleptic studies, benefits which accrued even to those subjects initially assigned to placebo. The purposes of the investigators are noble, not malicious.

3. Pharmacologic challenge studies have provided both safety data and preliminary findings of clinical improvement in perhaps a third of participating patients. Transitory improvement in negative symptoms noted in these studies has provided the empirical observation upon which clinical trials of dopamine enhancing drugs are presently being conducted for the treatment of negative symptoms. This is our most promising lead to date for the treatment of the deficit, or negative symptom, aspects of schizophrenia.

Improbable as it seems, critics of these studies seem to think that they never should have been conducted and that they should be stopped on ethical and human protection grounds. Should the critics prevail, would anyone suffering from schizophrenia be better off in their rights, their safety, or their future?

UNDERSTANDING RESEARCH RISKS IN THE CONTEXT OF CLINICAL REALITY

Most patients with schizophrenia have relapses regardless of drug treatment. The exception, a single episode patient without recurrent psychosis, faces substantial risk and little benefit from lifelong antipsychotic drug treatment. The prophylactic use of antipsychotic drugs is effective in delaying onset of relapse, and over time cuts relapse rates by a little more than half compared to the relapse rates for placebo. When a relapse occurs during research-based care, it is not evident that the relapse was caused by the research. It is the change in risk associated with drug withdrawal undertaken for research purposes that poses an ethical and clinical concern. Much can be done to reduce this risk, to minimize adverse consequences of relapse, and to enhance benefits of other aspects of treatment. Guidelines for these procedures have been discussed elsewhere (Carpenter *et al.*, in press). Judgments regarding the ethics and clinical prudence of medication-free research require intimate knowledge of patient selection criteria, the informed consent process, and the study's capacity to minimize risk, enhance benefit, and effectively treat symptom exacerbation. This complex evaluation must then be weighed against a similar evaluation of standard care. It has been possible to make this comparative evaluation based on empirical data in random assignment studies, and this has been reassuring as medication-free periods did not result in a worse overall course for research patients (Gaebel *et al.*, 1993; Schooler *et al.*, 1993; Curson *et al.*, 1986; Carpenter *et al.*, 1987, 1990).

In most cases, schizophrenia is a chronic disease, and the clinical care provided often falls short of what clinical investigators consider optimal. Most patients are treated with excessive amounts of medication by clinicians using their best judgment (Bollini *et al.*, 1994; Baldessarini *et al.*, 1988, 1990; Schooler, 1991), and, as a consequence, a terrible price is paid in adverse effects. Ordinarily, outpatients are seen far too infrequently to provide rapid intervention at earliest sign of relapse. Further, assertive clinical outreach into the community, psychoeducation, and specific family and patient-based psychosocial treatments shown to be effective in model programs and clinical trials research are not routine in clinical practice. Recommended necessities, such as a 24-hour on-call familiar clinician who is available to deal with warning signs of relapse, are not often provided. As a result, patients often belatedly receive interventions in emergency rooms, jails, or crisis centers when severity of relapse is greater and hospitalization is more likely. Poor compliance with medication is a widespread problem, and the occurrence of patients relapsing off medication with neither close supervision or rapid intervention is common in nonresearch practice. In contrast to clinicians in standard practice, clinical investigators, who are concerned both with what they wish to accomplish for their patients and what new knowledge they hope to gain from subject participation, are in a position to orchestrate research-based clinical care in a manner in which substantial benefits offset the risk elements associated with research procedures.

In maintenance and prophylactic phases, relapse is the one proven risk of medication withdrawal. Since definitions vary widely across studies, clarity as to what constitutes a relapse within each study is essential. In early studies testing the prophylactic efficacy of antipsychotic drugs, relapse in placebo-controlled studies usually signified a significant worsening of psychosis associated with rehospitalization. Now that an imperfect, but effective, prophylactic treatment has been established, relapse is often defined as a milder worsening that is usually responsive to outpatient intervention. Today, most relapses in research studies are not associated with rehospitalization, but rather with ratings of symptom changes. Given that the term *relapse* can cover a range from mild increase in anxiety and difficulty sleeping to severe psychosis and rehospitalization, any criticism of relapse risk associated with medication-free research must clearly specify how *relapse* is defined. Communication will be enhanced if this term is replaced by the term *symptom exacerbation*. This latter terminology will require specification of what symptoms have exacerbated, and to what extent. It will be evident that exacerbation is a continuum ranging from slight to severe. The risks of any particular research protocol must be judged accordingly.

There has also been the criticism that patients are abruptly dropped from research protocols without provisions for continued care (Irving and Shamoo, 1995).

No documentation has been forthcoming to validate this assertion, and it appears to be a misunderstanding of the terminology of *study attrition*. Most schizophrenia studies have very real attrition problems. Patients often wish to withdraw from studies for any number of reasons. In addition, side effects, inadequate response to treatment, and lack of compliance with protocol requirements may lead to the termination of study involvement of some patients. The fact that a patient either withdraws from, or is terminated from a research study, does not indicate that he/she is removed from clinical care. The usual procedure in such a case is to simply shift the patient to routine clinical care in the same setting. Some specialized research settings, such as the NIH and the Maryland Psychiatric Research Center (MPRC), finance routine clinical care as part of the institution's research agenda. In those instances in which patients no longer wish to be involved in research protocols, they are usually transferred to a standard care setting, allowing sufficient time to optimize routine care before transfer and providing patients adequate time to consider their options. At the MPRC, we have provided an enormous amount of free clinical care over the years to patients who were not participating in protocols. In my experience, I have encountered no cases in which clinical care was denied and the patient was abandoned due to his or her decision not to participate in research. This astonishing criticism did not seem credible, and I sought information on this point from authors of the forty-one studies that Shamoo and colleagues cite as a basis (Irving and Shamoo, 1995; Keay, 1995). Not a single patient was dropped from clinical care because he/she declined research participation, nor had any of the investigators received any request for information from this group of critics.

Despite expressed concerns regarding the ethics and safety of human medical experimentation, research-based clinical care is often perceived as a desired choice by patients and families. Expectations associated with the provision of clinical care by investigators include: exceptional diagnostic and therapeutic expertise, familiarity with current knowledge and recent advances, interest in difficult to treat cases, adequate resources, access to experimental treatments not yet generally available, development of a depth of knowledge regarding disease mechanisms, and an alternative approach to meet the financial burdens of high quality care. Not all of these advantages are found in all studies, but they underline Levine's view (1995) that the public seeks participation in research anticipating superior clinical care and therapeutic opportunity. In an illness syndrome such as the schizophrenias, the very best care is not curative and offers only partial relief from some aspects of the illness. While expectation of, and esteem for, clinical investigators may be high, so is experienced disappointment. This is a terrible disease, and research patients tend to be drawn from those patients who obtain unsatisfactory results with ordinary care. Scientists seek better therapeutics for this illness, but do not guarantee that experimental treatment will prove to be superior.

The material presented above supports the supposition that research-based care (in general) does not place patients at greater risk for harm than does standard clinical care (in general), and may actually offer substantial advantages. Although this is reassuring with respect to the existence of any widespread human protection problem in current research practice in schizophrenia, it does not address other key issues. How, for example, in the presence of favorable expectations, can investigators best assure that patients and significant others adequately appreciate the actual risks and benefits associated with a specific protocol? And how can these best be weighted in relation to alternative treatment arrangements? How can the field assure, and society have confidence, that patient selection and clinical provisions follow responsible guidelines for risk reduction and off-setting benefits? What constitutes reasonable documentation that autonomy is respected, and that the doctor-patient dialogue surrounding the informed consent document enhances the quality of information and patient autonomy?

A CHALLENGE TO THE ETHICS COMMUNITY

Recent personal attacks on investigators and general accusations and criticism leveled against those responsible for the ethical conduct of schizophrenia research have been sharp, public, and, in my view, have been offered without convincing documentation or adequate appreciation of the complexity of the issues involved. Clinical investigators share with other members of society who are concerned with these issues a commitment to the ethical conduct of research and to continually evolving guidelines and regulations to assure patient protection from harm and loss of autonomy. Criticism based on an adequate understanding of schizophrenia and on a clear view as to study purposes, methods, and actual effects on subjects is certain to be constructive, influential, and of central importance in determining not only which aspects of current research practice can be improved, but also how this can best be done.

I strongly believe that much of the discussion leading up to the Baltimore Conference failed to meet an acceptable standard of constructive and informed criticism. At the Conference, individuals made *ad hominem* attacks on specific investigators and institutions, with unseemly presentation of personal photographs of investigators being denounced, while denying the accused release of confidentiality. This denied the accused an adequate rebuttal, and deprives all participants of the information essential to understanding. Statements asserting unethical conduct were repeatedly asserted regarding the UCLA study and a Mount Sinai study, despite exonerating formal reviews by properly constituted bodies that had concluded that no unethical conduct was involved. Adverse events commonplace in schizophrenia treatment (i.e., relapse, suicide, pain, and suffering) were alleged to have been *caused* by research, and even were the intent of research, sometimes with no documentation that the events actually occurred or that they were, in fact, attributable to the research.

Fortunately, the above views were exposed to sharp rebuttal from clinical investigators, clinicians, concerned consumers, family members, and IRB lay members. Regrettably, the leading spokespersons in ethics and the law concerned with these issues remained silent. This silence is alarming, particularly in view of the fact that several of these professionals have voiced criticism of the aforementioned studies in the media and law reviews (Katz, 1993), despite their awareness that their information was incomplete and one-sided. To counter this one-sided discourse, I raise the following questions for the ethics and legal communities to consider:

1. In a forum requiring open, full, accurate, and candid discussion, what role do you envision for those individuals who wish to make accusations while denying the accused release of privileged information to use in their defense? Such individuals have ready access to the media, talk shows, lay organizations, public officials, and courts of law. What standards do you propose for participation in serious consideration of ethical issues?
2. When damaging accusations are admittedly made without foundation in fact (i.e., "many people were killed in these studies"), is there one among you who is willing to denounce this level of discourse? Is there a segment in our society concerned with legal and ethical issues of human experimentation in which there are leaders willing to articulate standards of evidence, documentation, and proper assessment of cause and effect in accusations of unethical conduct? How does the field of professional ethics defend its silence in the presence of virulent and unfounded accusations?
3. Can you articulate a standard of knowledge and understanding which you regard as essential for informed discussion? When you present your conclusions to an uncritical public, do your peers expect you to have cogent documentation and argument for their scrutiny? Are you yourself responsible for unethical conduct when you use inflammatory and misleading analogies? Do you think, for example, that a noble cause is advanced by telling the public that withdrawing a person with schizophrenia from antipsychotic drugs is equivalent to withholding immunosuppressant drugs during kidney transplant surgery?

FINAL COMMENT

I have prepared this material with a sense of urgency related to the vulnerable enterprise of schizophrenia research. Much could be accomplished with wiser public policy, increased resources, and better prepared clinicians and communities for dealing with schizophrenia. But I do not know anyone who believes we have either a cure or prevention for schizophrenia, nor do I know any informed person who fails to recognize the extensive morbidity and mortality associated with this disease despite the best clinical care available.

Medical research represents society's best and only real hope for progress in dealing with this dread disease. Yet, there is an incredible crisis in research funding, which is far below the standard federal funding for other major diseases; and funding from private sources such as NARSAD is small compared to the foundation and philanthropic support for cancer and heart disease. At present, about ten percent of peer reviewed grants having scientific merit are funded by NIMH, and clinical research is under extreme scrutiny because of the relatively high costs involved.

Regardless of the above circumstances, it is imperative that clinical research be ethical and conducted with informed consent, that patient autonomy be protected, and that risks be minimized and viewed in proportion to potential benefit. Present practices appear to work rather well in this regard. As the Executive Director of the National Alliance for the Mentally Ill indicated during the Baltimore Conference, there is no evidence of widespread abuse of patient rights or of harm to research subjects. If this is the case, then it follows that efforts to improve procedures should be considered dispassionately, employing feasible modifications and guidelines that would enhance the conduct of science without significantly undermining research opportunities or jeopardizing research funding. Impetuous and ill-advised changes imposed without due deliberation would both radically alter currently sound practices and lead to any number of negative consequences. Research costs would necessarily increase and the attainment of new knowledge would decrease.

As director of a research center focused on schizophrenia, I can report from personal experience that an ambience of accusation based on ethical issues without documentation of legitimacy is dangerous. We treat many inpatients and outpatients and, as is true in any clinical setting, there are adverse events. My concern in the present climate is that an adverse event will be presumed to be research "caused," devoid of any consideration of actual circumstances. What concerns me here, and what I think needs to concern society, is that clinical research in schizophrenia is quite vulnerable if confronted with irresponsible accusations of cause and effect linking separated and distinct events. Judging from the well-publicized and continued assertions that the UCLA researchers were responsible for a suicide that occurred over two years after protocol participation, we are acutely aware that critics are prepared to accept the simple occurrence of an adverse event no matter how remote, as sufficient evidence that it was *caused* by research. Imagine the impact on cancer research if the death of a study participant was presumed to be *caused* by the research. Furthermore, without release from confidentiality being granted, the oncologists would not even be able to be informative regarding the clinical care circumstances associated with the death.

Current experience shows that two sides of the story are not necessary to recruit the interest and criticism of the media and ethics community. A one-sided presentation seems far more likely to capture the interest and attention of the general public. And, unfortunately, clinical investigators and their institutions are too readily depicted as self-serving and defensive if they are vigorous in rebuttal. In view of the devastation that can be caused to legitimate research by what have become one-sided, poorly documented exposes, it is crucial, I believe, that patients, consumer advocates, non-research clinicians, and members of the ethics establishment vigorously address the issues raised in this presentation.

ACKNOWLEDGEMENT

This work was partially supported by NIMH Grant MHCRC 40279.

REFERENCES

Angrist, B., Peselow, E., Rotrosen, J., and Gershon, S. (1981) Relationship between responses to dopamine agonists, psychopathology, neuroleptic maintenance in schizophrenic subjects; in *Recent Advances in Neuropsycho-pharmacology* (*Advances in the Biosciences*). eds. B. Angrist, G. Burrows, M.

Lander, O. Lingjaerde, G. Sedvall, and D. Wheatley. New York: Pergamon Press, Volume 31, pp. 49–54.

Angrist, B., Peselow, E., Rubinstein, M., *et al.* (1975) Amphetamine response and relapse risk after depot neuroleptic discontinuation. *Psychopharmacology* 85:277–283.

Baldessarini, J.R., Cohen, B.M., and Teicher, M. (1988) Significance of neuroleptic dose and plasma level in the pharmacological treatment of psychoses. *Archives of General Psychiatry* 45:79–91.

Baldessarini, R.J., Cohen, B.M., and Teicher, M.H. (1990) Pharmacologic treatment; in *Schizophrenia: Treatment of Acute Psychotic Episodes.* eds. S.T. Levy and P.T. Minan. Washington, DC: American Psychiatric Press, pp. 61–118.

Bollini, P., Pampallona, S., Orza, M.J., Adams, M.E. and Chalmers, T.C. (1994) Antipsychotic drugs: is more worse? A meta-analysis of the published randomized control trials. *Psychological Medicine* 24:307–316.

Carpenter, W.T. Jr., Heinrichs, D.W., and Hanlon, T.E. (1987) A comparative trial of pharmacologic strategies in schizophrenia. *American Journal of Psychiatry* 144:1466–1470.

Carpenter, W.T. Jr., McGlashan, T.H., and Strauss, J.S. (1977) The treatment of acute schizophrenia without drugs: An investigation of some current assumptions. *American Journal of Psychiatry* 134(1):14–20.

Carpenter, W.T. Jr., Hanlon, T.E., Heinrichs, D.W., Summerfelt, A.T., Kirkpatrick, B., Levine, J., and Buchanan, R.W. (1990) Continuous versus targeted medication in schizophrenic outpatients: outcome results. *American Journal of Psychiatry* 147:1138–1148.

Carpenter, W.T. Jr., Schooler, N.R., and Kane, J.M. The ethics and deemed necessity of medication-free research in schizophrenia. *Archives of General Psychiatry* in press.

Curson, D.A., Hirsch, S.R., Platt, S.D., Bamber, R.W., and Barnes, T.R.E. (1986) Does short term placebo treatment of chronic schizophrenia produce long term harm? *British Medical Journal* 293:726–728.

Davidson, M., Keefe, R.S.E., Mohs, R.C., Siever, L.J., Losonczy, M.F., Horvath, T.B., and Davis, K.L. (1987) L-dopa challenge and relapse in schizophrenia. *American Journal of Psychiatry* 144:934–938.

Gaebel, W., Frick, U., Kopcke, W., Linden, M., Muller, P., Muller-Spahn, F., Pietzcker, A., and Tegeler, J. (1993) Early neuroleptic intervention in schizophrenia: are prodromal symptoms valid predictors of relapse? *British Journal of Psychiatry* 163(21):8–12.

Gilbert, P.L., Harris, M.J., McAdams, L.A., and Jeste, D.V. (1995) Neuroleptic withdrawal in schizophrenic patients: A review of the literature. *Archives of General Psychiatry* 52:173–188.

Hasoner, V. (1994) What is ethical? What is not? Where do you draw the line? *Journal of the California Alliance for the Mentally Ill* 5(1):4–6.

Hilts, P.J. (March 10, 1994a) Agency faults a U.C.L.A. study for suffering of mental patients. *New York Times.*

Hilts, P.J. (May 24, 1994b) Medical experts testify on tests done without consent. *New York Times.*

Irving D.N. (1994) Psychiatric research: reality check. *Journal of the California Alliance for the Mentally Ill* 5(1):42–44.

Irving, D.N. and Shamoo, A.E. (January 7–9, 1995) Washouts/relapses in patients participating in neurobiological research studies in schizophrenia. Background paper provided for the *First National Conference on Ethics in Neurobiological Research with Human Subjects, Baltimore, Maryland.*

Janowsky, D.S. and Davis, J.M. (1976) Methylphenidate, dextroamphetamine and lefamphetamine: effects on schizophrenic symptoms. *Archives of General Psychiatry* 33:304–308.

Janowsky, D.S., El-Yousef, K., Davis, J.M., and Sekerke, H.J. (1973) Provocation of schizophrenic symptoms by intravenous administration of methylphenidate. *Archives of General Psychiatry* 28:185–191.

Katz, J. (1993) Human experimentation and human rights. *St. Louis University Law Journal* 38:7–54.

Keay, T. (January 7–9, 1995) Approximating ethical research consent. *First National Conference on Ethics in Neurobiological Research with Human Subjects, Baltimore, Maryland.*

Levine, R.J. (January 7–9, 1995) Proposed regulations for research involving those institutionalized as mentally infirmed. *First National Conference on Ethics in Neurobiological Research with Human Subjects, Baltimore, Maryland.*

Lieberman, J.A., Kane, J.M., Gadaleta, D., Brenner, R., Lesser, M.S., and Kinon, B. (1984) Methylphenidate challenge as a predictor of relapse in schizophrenia. *American Journal of Psychiatry* 141:633–638.

Lieberman, J.A. (1993) Prediction of outcome in first-episode schizophrenia. *Journal of Clinical Psychiatry* 54:13–17.

Office for Protection from Research Risks, Division of Human Subject Protections. (May 11, 1994) *Evaluation of Human Subject Protections in Schizophrenia Research Conducted by the University of California Los Angeles.*

Rabiner, C.J., Wegner, J.T., and Kane, J.M. (1986) Outcome study of first-episode psychosis, I: Relapse rates after 1 year. *American Journal of Psychiatry* 143:1155–1158.

Schooler, N.R., Keith, S.J., Severe, J.B., and Matthews, S.M. (1993) Treatment strategies in schizophrenia: effects of dosage reduction and family management on outcome. (Abstract) *Schizophrenia Research* 9:260.

Schooler, N.R. (1991) Maintenance medication for schizophrenia: strategies for dose reduction. *Schizophrenia Bulletin* 17:311–324.

Scottish Schizophrenia Research Group. (1992) The Scottish first episode schizophrenia study. VIII. Five-year follow-up: clinical and psychosocial findings. *British Journal of Psychiatry* 161:496–500.

Shamoo, A. (December 12–16 1994a) Ethical considerations in medication-free research on the mentally ill. *33rd Annual Meeting ACNP*, San Juan, Puerto Rico.

Shamoo, A.E. (1994b) Our responsibilities toward persons with mental illness as human subjects in research. *Journal of the California Alliance for the Mentally Ill* 5(1):14–16.

Sharav, V.H. (January 7–9, 1995) Independent family advocates challenge the fraternity of silence. *First National Conference on Ethics in Neurobiological Research with Human Subjects, Baltimore, Maryland.*

Tamminga, C.A., Schaffer, M.H., Smith, R.C., and Davis J.M. (1978) Schizophrenic symptoms improve with apomorphine. *Science* 200(4341):567–568.

van Kammen, D.P., Docherty, J.P., and Bunney, W.E. Jr. (1981) Acute amphetamine response predicts antidepressant and antipsychotic responses to lithium carbonate in schizophrenic patients. *Psychiatry Research* 4:313.

van Kammen, D.P., Docherty, J.P., and Bunney, W.E. Jr. (1982a) Prediction of early relapse after pimozide discontinuation by response to d-amphetamine during pimozide treatment. *Biological Psychiatry* 17(2):233–242.

van Kammen, D.P., Bunney, W.E. Jr, Docherty, J.P., Marder, S.R., Ebert, M.H., Goodwin, F.K., Rosenblatt, J.E., and Rayner, J.N. (1982b) d-Amphetamine-induced heterogeneous changes in psychotic behavior in schizophrenia. *American Journal Psychiatry* 139(8):991–997.

van Kammen, D.P., Bunney, W.E. Jr, Docherty, J.P., Jimerson, D.C., Post, R.M., Siris, S., Ebert, M., and Gillin, J.C. (1977) Amphetamine-induced catecholamine activation in schizophrenia and depression: behavioral and physiological effects; in *Advances Biochemical Psychopharmacology*. eds. E. Costa and G.L. Gessa. New York: Raven Press, Volume 16, pp. 655–659.

Watt, D.C., Katz, K., and Shepherd, M. (1983) The natural history of schizophrenia: a five-year prospective follow-up of a representative sample of schizophrenics by means of a standardized clinical and social assessment. *Psychological Medicine* 13(3):663–670.

Weisburd, D. (ed). (1994) *Journal of the California Alliance for the Mentally Ill* 5(1).

Informed Consent with Cognitively Impaired Patients: An NIMH Perspective on the Durable Power of Attorney

Trey Sunderland[†‡] and Ruth Dukoff[†]

[†]Section on Geriatric Psychiatry, Laboratory of Clinical Science, National Institute of Mental Health, Building 10, Room 3D41, Bethesda, MD 20892
[‡]Chairman, National Institute of Mental Health's Institutional Review Board

INTRODUCTION

Can someone with a cognitive impairment possibly give informed consent for a research project? Furthermore, if we allow cognitively impaired individuals to enter research protocols, are we not placing already vulnerable people at risk of potential harm? These are questions any institutional review board (IRB) must face when addressing the inclusion of cognitively impaired subjects in research protocols. And while definitive answers to these questions are elusive, there are systematic approaches which can be applied to increase the safeguards for cognitively impaired subjects and offer some assurance that adequate thought has been given to the risks involved in any individual research project. The following discussion presents the approach and perspective taken within the intramural research program at the National Institute of Mental Health. It represents an attempt to deal with the vexing problems of assuring proper informed consent in impaired subjects and is offered as a working model for other institutions.

RESEARCH WITH COGNITIVELY IMPAIRED SUBJECTS

When considering research with cognitively impaired subjects, it must first be acknowledged that much previous thought has been given to this issue (Melnick et al., 1984; High, 1992; DeRenzo, 1994). For the purposes of this discussion, we will concentrate on five basic questions (Table I). Each of these questions raises different medical and/or ethical issues. While we will approach these issues separately, it should be clear from the outset that they all reflect on different facets of the same informed consent process. Definitive answers to each question may not be pos-

This article will also appear in the journal: *Accountability in Research*, Volume 4, pp. 217–226 (1996).

TABLE I   Five basic questions surrounding the informed consent
process with cognitively impaired subjects

(1) What constitutes cognitive impairment?
(2) Who determines whether someone is cognitively impaired?
(3) What medical diagnoses are associated with cognitive impairment?
(4) Is cognitive impairment a transient or permanent condition?
(5) Can cognitively impaired subjects participate in *this* research project?

sible, but the process of addressing each question is vital to provide adequate protection of cognitively impaired research subjects.

What Constitutes Cognitive Impairment?

According to the American Psychiatric Association's *Diagnostic and Statistical Manual* (American Psychiatric Association, 1994), cognitive impairment can be manifested by any one of the following disturbances: a failure to recognize or identify objects despite intact sensory function (agnosia), an inability to carry out a motor task despite intact comprehension and motor abilities (apraxia), language difficulties (aphasia) or demonstrating a disturbance in planning, sequencing, abstracting or organizing (executive ability dysfunction). More often, cognitive impairment is indicated by trouble learning new information or recalling previously learned knowledge (memory impairment). These difficulties can be seen in isolation or together, and the range of severity is tremendous, depending on the underlying cause of cognitive impairment. This last point is extremely important, because cognitive impairment in general, and the aforementioned specific examples of cognitive impairment in particular, are simply descriptors and should not be thought of as medical diagnoses themselves. For example, the ataxia following a mild stroke may not interfere significantly with one's ability to give informed consent for a research protocol, but the same apraxia along with memory impairment and aphasia of a progressive dementing process such as Alzheimer's disease would have very different implications. As a result, these symptoms must all be viewed in the medical context of the whole individual and not as specific determinants of competence.

For legal and financial issues, the courts usually are the arbiter, often with appropriate medical consultation, and power of attorney and legal guardianship are the major options. But within the medical research community, there have been no quantitative standards for measuring cognitive impairment with respect to informed consent issues. Practically speaking, when cognitive impairments are severe, this question is superfluous, because the impairments are obvious to all observers, and informed consent is considered unlikely in the absence of advanced directives. However, in the early or subtle stages of cognitive impairments, the answer is less certain. Traditionally, assessments of cognitive status are made by neurologists, psychiatrists and neuropsychologists. It is generally the neuropsychologists who generate the most thorough and quantifiable profile of cognition. However, these profiles are usually interpreted in the medial context by the physician involved to help determine the prognosis with respect to any specific patient and his/her cognitive impairments.

What Medical Diagnoses Are Associated with Cognitive Impairment?

Since we have already asserted that the cognitive impairments defined above are symptoms of underlying medical conditions, then we might further suggest that there are many causes of cognitive impairments. Indeed, the list of illnesses associated with such impairments is long, including all the forms of dementia, delirium and amnesia (McKhann *et al.*, 1984). While a dementia such as Alzheimer's disease or vascular dementia is by definition irreversible, the cognitive impairments of a delirium or amnesia can be reversible. It is not uncommon, for example, to find a cardiac patient in a temporary delirium following bypass surgery while recuperating in the intensive care unit of a hospital. During that period of delirium, cognitive impairments are obvious. Similarly, there are various psychiatric conditions such as schizophrenia, manic-depressive disorder and delusional depression which can be accompanied by transient cognitive impairments. In effect, any person can become temporarily impaired while under the stress of an acute hospitalization, secondary to multiple medications, or associated with surgery. Underlying medical conditions such as alcoholism or Alzheimer's disease which affect the brains' function certainly predispose one to cognitive impairments, but they are not the only causes of such impairments. The fact that temporary cognitive impairments can affect anyone clearly suggests that diagnosis is not the only factor which should be involved in our deliberations.

Is Cognitive Impairment a Transient or Permanent Condition?

Cognitive impairments are usually considered in a static time framework. Specifically, the testing to determine whether one is cognitively impaired often takes place over a few minutes or hours, whereas the underlying conditions being tested is often quite dynamic. For example, an individual who overdosed on a medication can be overtly delirious at one test point and then be fully functional the next morning with no evidence of residual cognitive impairment. An Alzheimer patient may show little day-to-day variability in test results but will demonstrate definite deterioration when followed for months and years with the same test battery. On the other hand, patients with one single stroke associated with aphasia or another cognitive impairment may remain remarkably stable for years. As a result of this dramatic variability, cognitive impairments must be considered in the medical context in which they are found. Decisions regarding informed consent are therefore influenced by the expected duration, severity and possible progression of symptoms, not just the presence of symptoms at any one point in time.

How Can Cognitively Impaired Subjects Participate in This Research Project?

After all the general discussion about different types of cognitive impairment, degrees of impairment, and the transient or permanent nature of cognitive decline, the IRB's must still make concrete decisions about the informed consent process for individual research projects. Given the enormous variety of research questions being addressed at different institutions, it may be surprising to learn that standard guidelines for the inclusion of cognitively impaired subjects in research do not presently exist. Whereas minor children are protected under specific federal regulations (Melnick *et al.*, 1984; High, 1992; DeRenzo, 1994), there are no such national protections for the cognitively impaired adults and seniors. Each institution is left to establish its own IRB guidelines within the general federal regulations. In the absence of rigid standards, most IRB's carefully review the relative risk of a research protocol and compare it to the possible benefit of the project. For example, a simple diagnostic procedure such as a blood test or a neuropsychological battery would not be considered much of a risk. Similarly, a research medication that had the potential for benefit and a history of little toxicity in other patient groups might be considered a minimal risk to a cognitively impaired subject. In these situations, next-of-kin approval has traditionally been acceptable. However, when the diagnostic procedures become more invasive (i.e., a positron emission tomography scan involving radiation exposure) or the medications are newer (i.e., little or no past experience in humans), then the relative risk of the research is greater, and the IRB must have alternative pathways to ensure protection of the cognitively impaired subject.

SURROGATE DISCUSSION MAKING FOR IMPAIRED SUBJECTS

At the National Institutes of Health, there has been a long history of research with cognitively impaired subjects. That research has included subjects with Alzheimer's disease, Down's syndrome, acquired immune deficiency syndrome, Parkinson's disease and many other conditions, including functional disorders such as schizophrenia. Traditionally, the informed consent was obtained from the individual patients as much as possible along with the next-of-kin relatives when appropriate. By the mid-1980s, however, that informal policy was no longer considered adequate. Unfortunately, state and federal law did not specify how informed consent should be achieved for research purposes under circumstances of cognitive impairment. In general, federal regulations require informed consent from an individual or that person's "legally authorized representative" as in cases with legal guardianship or with minors. For routine medical care, state law is clear in this issue and there are various legal options, but the problem when once considered entry into medical research. In the absence of state law with provision for research with cognitively impaired subjects, the NIH set out to establish its own guidelines for use at the Warren G. Magnuson Clinical Center where most intramural NIH human research is performed.

The Use of Durable Power of Attorney

In 1985, the NIH introduced a trial policy for obtaining consent with impaired human subjects (Fletcher *et al.*, 1985). Rather than relying on the time-consuming and limited legal guardianship approach, the NIH proposed the use of a durable power of attorney (DPA) for research purposes. Historically, the DPA has been employed by the court with regards to financial matters. More recently, the state of

Maryland had extended the DPA for use with routine health care matters (Maryland General Assembly, 1984). It was the intent of the NIH policy to adapt the DPA to include surrogate discussion making with respect to medical research.

Since its trial introduction in 1985, the DPA has been formally adopted at the NIH as the procedure for doing research with cognitively impaired subjects (Fletcher and Wichman, 1987). To help ensure the protection of individuals, several important safeguards were built in to the application of a DPA (Table II). First, it is essential that the patient be able to understand the meaning of a DPA before it can be assigned. It is therefore assumed that the individual's cognitive impairment will not be so severe as to preclude this level of understanding. Second, the patient him or herself must choose their own designee as a DPA. This stipulation differs from the routine policy with next-of-kin consent and allows the individual greater flexibility of choice. Third, the patient must have discussed his/her wishes with the DPA designee before it can be enacted. The process must be interactive and mutually agreeable. Finally, the person named the DPA must understand the responsibilities of a DPA and be willing to make decisions if called upon. The process of choosing a DPA is therefore not an automatic assignment. The NIH must also be assured that this individual has the ability and determination to carry out the responsibilities.

If a patient has the capacities to understand and assign a DPA, why is it needed in the first place? Clinically, it is not unusual for someone with cognitive impairments to understand the above process and be able to pick some trustworthy individual to act as DPA but at the same time be unable to completely understand the complicated procedures, risks and benefits associated with an intricate research protocol. Furthermore, the DPA is an instrument designed to be assigned *before* it is actually needed. In many cases, cognitive impairments are progressive such that the DPA becomes increasingly involved in the decision-making as the illness advances. This is exactly the situation the DPA is designed to address, so that the wishes of the individual can be carried out with the assistance of a surrogate as the cognitive impairment increases.

In order to further assure the proper use of the DPA, the NIH has put forth several other safeguards including a requirement that the IRB review each protocol employing the DPA more carefully and that all investigators follow the general guidelines as to when and how the DPA can be assigned and enacted. For instance, when an IRB reviews a protocol which may involve patients with DPA's, the investigators must specify in their protocols what additional safeguards have been

TABLE II   Four essential criteria regarding the use
of a durable power of attorney (DPA) for research purposes
at the National Institutes of Health

(1) The patient understands the meaning of a DPA.
(2) The patient names someone of his/her choice.
(3) The patient has discussed his/her wishes with the DPA.
(4) The assigned DPA understands the responsibilities of a DPA *and* is willing to make decisions if the patient becomes too impaired.

included to protect cognitively impaired subjects. To help investigators determine when it is appropriate to assign and use the DPA in a research setting, a series of 8 case examples have been developed to serve as guidelines (Table III). These cases represent eight clinical situations where research is contemplated. The key variables are the underlying competence of the patient and the degree of risk presented by the research protocol.

The level of cognitive impairment is by now a familiar term in this discussion, but the degree of risk is another variable which is determined by the IRB according to federal guidelines set by law [45 Code of Federal Regulations (CFR) 46]. There are four levels of risk: minimal and more than minimal risk each with or without potential for direct benefit. Examples were cited earlier in the discussion where a routine medical exam would usually constitute minimal risk in a research protocol while the administration of a new medication with potentially toxic side effects would pose more than minimal risk but with the potential of direct benefit. To help the investigators with guidelines for informed consent, these attributes of a protocol are set by the IRB long before any cognitively impaired subject would be invited to join in the research.

As can be seen in Table III, either increasing the level of risk or decreasing the capacity to provide informed consent dictates a higher level of review and eventually exclusion from research entirely (Case 8). To help in this process, the NIH has also provided for consultations with a bioethicist when questions arise. In fact, a bioethics consultation is required once the capacity to give informed consent is decreased or the level of research risk is increased beyond minimal. This consultation is intended to assure objectivity in the process of assigning a DPA and helps protect both the patient and investigator when any questions arise. Once a subject has become incapable of giving informed consent *and* the research is beyond minimal risk, a *DPA can no longer be initiated*, even with a bioethics consultation. At that point, either the courts must be involved or research cannot be considered.

Recent Experience with DPA at the NIMH

Over the last five years, the NIMH's Section on Geriatric Psychiatry has had experience with 82 patients and their families with respect to the DPA. The patients

TABLE III   Eight case examples for possible use of the DPA at the NIH

Impairment	Research Risk	Informed Consent Action
1. Capable	Minimal risk	Assign DPA for later use
2. Incapable	Minimal risk	Ethics consult and kin consent
3. Capable	Risk but possible benefit	Ethics consult and DPA
4. Capable	Risk but no benefit	Ethics consult and DPA
5. Incapable	Risk but possible benefit	Ethics consult and guardianship
6. Incapable	Risk but no benefit	Ethics consult and guardianship
7. Incapable	Risk but possible benefit	No intact family: emergency only
8. Incapable	Risk but no benefit	NO RESEARCH POSSIBLE

have for the most part been mild-to-moderately impaired individuals who are still living independently with their families. In general, they are highly-educated, mostly in the "young-old" age group (65–75 years old), relatively early in their onset of dementia and more often female (see Table IV). Their dementia ratings have proved them to be mostly mildly impaired, and they have all had the capacity to assign a DPA for research purposes, at least when they were first enrolled in the research program (Dukoff and Sunderland, In preparation).

Of this group, 35 individuals have returned for multiple additional studies between 1989–1994. Whereas each subject was able to assign a DPA and/or participate in the informed consent process at the time of the first research project, many of these same individuals had progressed significantly by the time of their most recent admission to the research program (Table V). The actual progression of the dementia is perhaps best demonstrated with the help of the Clinical Dementia Rating scale, a general measure of dementia severity (Hughes *et al.*, 1982). As can be plainly seen in Figure 1, many of the individuals progressed to higher levels of severity, In fact, almost one third of the subjects were now classified as severely demented, precluding them from assigning a DPA at that time or initiating research at the NIH because they could most likely not provide informed consent. Yet, since each subject had already assigned a DPA, these individuals could continue in research protocols with the informed consent of the DPA and the consent (if available) and/or assent of the individual.

There is no formal "starting point" for the enactment of a DPA, only a formal assignment at the time of document signing when the subject is still sufficiently intact cognitively. In this way, the actual execution of the DPA is seamlessly woven into the ongoing research participation of the cognitively impaired individual. At no time is he or she formally told that the DPA has taken precedence over their own participation in the informed consent process (even when/if the subject only signs the informed consent document with an "X"). Furthermore, the patients' assent at the initiation of a protocol and throughout any research project is always solicited. The importance of patient assent cannot be overemphasized. While there is no formal assent document with cognitively impaired adult subjects, as is required with minors participating in research authorized by their parents or guardian, the patients' ongoing assent is essential. In fact, no NIMH research is allowed to go forward if assent is withdrawn, irrespective of the level of dementia severity or the wishes of the DPA.

TABLE IV   Description of Alzheimer patients
using DPA at the NIMH (1989–1994)

Category	Mean ± SD
Subjects	82
Gender (Male/Female)	32M/50F
Onset (years)	65.6 ± 9.2
Education (years)	14.6 ± 3.1
Clinical Dementia Rating (1–3)	1.4 ± 0.9

TABLE V   Description of Alzheimer patients with multiple
hospital admissions who assigned a DPA (1989–1994)

Category	First Admission	Last Admission
Age (years)	66.4 ± 8.2	69.3 ± 8.4
Admission Number	—	5.0 ± 2.2
Duration (years)	3.2 ± 1.8	6.1 ± 2.5
Clinical Dementia Rating (1–3)	1.2 ± 0.5	2.2 ± 0.6

Possible Expansion of DPA to Other Conditions

Alzheimer's disease perhaps represents the classic example of how a DPA can be used in a research setting because there is an easily recognizable cognitive impairment and the expectation of profusion. But what about other, less obvious conditions where the cognitive impairments are more subtle or less predictable? How,

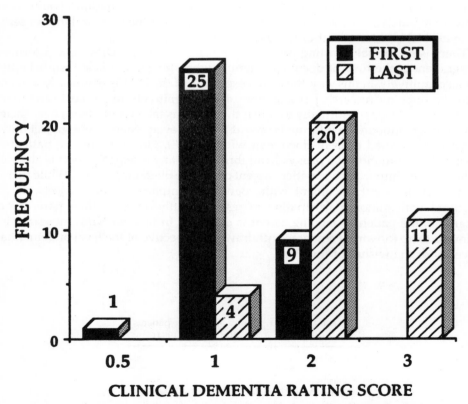

Figure 1.   Frequency distribution of clinical dementia rating scores during the first and last admissions of Alzheimer patients with DPA.

for instance, should the NIH or any other institution handle the variable cognitive impairments associated with metabolic conditions such as kidney failure, the stress-related cognitive dysfunction of postoperative psychosis, the delirium caused by pharmacologic agents or the thought disorders commonly associated with functional disorders such as schizophrenia? Each of these conditions is worthy of much research study, but each condition can be associated with significant cognitive impairments which *might* contaminate the informed consent process.

Rather than attempting to answer these questions individually and become locked into interminable discussions about degrees and duration of cognitive impairment, let us assume that significant cognitive impairments can occur in these conditions. In addition, let us reason that there is an element of unpredictability in the timing of these cognitive impairments. If these two assumptions are true, then it follows that trying to arbitrarily differentiate amongst these conditions as to which one should require a DPA and which one should not would inevitably lead to the introduction of bias and perhaps even stigma (in the case of the mental disorders).

Given the simplicity of the DPA process with the Alzheimer patients and the general acceptance of the process with families, it might make more sense to include all research patients in the DPA model. As part of the admission process, patients would be expected to assign a DPA either because of a pre-existing cognitive impairment or in the event of a treatment-emergent cognitive impairment. Just as Alzheimer patients can revoke the DPA at any time or refuse to assent to any procedure, so would every other subject maintain those patient rights. Instead of being in the awkward ethical position of deciding whether to continue a research protocol *after* a previously intact subject has become cognitively impaired, the pre-arranged DPA will be available to help guide researchers as to the wishes of the individual, irrespective of the medical cause of the subsequent cognitive impairment.

Does the DPA Put Cognitively Impaired People at Risk?

In any research project, there are always inherent risks involved, sometimes as small as an evaluation form or as large as an experimental surgical procedure. It is the responsibility of the principal investigator and the Institutional Review Board to evaluate the risks and potential benefits of the research and to assure that there is a reasonable balance before proceeding. The informed consent document is the external or public record of that balance, and it should reflect in common language all important aspects of a research project for any individual subject or the assigned durable power of attorney. While the final informed consent is codified in a written document, the informal consent process is multifactorial and should involve the common-sense good judgment of subject, family, researcher, and the entire medical staff as well. Each person is responsible for minimizing the risk as much as possible, and each person has the opportunity to stop the process at any time if there is a suspicion of wrong-doing, coercion or unnecessary risk. In so doing, all research subjects have multiple levels of protection.

CONCLUSION

At the NIH, the DPA is used as an instrument to protect the wishes of the cognitively impaired individual in a research setting. With subjects who have mild cognitive impairments, the person assigned as DPA is initially present during the informed consent process to observe the subjects as they chose amongst a variety of research paths. When the cognitive impairments have increased, it is the job of the previously-selected DPA to evaluate and choose which research tracks the subject might have wanted for themselves if they were less impaired. In this way, the more-impaired subjects can still profit from the potential benefits of the research and take pride in feeling that they are contributing to an ongoing research endeavor. For an Alzheimer patient who may otherwise feel themselves to be a burden to their family, the enhanced self-esteem associated with continued research involvement is of palpable value. Throughout this process, the individual maintains the right to refuse assent to any procedure. When used properly, the DPA extends the choice of an individual beyond the time that cognitive impairments robs him/her of the capacity to give informed consent.

REFERENCES

American Psychiatric Association. (1994) *Diagnostic and Statistical Manual of Mental Disorders*, Fourth Edition. Washington, D.C., American Psychiatric Association.

DeRenzo, E. (1994) Surrogate decision making for severely cognitively impaired research subjects: the continuing debate. *Cambridge Quarterly of Healthcare Ethics* 3:539–548.

Dukoff, R. and Sunderland, T. (in preparation) Clinical research with Alzheimer's disease patients and the Durable Power of Attorney.

Fletcher, J.C., Dommel, F.W.J., and Cowell, D.D. (1985) A trial policy for the intramural programs of the National Institute of Health: consent to research with impaired human subjects. *IRM* 7:1–6.

Fletcher, J.C. and Wichman, A. (1987) A new consent policy for research with impaired human subjects. *Psychopharmacology Bulletin* 23:382–385.

High, D.M. (1992) Research with Alzheimer's disease subjects: informed consent and proxy decision making. *Journal of the American Geriatric Society* 40:950–957.

Hughes, C.P., Berg, L., Danziger, W.L., Coben, L.A., and Martin, R.L. (1982) A new clinical scale for the staging of dementia. *British Journal of Psychiatry* 140:566–572.

Maryland General Assembly. (1984) Health General Article, Annotated Code of Maryland, Chapters 591 and 540 of the Acts of 1984. Section 20-107.

McKhann, G., Drachman, D., Folstein, M., Katzman, R., Price, D., and Stadlan, E.M. (1984) Clinical diagnosis of Alzheimer's disease: report of the NINCDS-ADRDA Work Group under the auspices of Department of Health and Human Services Task Force on Alzheimer's Disease. *Neurology* 34:939–944.

Melnick, V.L., Dubler, N.N., Weisbard, A., and Butler, R.N. (1984) Clinical research in senile dementia of the Alzheimer type: suggested guidelines addressing the ethical and legal issues. *Journal of the American Geriatric Society* 32:531–536.

A Communal Model for Presumed Consent for Research on the Neurologically Vulnerable

David C. Thomasma

Loyola University of Chicago, Medical Humanities Program, 2160 South First Avenue, Maywood, IL 60153

The origin of modern medical ethics lies in the effort to control and to regulate research on human subjects. This international effort flowing from the Nuremberg Trials through the Helsinki Accords, and nationally, through the legislation governing Institutional Review Boards through the National Institutes of Health, has been acceptable for most research, but not always completely successful. Most recently, efforts to conduct research on schizophrenics have raised serious questions about harms to the subjects, and their families and loved ones. These harms and the risks to subjects have been detailed at this conference already. No need to repeat them here. So too have proposals been offered for optimizing and regulating this kind of research. What I will contribute is an exploration of a communal model for presuming consent regarding subjects who have impaired neurological function.

I will state my conclusion first. For those who are most vulnerable in a modern, scientific society, greater care must be taken for their vulnerability. In effect this means that no action can be taken that places these beings and persons at greater risk than others of being objectified and manipulated as if they were without subjectivity and personhood. For the majority of us consent is the only way that such "use" of persons can occur, since we are able to agree ahead of time to alienate our subjectivity from an objective body to be studied, either in hopes of obtaining some benefit, or even without such hopes. Further we consent in such circumstances to suspend the normal stages of a doctor-patient relationship in favor of altruistic ends both the subject and the researcher presumably employ.

THE COMMUNAL MODEL

I propose that research on the neurologically impaired can be conducted if there is little to no risk to subjects using already existing criteria of surrogate decision making and IRB requirements and oversight already in place. Surrogate decision

This article will also appear in the journal: *Accountability in Research*, Volume 4, pp. 227–239 (1996).

makers can offer the consent if the subject is currently incapable of offering it. However, in cases where moderate and even unknown risk is involved, I propose that a communal model for providing consent be employed. This model expands on methods we already use for assessing the ethics of research, namely, the consensus panel method, and combines it with greater oversight by the IRB on the progress of the research.

Nonetheless, the only research that can ethically be conducted on the neurologically impaired is that which either benefits them directly (direct benefit research), or benefits the class of such beings directly (class benefit research, direct type), themselves only indirectly, but poses either no or only very mild risk. Of course, even for direct benefit or class benefit research to be conducted some consent must be obtained. In this regard, a committee of research consent according to the communal model should both provide that consent and monitor the research progress. This research consent committee should include but not be limited to the legally authorized surrogates. It should have at least one member of the class of subjects who have the disease upon which the research is being proposed. Preferably, this person or these persons would have undergone similar research in the past. Representatives of advocacy groups should also be included.

For high risk but potentially high benefit research, the neurologically impaired cannot be considered suitable subjects, even if down the line they might profit as a class from this research. A general principle should be that for serious disease, such as schizophrenia, a relapse must also be considered a serious harm, and its potential, even in direct benefit research would preclude the ability of any committee to subject a person to such harm. Following an intermediate path such as I propose, society can benefit from research, and still protect the vulnerable from harm. This doctrine of protecting from harm has served us well, not only by protecting all individuals in medical research, but especially those who might be the most vulnerable to manipulation. Notwithstanding the moderate view I have proposed, it still involves a communitarian rather than libertarian component, namely, permitting a committee representing the community of care to replace a seriously ill person's ability to consent to moderate risk research.

The rest of the paper constructs steps of an argument in support of this approach. The communal model is proposed to meet the complexity of a specific type of incompetency, a variable one, for adolescent and adult schizophrenics, but it can be applied to other forms of neurological impairment as well. In particular, it will help to more fully examine the need for informed consent to detect any ways in which we might be able to override its requirement for a higher good.

PRESUMPTION *A PRIORI*

An *a priori* regarding research could be that we can presume the subject is neutral toward either no risk or moderate risk, direct benefit research. That is to say, we cannot be certain that the subject would, if able, consent to or deny participation in such research. Thus, we would not violate necessarily their best interests in deciding to enter them in studies from which they might benefit. For this reason, the normal process of relying upon a surrogate decision maker, most often a parent or spouse, to decide about direct benefit research would be appropriate.

With respect to research that holds less direct benefit, but might benefit the class of patients who have the disease of the potential subject, I propose that we rely upon the surrogates to decide if the risk is minimal. If the risk increases to moderate, however, the communal model should be employed. If the risk is high, and the disease seriously impairs the person, then the presumption of best interests would deny the possibility of entering the person on the study. No decision making model need to be employed, since by public policy the surrogates could not be approached with a request in this instance.

Hence, based on the benefit presumption *a priori* I have proposed, the communal model functions in cases of potential but intermediate harm between no risk of harm and serious risk of harm. For no risk, surrogates decide. For serious risk, public policy denies access to vulnerable subjects. In the middle are research studies that would be decided on the basis of a committee representing the community's care for the individual, composed of those who know and love the individual, and those who are touched by the disease most directly or are advocates for such persons.

VARIABILITY OF COMPETENCY IN SUBJECTS

A second reason for the communal model comes from the variability of competency.[1] There is a wide range of persons who lack the capacity to consent, either by reason of age, maturity, or disability. Among those impaired by reason of age are embryos, fetuses, and children until at least the age of reason and beyond. Among those impaired by reason of maturity are those who have suffered a stroke or suffer from sufficient Alzheimer's to limit or impede their capacity to consent. Disability, too, can cover a wide range of lives, from those whose pain overrides their ability to reason things through, to various genetic or organic physiological impairments. For each of these classes there are special considerations. For example, there is a moral difference between those who have never constructed a biography (e.g., the newborn), and those who have (e.g., a grandmother after a severe stroke). In the former case, we must exercise greater efforts to protect from harm precisely because we do not know the value-system of the subject, while in the latter instances, having in hand some statements or life-histories regarding values, we can take those into account in our considerations about research.

Those whose disability is neurological impairment face a particularly difficult process if they are to participate in neurobiological or any other kind of research. The normal standards of informed consent do not seem applicable, particularly since informed consent doctrine requires on the part of the participant at the very least a capacity to process the information and to apply it to one's own circumstances. Neurologically impaired individuals may either understand but be unable to foresee the consequences for themselves, or more often, be unable either to process information or apply it, especially to their future capacities.

The class of neurologically impaired might include both those who are temporarily so because of mood swings or personality disorders, and those who are permanently so because of permanent damage to the brain. The ethics of conducting research on Alzheimer's patients, while an analogue of the ethics of research on schizophrenics, does not fit it exactly, since schizophrenia is capable of being controlled and patients may become competent during periods of remission, while Alzheimer's is a progressive and fatal condition. This point need not be belabored. But it does indicate a difference in the way we ought to protect the vulnerable from harm.

In permanent conditions of incapacity, full-scale paternalism requires extreme caution about subjecting individuals to research, even that which might benefit them directly. In temporary conditions, however, greater reliance on individual wishes while competent must occur. Notwithstanding our commitment to self-determination, the cautions for schizophrenics ought to include the communal model already adumbrated. Thus the person's autonomy with respect to medical research is modified to some extent by the consent and oversight committee, regarded as a sort of "committee of the whole community."

ASSESSMENT OF DEEPER RISKS

A third argument supporting the communal model comes from a reflection about the deeper risks of relationship alienation. We are all vulnerable when it comes to modern, scientific research within the context of a medical establishment. The normal one-to-one doctor-patient relationship is disrupted in favor of gaining new knowledge. Nonetheless, one must not forget that health professionals' "act of profession," is a declaration of commitment, the act of "consecration" to use Cushing's word, to a way of life that is not ordinary. In that act health professionals promise that they will not place their own interests first,[2] that they will not exploit the vulnerability of those they serve, that they will honor the trust illness forces on those who are ill.[3] This necessity for a higher standard impelled Plato to use medicine as a paradigm for the ethical use of knowledge. Medicine was for him a *tekne*, a craft and art to be sure, but a craft with a very significant difference from all the others.

The normal doctor-patient relationship proceeds through at least three separate steps.[4,5] The first is one of compassion or immediacy. This is a personal stage in which both patient and physician respond to the disability, pain, or suffering of the patient on an empathetic level. Becoming a research subject changes this immediate bond between doctor and patient. The second stage is one of alienation, during which the help to be offered by the physician requires an objectification of the body of the patient, a categorization, a necessary but somewhat Cartesian step that is normally synthesized in the third step. In a research model of a subject-researcher relationship, the second stage of the doctor-patient relationship becomes the standard. The third stage is a personal synthesis stage, where the physician returns to the particularities of the patient as a person, in the context of his or her own values.[6] What is most distressing to the neurologically impaired who have been research subjects is a lack of sensitivity to this final, synthetic stage in the research model. Perhaps expecting such concern for values is too much to ask of a research model?[7]

All medical research, then, involves greater risks than those normally detailed in the consent form regarding the protocol itself. There is a risk of alienation, both from oneself and from the physician. If one is incapable, not only of understanding the nature of the research and its consequences, but also of ascertaining, of processing the fundamental changes in character of the physician-researcher and the relationship with that professional, then the risks of the research are too high to permit the neurologically impaired to participate. I submit that this is what happened in the studies about schizophrenia. The subjects and their families found themselves facing the terror of relapse without having fully comprehended its possibility and the damage it would do to relationships within the family, in society, and with the physician-researchers.

That being said, it is important also to note that modern biomedical research advances in very tiny steps. Most often the possible gain for individual subjects is almost minuscule. Rarely are there major leaps in knowledge provided by research protocols. This makes research on all classes of neurologically impaired individuals highly burdensome on those individuals, not to speak of the imbalance in the risk/benefit ratio. Knowing this ahead of time, even when persons might be in remission from their disease, it is important to circumscribe their autonomy with the deliberations of a committee that can assess the dangers and risks to values, and not just to the physiology of the subject. In this way, the results of the research, its impact on family and society, can also be brought to the table. Should consent be offered, this same care committee can monitor the progress of the research very closely and judge its impact on those same values, even if the subject has no physical side-effects from the drugs or treatment arm. This monitoring is important, too, since not all risks and impacts can be predicted, especially when we consider the broader risks I have suggested in this section.

ANOTHER LOOK AT AUTONOMY

A fourth argument concerns not only broadening our concerns about risks beyond physiological ones, but broadening our respect for the neurologically impaired beyond a concern for autonomy itself.

As is well known, the principle of autonomy lies behind the doctrine of informed consent. We are now able to examine its role a little more carefully. Throughout the world, there has been an increased emphasis on autonomy, not only in research, but also in clinical practice. The standard for information given a patient is either what the patient requires in order to participate, or what the average, prudent patient, might wish to know, rather than what the medical profession thinks is appropriate. This shift to the patient is based, not only on autonomy, but also on a deeper commitment to respect persons. In this view, the best interests of the patient cannot be formulated without reference to that patient's own desires and preferences.

James Gustafson has said: "A person has a right to determine his or her own destiny ... capacity for self-determination is what makes humans distinctive as a species."[8] It is a testimony to the strength of our commitment to this belief that concern over the treatment of incompetent patients has now gained so much attention.[9,10] If a person has a right to determine his or her own destiny, what must we make of incompetent patients? Are they no longer persons? Have they forfeited the right of self-determination because they are now incapable of sometimes expressing it? What should be our duties towards them?

My response is that autonomy and beneficence are closely linked: one cannot ignore one for the sake of the other. Tom Beauchamp has pointed out that competency and autonomy are quite separate concepts, with quite separate grammars and considerations.[11] Customary arguments about making decisions for incompetent patients draw too close a link between the two concepts. As a result, far too much attention is paid to what patients might wish, or formerly wanted, and not enough to medical indications themselves. That is, too much attention is paid to autonomy and consent-related concerns. Quality of life judgments, which employ criteria such as Substituted Judgment or Reasonable Persons, are tinged with a preoccupation about consent, or better, decision-making, as the basis of our commitment to respect the dignity of human beings. Self-determination or autonomy need not be the only basis of personal dignity. There are many other features.

Examples of features of human dignity not necessarily connected with autonomy and consent are first, other forms of freedom, and second, other dimensions of human existence. There are at least five sorts of freedom, only one of which may properly be called autonomy, and yet none of which could be lacking without serious impairment occurring to human dignity.[12] One example can suffice. If autonomy or freedom is too closely identified with freedom of choice (decision-making), one neglects therein the primal freedom of creating new choices or of committing oneself to a single choice.

Similarly, human *dignity* should not be totally identified with autonomy. Loss of autonomy, almost all physicians, ethicists, and lawyers concede, does not mean a person loses status as a person, and no longer deserves respect. Thus Wilfred Gaylin, perhaps acknowledging an earlier 1972 sketch by Joseph Fletcher regarding indicators of humanhood,[13] developed an argument that human dignity should encompass such elements beyond autonomy as responsibility, past achievements, fundamental genetic human nature, a capacity for technology, and freedom from instinctual fixation. Gaylin notes: "Modern developments in medicine raised questions about individuals whose autonomy was limited, and yet who commanded a sense of special worth—the infant and the child, the senile, comatose, or retarded person. The conflation of dignity into autonomy threatened their position in the moral world, and compromised our treatment of them."[14]

Generally speaking, then, the principle of autonomy can be overridden for reasons of beneficence when doctors and surrogates, such as parents, believe that a treatment or research protocol would be in the best interests of a minor or offspring who is neurologically impaired. Thus, minors may assent for a procedure but cannot refuse one that the parents and doctor think is necessary. By analogy, I suggest, human dignity may still be preserved when surrogates consent "parentally" to entering a neurologically impaired person on a research study. Yet the parentalism involved in this surrogate consent, as I have said, requires that the research study directly benefit the subject/patient and present little or no risk. If the risk of harms increases from this level, greater protections must be employed.

There is an important objection to this line of reasoning. The objection comes from a sophisticated analysis of autonomy and incompetence provided by Robert Veatch. According to Veatch, the current preferences of even legally and psychologically incompetent patients should generally prevail. The same commitment of respect, according to Veatch, must be made to those who previously expressed their wishes, although the status of those wishes can only be legally confirmed through living wills or a mechanism like the durable power of attorney. Only in this instance does the legal doctrine of substituted judgment make much sense,

Veatch argues, as it then would be a process of a guardian determining what the expressed wishes of the patient actually entailed.

Veatch includes two other categories of patients: those who were never competent, and have no relatives or other agents to step into the guardian role. The third class is that of incompetent patients who do have guardians or family members able and willing to act as guardians. For Veatch, both of these categories include patients who may once have been competent, but failed to make clear their wishes about the situations which eventually befell them. He argues that decisions can be made for such patients without appeal to subjective, substituted judgment criteria, that is, by substituting an earlier judgment they themselves made for a missing one at this time.

He confronts the fact that when patients have never expressed prior wishes, the principle of autonomy has no further merit. It makes no sense to try to respect a person's autonomy (read, "decision-making") when they have left no clues about what they would wish. Instead, and this is the danger in his position, one no longer aims at the best interests of the patient. Veatch holds these as now indeterminable. Instead, as he says, "the goal is not to serve the patient's best interests, but to honor his wishes out of respect for him."[15] But how can wishes not made be "honored?"

The willingness to abandon a search for the patient's best interests seems to lie in an identification of respecting persons with honoring their wishes, and in a repugnance for substituting one's own judgment about the quality of life for a missing capacity. This is an important, even necessary component to the whole notion of respect for persons. Consequently, there is much to admire in Veatch's position. But respecting wishes is a necessary but not a sufficient way to respect persons, especially in the absence of a decision-making capacity, precisely what is missing in incompetent subjects. Keeping faith with such patients, then, requires some other kind of judgment.

There is a way to keep that faith, and not abandon it, as Veatch's reasoning seems to have forced him to do. As noted, the flaw in that reasoning concentrates respect for persons on decision-making capacity. A better approach broadens our concerns for respecting persons to the locus of their life, the context of care in which they thrive best. Persons responsible for maintaining this support are then able to judge best about the impacts of research on that environment. Recall that that environment is only tentatively stable and reached after a long and painful struggle.

Note that quality of life judgments themselves are not the nemesis of care we often take them to be. The real threat is their subjectivity.[16] One may make quality of life judgments in an objective manner by assessing physiological and social function and comparing this to the subject's perception (if moderately competent) or to other more "normal" categories of function. No matter how many objective criteria are employed, however, at some point the committee will have to make assessments from the standpoint of their own values as well.

PARENTALISM AND KEEPING FAITH WITH THE INCOMPETENT SUBJECT

A fifth argument turns our attention to parentalism itself, for it need not be considered so evil in the context of the neurologically impaired as it might be in other contexts of greater capacity for decision making.

In light of the increasing emphasis on autonomy in American bioethics and public policy, paternalism has acquired a bad reputation. Most often this is deserved. However, if one extracts from the concept the goal of acting in the best interests of others, namely the underlying beneficence inherent in the concept, then we can quite properly advocate a kind of parentalism, a beneficent "looking out" for even competent persons who may not foresee the consequences of their choices as clearly as we would want.[17] I prefer this point of view to that of abandoning the individual subjects to their own resources, given not only the history of questionable research on the most vulnerable populations in society, but also the complexities of alienation as I have already described them that occur within all medical research contexts. In this way we "keep faith" with either temporarily or permanently incompetent subjects.[18]

By now it is well-recognized that by "incompetent" is meant an inability to execute decisions about one's care, whether the cause is senility, retardation, a stroke, mental illness, genetic disease, or any temporary or permanent incapacity. This statement is merely a negative elaboration on the common definition of competence as a capacity for performance of a specific task. As Pellegrino defines it with respect to medicine, a "competent person possesses the capacity to make an explicit, reasoned, and intentional choice among alternatives."[19] This requires the capacity to receive information, perceive the relation of the information to one's own predicament, integrate it and calculate risks and benefits, make a choice among options, and convey or communicate that choice in some manner. Unless otherwise the case, one ought to assume that a person is competent.[20] But incompetence can occur in any one of the categories Pellegrino mentioned, impairing that person's decision-making ability. It is for this reason that the wishes of patients and even their consent are sometimes not solicited nor honored in medicine.[21,22]

By "keeping faith" with incompetent subjects, I mean acting for their ultimate benefit insofar as we may with our best intentions. Thus, the communal model is a form of parentalism that permits a small group of interested and committed persons to presume the consent of a neurologically impaired person with respect to research that may have moderate risk to the subject, or may only potentially benefit the class of patients to which the potential subject belongs, but not benefit him or her directly. Keeping faith with such patients is equivalent to acting beneficently toward them.[23] As I put this earlier, there is just as much evidence that an individual may wish to participate in research on a disease one possesses as that individual may *not* wish to participate. Independent but caring judgments about the risks in the protocol and in the health professional relationship is necessary to help assess the good for the individual and for his or her class of patients.

Acting on behalf of the incapacitated persons is quite difficult. If medicine's object is to protect or even restore autonomy and self-determination,[24] how can one protect this value while intervening in such a way that might disrupt that autonomy, neglect or ignore it, or even, possibly impede its develop through the relapse of a schizophrenic episode, or perhaps even destroy it permanently? Put another way, how can one maintain beneficence in face of uncertain benefit? The usual method is to calculate beneficence on the basis of the wishes of the patient, thus identifying what is best for the patient with his or her previously expressed wishes.

However, in dealing with incompetent patients, it is not always the case that one's best interests are served by a respect for prior wishes, or an effort to determine what someone would wish were that person now competent. Indeed, there is a standard therapeutic assumption that can be applied to the communal model: The physician is justified in treating the disorder that renders the patient incompetent.[25] By analogy, the committee can place the patient on a study that may remove impediments to competence (e.g., decrease the chances of relapse). There is nothing essentially sacrosanct about respecting prior wishes of patients who were once competent. Because of the new situation that was perhaps not foreseen, these prior wishes themselves may now be suspect. Morreim argues that respecting prior wishes of the incompetent may not be the best way to respect autonomy itself.[26] Social pressure to respect prior wishes for utilitarian reasons, given current and future economic pressures, may lead to rapid acceptance of a refusal of therapy (by prior wish). This may come to be the preferred social outcome.[27] If so, either one of two things must give. Either we will no longer seek the best interests of patients (but rather that of society), or we will sacrifice quality of care for inappropriately applied prior wishes.

Two traditional attempts to avoid capricious decision-making in quality of life judgments should now be examined, since the committee would most certainly attempt to employ them in deciding the appropriateness of entering a neurologically impaired person on a study. In general, both often are used to try to decide on treatment plans for incompetent patients. They are Substituted Judgment and Reasonable Persons Standards.

First, Substituted Judgment has been employed in an effort to decide what an incompetent patient would now wish done. I agree with Veatch who argues that this criterion has only very limited applicability.[28] There are at least four kinds of Substituted Judgment. In the first, all that is substituted is a "wish." This is an objective judgment about current therapy based on what a patient once said about it in the past. A second more subjective and therefore problematic form is one in which we extrapolate from a patient's probable choices in the past to the present treatment question. In this form we truly substitute for present decisions an interpreted past history.[29] In the third more dangerous form, we actually substitute our judgment for the patient's (usually when we have no evidence to use either of the first two forms); in effect, we ask "what would I do if I were in the patient's shoes?" In the final and most dangerous kind, we pass judgment on another's life, holding it either worth or not worth living.

The point when Substituted Judgment becomes a violation of respect for persons occurs when prior wishes cannot be easily determined or facilely applied to the current medical situation, or the crisis itself was actually never foreseen by the subject, when competent, if ever, or not properly contemplated by the surrogates. In this respect John Robertson notes: "Until recently the courts, policy-makers and others have shrunk from pronouncing such stark judgments on the relative worth of incompetent persons. Yet such judgments have always been implicitly made, if only in the manipulation of purportedly patient-centered tests such as substituted judgment to reach the result of nontreatment."[30]

In the end Substituted Judgment can fail on several counts. First, it can become capricious if the current wishes of the patient are not known. It is too subjective if it is based totally on another individual's judgment of what is best for a patient. Finally, it is prone to personal and cultural miscalculations about the quality of life of another human being. Our political, religious, and philosophical traditions require avoiding any temptation to impose a value on another person. To do so violates a belief in the intrinsic dignity of human beings.[31] Yet those same traditions also require we care for the most vulnerable in society. The only way to circumvent the danger of invalid imposition of values and yet provide for those who depend on us for their care, is to broaden the base of decision making. This is partially accomplished through the communal model, although the whole of society must become involved in restricting high risk neurological research.

What about Reasonable Persons then? Is that what I am proposing? Unfortunately, the Reasonable Persons criterion is also flawed in a significant way. It represents an attempt to develop an alternative to subjective judgments about the quality of life of another person by appealing to external authorities. These authorities are the "reasonable persons," the same people whom one would consult in any difficult matters of prudential judgment. According to this view, one should not make difficult moral decisions about incompetent patients without first checking with what a body of rational and wise persons might suggest.

At first blush this standard, proposed by Richard McCormick regarding treatment decisions about defective newborns, seems a way out of the quality of life dilemma. One does not impose one's own necessarily limited view of what would be best for the patient. Instead, reasonable individuals are consulted. In the end, however, one merely substitutes for one individual's judgment a community of authorities who also make subjective judgments about another more vulnerable human life. That is why I propose that the "authorities," if they can be so called, should be those who care for the subject and provide his or her context of value. Primarily this would be the family members making decisions together for their loved one, accompanied by significant other persons, precisely because the decision to join a study would affect all of them as well as the impaired subject.[32]

A stronger argument against Substituted Judgment and Reasonable Persons can also be formulated. Both doctrines employ a device to determine what a person would desire. But the incompetent person never was nor will be like the rest of us. Determining possible preferences is impossible. Hence, the attempt to determine the patient's best interests falters. This is why Veatch was willing to abandon the best interests standard. Instead, he proposed under his "limited familial autonomy" principle, that families (or guardians) make a choice from a set of reasonable alternatives. They would not be saddled with a futile attempt to determine the most beneficent course of treatment.[33] I am now applying his idea to research studies as well. An important suggestion, Veatch's approach would clearly help avoid quality of life judgments. But does it properly respect the human dignity of such patients? While protecting society and the patient from unreasonable (and no doubt costly) interventions, as I have suggested it nevertheless entails an abandonment, at least in its pristine form, of the ethic of beneficence at the heart of medicine.[34]

CONCLUSION: PROTECTING THE VULNERABLE FROM HARM

A final argument stems from the duty to protect the vulnerable from harm.[35] Protecting the vulnerable from harm is a virtue of modern scientific and medical research. The Nazi experience was the first time in human history that a powerful state marshalled the forces of medicine to its own ends. Those ends were advancement of the state's interests in the world and the "hygiene" of its populace so that the war effort could be most efficient. As a result, individuals were valued for their contribution to the state. Those who could not so contribute were regarded as "ballast existence" on the state and proper candidates for euthanasia.[36]

To counter both the manipulation of human beings by medicine and the state, and the very power of those forces over individuals, the requirement of informed consent was born from the Nuremberg trials after the war and subsequent revelations about research in the United States like the Tuskegee Syphilis Study. Recall that informed consent is only a minimal way to protect individuals and honor their personhood. Essentially it defines the borders beyond which we cannot go without the participation of the individual. Without consent we cannot reduce a person to an object for any purpose. This is not the same as honoring and promoting individual worth, no matter what the circumstance of their neurological condition.

Recent revelations about our own government and scientific community conducting radiation research without consent on the retarded and pregnant women in the 1950's and 1960's demonstrates that the Nazis did not have a lock on the cavalier treatment of vulnerable populations. Boundaries must be established on the power of the state and on the power of science over individuals in our society.

Thus, what I have called the virtue of informed consent is a negative virtue; it sets requirements limiting the power of others over vulnerable individuals. A more positive virtue would be honoring the individual to a greater extent than we honor those who can contribute to society. This honor consists in ensuring that such persons are not treated as objects.

The vulnerability of impaired persons lies not so much in their loss of capacity, compared to "normal" people, but in the sheer force of objectification found in modern life, by which standard and accustomed ways in which we deal with one another as objects (salespersons, garbage men, etc.), is immeasurably increased due to loss of interactive ability. Those persons who suffer progressive and irreversible cognitive decline might at one time been mentally intact. In that case, participation in neurological research might be based on presumed consent. This can be extrapolated either directly from an advance directive that would be obtained in the earlier stages of the disease, or indirectly, through surrogates articulating the previous wishes or value system of the patient.

Those persons who suffer from fixed neurological disorders and mental incapacity, and who have never been competent to execute an advance directive, might still have a set of values that surrogates can express, but these are most often very concrete concerns about daily living, e.g., that Jimmy liked to swim or to eat, rather than more abstract values like Jane always wanted to contribute to society. From the former set of values, little can be extrapolated about medical research. From the latter, we might be able to argue that presumed consent for research would follow from a value of contributing to society.

In order to avoid any scenario of disrespect for the neurologically impaired and their families, and simultaneously avoid having consensus meetings with thousands of points of view represented, we must turn to certain *a prioris* that should be on the table during the discussion by the committee of care. A prime candidate for such an *a priori* should be what I call the Principle of Dominion. Instead of trying to justify its presence during the discussion on conceptual grounds, it should be justified on experiential ones. To wit: the experience of Western Civilization is such that it has erred grievously on the side of domination and control over all that it has objectified, including animals and human beings.

Thus, the Principle of Dominion would state that for every intervention into a natural process, special care must be taken not to objectify or manipulate for our own ends, however noble, the person for whom the intervention is contemplated. Since that individual cannot make judgments for him- or herself, analyses of the consequences on his or her life, and on the lives of others, must be an intrinsic part of the decision to intervene. Under this principle it would be impossible to justify non-therapeutic research on neurologically impaired individuals. Some direct benefit must be contemplated and designed into the protocol.

NOTES

1. Drane, J. (1985) The many faces of competency. *Hast. Ctr. Rep.* 15(2):17–21.
2. Pellegrino, E.D., and Thomasma, D.C. (1993) *The Virtues in Medical Practice.* New York: Oxford University Press.
3. Kass, L. (1983) Professing ethically. *J. Am. Med. Assoc.* March 11, 249(10).
4. Von Gebsattel, E. (1996) The meaning of medical practice. *J. Theor. Med.* 16: forthcoming. Tr. from the German by Welie.
5. Welie, J. (1995) Viktor Emil von Gebsattel on the doctor-patient relationship. *Theor. Med.* 16: forthcoming.
6. Pellegrino, E.D., and Thomasma, D.C. (1988) *For the Patient's Good: The Restoration of Beneficence in Health Care.* New York: Oxford University Press.
7. Thomasma, D.C. (1992, 1993) Models of the doctor-patient relationship and the ethics committee: part one. *Cambridge Quarterly of Healthcare Ethics* 1:11–31; Part two. *Cambridge Quarterly of Healthcare Ethics* 3:10–26.
8. Gustafson, J. (1982) Ain't nobody gonna cut on my head. In Levine R. and Veatch, R. (eds.): *Cases in Bioethics.* New York: Hastings Center, p. 37.
9. Monagle, J.F., and Thomasma, D.C. (eds.) (1988) *Medical Ethics: A Guide for Health Care Professionals.* Rockville, MD: Aspen Publishers, Inc.
10. Robertson, J. (1988) The geography of competency. In Engelhardt, H.T., Jr., Cutter, M.A., and Shelp, E. (eds.): *When Are Competent Patients Incompetent?: A Study of Informal Competency Determinations in Primary Care.* Dordrecht/Boston: Kluwer Academic Publishers.
11. Beauchamp, T. (1988) Competent judgments of the competence to consent. In Engelhardt, H.T., Jr., Cutter, M.A., and Shelp, E. (eds.): *When Are Competent Patients Incompetent?: A Study of Informal Competency Determinations in Primary Care.* Dordrecht/Boston: Kluwer Academic Publishers.
12. Thomasma, D.C. (1984) Freedom, dependency, and the care of the very old. *J. Am. Ger. Soc.* 32: 906–914.
13. Fletcher, J. (1972) Indicators of humanhood. *Hast. Ctr. Rep.* 2:1–4.
14. Gaylin, W. (1984) In defense of the dignity of being human. *Hast. Ctr. Rep.* 14:8–22.
15. Veatch, R. (1984) An ethical framework for terminal care decision: a new classification of patients. *J. Am. Ger. Soc.* 32:665–669; quote: 667.
16. Gutheil, T., and Appelbaum, P. (1983) Substituted judgment: best interests in disguise. *Hast. Ctr. Rep.* 13:8–11.
17. Blustein, I. (1993) The family in medical decision-making. *Hast. Ctr. Rep.* 23:6–13.
18. Brock, D. (1991) Surrogate decision-making for incompetent adults: an ethical framework. *Mt. Sinai J. Med.* 58:388–392.
19. Pellegrino, E. (1988) Informal judgments of incompetence: the patient, the family and the physician. In Engelhardt, Jr., H.T., Jr., Cutter, M.A., and Shelp, E. (eds.): *When Are Competent Patients Incompe-*

tent?: A Study of Informal Competency Determinations in Primary Care. Dordrecht/Boston: Kluwer Academic Publishers.

20. Pincoffs, E. (1988) Judgments of incompetence and their moral presuppositions. In Engelhardt, H.T., Jr., Cutter, M.A., and Shelp, E. (eds.): *When Are Competent Patients Incompetent?: A Study of Informal Competency Determinations in Primary Care.* Dordrecht/Boston: Kluwer Academic Publishers.

21. Knight, J. (1988) Judgments of incompetence and their moral presuppositions. In Engelhardt, H.T., Jr., Cutter, M.A., and Shelp, E. (eds.): *When Are Competent Patients Incompetent?: A Study of Informal Competency Determinations in Primary Care.* Dordrecht/Boston: Kluwer Academic Publishers.

22. Meisel, A. (1979) The exceptions to the informed consent doctrine: striking a balance between competing values in medical decision-making. *Wisc Law Rev.* 1979:413–488.

23. Shelp, E. (1982) *Beneficence in Health Care.* Dordrecht/Boston: D. Reidel Co.

24. Cassell, E. (1977) The function of medicine. *Hast. Ctr. Rep.* 16–19.

25. Thomasma, D.C., and Pellegrino, E. (1987) The role of the family and physicians in decisions for incompetent patients. *Theor. Med.* 8:283.

26. Morreim, E.H. (1988) Different notions of autonomy and competence: their implications for medicine and public policy. In Engelhardt, H.T., Jr., Cutter, M.A., and Shelp, E. (eds.): *When Are Competent Patients Incompetent?: A Study of Informal Competency Determinations in Primary Care.* Dordrecht/Boston: Kluwer Academic Publishers.

27. Abernathy, V. (1988) Judgments about patient competence: cultural and economic antecedents. In Engelhardt, H.T., Jr., Cutter, M.A., and Shelp, E. (eds.): *When Are Competent Patients Incompetent?: A Study of Informal Competency Determinations in Primary Care.* Dordrecht/Boston: Kluwer Academic Publishers.

28. Veatch, R. (1984) An ethical framework for terminal care decisions: a new classification of patients. *J. Am. Ger. Soc.* 32:665–669.

29. Beauchamp, T., and McCullough, L. (1984) *Medical Ethics: The Moral Responsibilities of Physicians.* Englewood Cliffs, N.J.: Prentice-Hall, Inc.

30. Robertson, J. (1988). The geography of competency. In Engelhardt, H.T., Jr., Cutter, M.A., and Shelp, E. (eds.): *When Are Competent Patients Incompetent?: A Study of Informal Competency Determinations in Primary Care.* Dordrecht/Boston: Kluwer Academic Publishers.

31. Thomasma, D.C. (1990) *Human Life in the Balance.* Louisville, KY: Westminster Press.

32. Hardwig proposes to "reconstruct medical ethics in light of family interests." See Hardwig, J. (1990) What about the family? *Hast. Ctr. Rep.* 20(2):5–10; quote: 10.

33. Veatch, R. (1984) An ethical framework for terminal care decisions: a new classification of patients. *J. Am. Ger. Soc.* 32:665–669.

34. Pellegrino, E.D., and Thomasma, D.C. (1988) *For the Patient's Good: The Restoration of Beneficence in Health Care.* New York: Oxford University Press.

35. Thomasma, D.C. (1990) The ethics of caring for vulnerable individuals. In *Reflections on Ethics.* Washington, DC: American Speech-Language-Hearing Assoc. 39–45. Reprint in National Student Speech Language Hearing Association Journal 1992–93; 20:122–124.

36. Thomasma, D.C. (1994) Euthanasia as power and empowerment. In Bland, R., and Bonnicksen, A. (eds.): *Medicine Unbound: The Human Body and the Limits of Medical Intervention.* New York: Columbia University Press. pp. 210–227.

Patients' Competence to Consent to Neurobiological Research

Paul S. Appelbaum

Department of Psychiatry, University of Massachusetts Medical Center, Worcester, MA 01655

Progress has been made defining the standards for determining subjects' competence to consent to research: abilities to communicate a choice, understand relevant information, appreciate the nature of the situation and its consequences, and manipulate information rationally. Available data show clearly that persons with mental illness display a spectrum of decisionmaking abilities, with many performing well, but some doing quite poorly. More attention now is required to identifying the degree of capacity required for competent consent to projects with varying risk/benefit characteristics. Practical means are also required for screening for subjects with impaired capacities, attempting to improve their performance, and providing substituted consent when their deficiencies are intractable.

Keywords: Competence, informed consent, research, decisionmaking

Federal regulations and common law have combined to make informed consent a prerequisite to participation by potential subjects in research protocols (Appelbaum, Meisel, and Lidz, 1987; Code of Federal Regulations, 1991). This requirement rests on the belief that it is illegitimate to use persons for research purposes without obtaining a consent that reflects their interests and desires. Although exceptions may exist—e.g., when projects pose minimal risks and obtaining consent would make the projects nearly impossible to pursue—solicitation of consent is now taken for granted in almost all research settings.

For consent to be valid (that is, for it genuinely to reflect the interests and desires of the subject), several elements must be in place. Sufficient *disclosure* must occur for potential subjects to have available the information material to a decision regarding participation. This requirement is premised on the commonsense view that a decision taken without knowledge of the relevant facts will not likely vindicate the interests and desires of the person. Considerable effort is devoted in the federal regulations to identifying the categories of information that must be disclosed to accomplish this end, and institutional review boards typically devote the greatest part of their efforts to insuring that consent forms contain the detail they consider requisite.

Assuming adequate disclosure has occurred, potential subjects must be permit-

This article will also appear in the journal: *Accountability in Research*, Volume 4, pp. 241–251 (1996).

ted to make a decision free of coercion, undue influence, and unfair manipulation. That is, the decision they reach must be made in a *voluntary* manner, at least insofar as illegitimate pressure from treaters or members of the research team is excluded. This is aimed at insuring that the potential subject's desires take precedence in the decision, rather than the wishes of the researchers or treatment team.

Finally, potential subjects must be *competent* to make a decision. By competent, we mean that subjects must have sufficient cognitive capacities to make it possible for them to reach decisions in a rational fashion. This final element of a valid consent is derived from the view that some minimal degree of rationality in decision making is required if individual interests are to be protected; otherwise, decisions will be made in an essentially random way with regard to the interests of the subject.

It is this last element of informed consent—competence—that is my focus here. One might expect that issues related to competence might be a particular problem in neurobiological research with human subjects (Shamoo and Irving, 1993). Many of the target subjects of such research, after all, will have disorders that impair the very cognitive faculties on which they must rely for decision making with regard to research participation. These disorders may render some proportion of potential subjects incompetent to give consent for research, requiring that they not be entered into studies (even if they appear to be agreeing to participate) or that alternative mechanisms for authorizing their participation be developed.

My goal in this paper is to describe the standards that have evolved by which persons' competence to make decisions such as whether to participate in research can be determined; to review the data on the degree to which target populations of neurobiological research are likely to fail to meet such standards; and to suggest means of dealing with the impairments in competence-related functions that may exist in groups of potential research subjects.

STANDARDS FOR DETERMINING DECISIONMAKING COMPETENCE

Over the last twenty years, a period during which increased attention has been given to the concept of decisionmaking competence, a rough consensus has been reached as to the applicable standards. This consensus is embodied in case law, and the legal, medical, psychological, and ethical literature. Although none of the case law, and little of the literature, focuses on competence to make decisions with regard to research, there is a fair body of court decisions and writing related to competence to make treatment decisions (Appelbaum and Grisso, 1995; Cutter and Shelp, 1991).

There are several reasons why it is not unreasonable to apply the standards developed for competence to consent to treatment to questions of decisionmaking related to research. First, many research projects involve administration of treatment, making the treatment decision and the research decision one and the same. Second, the legal standards elaborated for competence to consent to treatment are closely related to standards for other decisionmaking tasks, such as making contracts (White and Denise, 1991), writing wills, and giving gifts, perhaps even for decisions related to criminal defense (Bonnie, 1993). Indeed, although this is not the place to elaborate the argument, it is probably the case that legal notions of decisionmaking competence are not specific to any category of decision, but encompass decision tasks in general.

When we turn from the legal literature to medical, psychological and bioethical writings, we find a similar commonality of approach. Commentators do not distinguish between competence for treatment and for research. Although the type of information to be revealed may differ in the two situations, and different procedures may be necessary to avoid infringing persons' voluntary choice, the capacities needed to make a meaningful decision are much the same. An analysis of commentaries on competence to consent to research, in fact, revealed identical standards to those used in the treatment setting (Appelbaum and Roth, 1982).

In general, four standards have been applied to determine decisionmaking competence (Appelbaum and Grisso, 1988; 1995). There is some variation across legal jurisdictions and among commentators as to which of these standards are endorsed, but almost all select one or more of these options. A good deal of the variation, moreover, may be due to the facts of a particular case (this is certainly true of the case law), to which analyses of competence are often highly reactive.

The first standard applied to assess competence asks whether the person has the *ability to communicate a choice*. This is the least stringent measure of decisionmaking adequacy, requiring only that the person offer evidence of their wishes. Few would quarrel with the conclusion that a person who, because of mental disorder, cannot reach a decision or make that decision known to the outside world, ought not to be afforded the power to guide his or her own affairs. Mute catatonic or severely depressed patients, patients with manic or catatonic excitement, and patients with severe psychotic thought disorder may fall into this category. Some patients who can evidence a choice but, because of illness, cannot maintain that choice for a sufficient period of time for it to be implemented may also be considered impaired on this dimension.

Ability to understand relevant information is the second and most common standard for decisionmaking competence. There is a commonsense feel to the contention that a person who cannot comprehend the information material to a decision is in no position to make that decision. In both treatment and research settings, the material information is that which must be disclosed according to the rules of informed consent. Impairments of intelligence, attention, and memory, whether due to organic or functional disorders, can affect this ability.

A third standard is the *ability to appreciate the nature of the situation and its likely consequences*. This standard differs from the ability to understand relevant information by insisting that persons be able to apply the information that they understand abstractly to their own situation. Thus, persons who understand that their physicians believe they are ill, but in the face of evidence that would persuade a reasonable person, deny that this is so; or who understand that treatment is recommended, but refuse to accept that it may be beneficial for them, may pass a test of understanding, but fail to meet the appreciation standard. Denial (often called "lack of insight"), delusions, and psychotic levels of distortion can impair appreciation.

Finally, the *ability to manipulate information rationally* is the fourth standard applied to determine decisionmaking competence. Relevant here is the person's ability to employ logical processes to compare the risks and benefits of treatment or research participation. What is examined here is not the outcome of the person's decision process, i.e., whether the person has made an objectively "good" choice. Rather it is the process that is considered, with an emphasis on whether it adequately allows the person's own values to come into play in weighing alternative courses of action.

As noted above, courts and commentators typically select several of these standards to formulate a compound standard of decisionmaking competence. Unless regulations, statutes, or case law in a particular jurisdiction indicate otherwise, it is not unreasonable to assume that each of these standards may be relevant to determining the competence of a potential research subject.

RESEARCH ON IMPAIRMENTS OF DECISIONMAKING COMPETENCE

How well do persons with mental disorders, who frequently will be recruited to be subjects in neurobiological research, perform on these standards of decisionmaking competence? The literature on patients' performance in the context of consent to research is sparse. Fortunately, the somewhat larger literature on consent to treatment can be called on here to round out the picture. Indeed, recent data provide the most comprehensive look we have had at the comparative performance of persons with mental illness on measures related to decisionmaking competence.

In a recent paper, my colleague Tom Grisso and I reviewed the empirical literature on the decisionmaking competence of persons with psychiatric disorders (Appelbaum and Grisso, 1995). The following draws on that review; the interested reader is referred to the original paper for additional details.

With regard to the ability of psychiatric patients to meet the standard of *communicating a choice*, an irony is evident: to participate in the very studies that attempt to assess this ability, patients must offer consent to researchers. Persons so impaired that they cannot even communicate their decisions are selectively excluded from further study. Nonetheless, one study of newly admitted patients in a community mental health center reported that 9% were mute or catatonic when approached to participate in a study of decisionmaking capacity (Appelbaum, Mirkin, and Bateman, 1981). Another study that presented inpatients, mostly with depression, with vignettes that required resolution of dilemmas noted that 18% of subjects were "incapable of deciding" (Radford, Mann, and Kalucy, 1986). These figures likely will vary according to the nature and degree of illness of study samples. However, this population is probably least problematic from the perspective of neurobiological researchers: potential subjects' inability to offer consent will be evident and consent will be sought elsewhere or subjects excluded from the study.

Ability to understand relevant information has been explored in numerous studies. The first generation of such work, however, generally involved querying psychiatric patients about their illnesses and treatment without ascertaining that the appropriate information had been communicated to them. Although performance in these studies was generally poor, it is difficult to know how much to make of the findings. A second set of studies examined patient understanding after information disclosure. Here, too, significant proportions of patients, sometimes more than 50%, had substantial difficulty understanding disclosed information.

Two studies by Loren Roth and his colleagues, which focused on understanding by psychiatric patients of information related to participation in research, exemplify the best of this work. In the first study, 41 patients with affective disorders (of whom only 5 were psychotic) were asked to participate in a study of their EEG patterns during sleep (Roth, Lidz, Meisel, Soloff, Kaufman, Spiker, and Foster, 1982). Approximately one-quarter of the subjects understood half or less of the disclosed information, only one-half understood more than two-thirds of the information, and a mere 5% understood 87% or more of the information. Four of 19 subjects whose consent discussions were videotaped were rated as probably incompetent by independent judges. Of interest is that results were essentially the same for a sample of 57 patients who were being asked to consent to ECT for treatment purposes.

The second study (Benson, Roth, Appelbaum, Lidz, and Winslade, 1988) examined understanding of 88 psychiatric patients recruited to participate in four different psychiatric research projects. Data analysis revealed that "[w]hile prospective research subjects generally demonstrated good understanding of the purpose of the written consent form and their right to refuse or withdraw from the study ... they frequently did not understand the psychiatric research project's purpose, or why they had been asked to participate in it. Subjects also often demonstrated poor understanding of important methodological aspects of the study, including the randomized and double-blind treatment assignment." (p. 469) These findings suggest that how well we rate the ability of psychiatric patients to understand information in research settings will depend on what types of information we consider it most important for them to comprehend. The study also found that patients with schizophrenia and those with high levels of psychopathology (measured by the Brief Psychiatric Rating Scale) performed more poorly than patients with borderline personality disorder and generally less impaired subjects.

The *ability to appreciate the nature of a situation and its likely consequences* has been subject to extensive exploration in psychiatric populations because of its close connection to core symptoms of mental disorders. Thus, inability to appreciate that one is ill (termed "lack of insight" or "denial of illness") has been identified as a key diagnostic feature of schizophrenia in two multi-national studies (Carpenter, Bartko, Carpenter, and Straus, 1976; Wilson, Ban and Guy, 1986), although rates of poor insight have varied from 27% (Wing, Monck, Brown, and Carstairs, 1964) to 97% (Carpenter *et al.*, 1976) of schizophrenic populations studied. Schizophrenic patients also have been reported to have difficulty appreciating their need for treatment (McEvoy, Apperson, Appelbaum, Ortlip, Brecosky, Hammill, Geller, and Roth, 1989). In contrast, depressed patients typically are assumed to be impaired in their ability to appreciate the potential value of treatment, because of the hopelessness that so often accompanies the disorder.

Appreciation problems have been demonstrated directly in connection with consent to research. Potential subjects with psychiatric disorders often fail to appreciate the nature of the research situation and its potential impact on their treatment. Although they may understand abstractly what is revealed in consent forms, subjects often fail to appreciate that research methods may sacrifice personalized decisionmaking about their care for the sake of producing generalizable data. Because of what my colleagues and I have referred to as "the therapeutic misconception," research subjects may lack appreciation of the implications of randomized assignment, double-blind procedures, placebo-control groups, and fixed treatment protocols (Appelbaum, Roth, Lidz, Benson and Winslade, 1987). Recent controversies raise the question as to whether the consequences of wash-out periods and drug discontinuation protocols might fall into this category as well (Willwerth, 1994). Whether psychiatric patients are particularly likely to misconceive the nature of research procedures, compared with medical patients in general, remains unclear.

Studies of psychiatric patients' *ability to manipulate information rationally* have been few, and often have relied on paradigms that bear little resemblance to decision making in clinical or research contexts. If one takes decision outcome as a proxy for rationality of decision process, two studies by Barbara Stanley and colleagues are relevant here. Looking at a group of psychiatric patients with mixed diagnoses who were asked whether they would participate in hypothetical research projects, Stanley et al. found that 40% of acutely hospitalized patients said they would agree to take part in high risk/low benefit projects, while up to 32% refused low risk/high benefit participation (Stanley, Stanley, Peselow, Wolkin, Deutsch, Platt, Speicher, Golash, and Kaufman, 1982). While this suggests poor performance in the weighing of risks and benefits, a similar study that compared psychiatric and medical inpatients found no differences between the groups (Stanley, Stanley, Lautin, Kane, and Schwartz, 1981).

The findings of the recently completed MacArthur Treatment Competence Study, which addressed all four standards of decisionmaking competence, are worth noting here as well (Appelbaum and Grisso, 1995; Grisso, Appelbaum, Mulvey, and Fletcher, 1995; Grisso and Appelbaum, 1995). In contrast to most previous studies, which demonstrated impairments in the performance of persons with psychiatric disorders but did not include comparison groups, the MacArthur study compared newly hospitalized patients with schizophrenia, major depression, and angina pectoris with community controls matched for age, gender, race, education, and occupation. Nearly 500 subjects were studied. Since, from the policy perspective, competence is not judged in isolation, but in comparison to how well most people would perform at the task, this approach provides crucial information for identifying impairments with potential policy significance.

Instruments intended to measure performance on each standard were designed. Few subjects (5%) had difficulty *evidencing a choice*, as would be expected in a study that required preexisting subject consent. On measures of *understanding*, based on disclosure of treatment-related information, schizophrenic subjects did significantly worse than depressed or angina subjects, who did not differ from each other. Concretely, 28% of schizophrenic subjects fell into a range we defined as impaired (below two standard deviations of the mean of all subjects combined), compared with only 5% of depressed subjects and 2.4% of controls.

Appreciation was measured both in regard to the presence of illness and the likely benefit of treatment. Again, schizophrenic subjects performed significantly worse than the two other groups (control subjects were not used with this instrument), although depressed patients, too, significantly underperformed compared with medical patients. Twenty-three percent of schizophrenic subjects fell into the impaired range, along with 12% of depressed subjects. As might be expected, schizophrenic subjects were more likely to deny the presence of illness than were depressed patients (35% vs. 4%), while similar segments of each diagnostic category denied the potential for effective treatment (13% vs. 14%).

Finally, testing of *rational manipulation of information* or "reasoning" showed results very similar to those for understanding, with schizophrenic subjects again performing significantly worse than other patient groups and their own controls. Depressed patients did significantly worse than angina patients and than their control group. In percentage terms, 24% of schizophrenic subjects, 8% of depressed subjects, and 2% of control subjects fell into the impaired range. Impaired performance for schizophrenic patients on understanding, appreciation, and reasoning was related to scores on measures of conceptual disorganization, unusual (i.e., delusional) thought content, and to a lesser extent, hallucinations.

In sum, existing data suggest that persons with psychiatric disorders, especially schizophrenia, manifest impairment of competence-related abilities with some frequency. On the other hand, the majority of schizophrenic and depressed patients in the MacArthur Treatment Competence Study performed in the same range on each individual measure as their non-ill controls. Thus, the glass can be seen as either half-empty or half-full. It should be noted, though, that the data likely underestimate the degree of impairment among psychiatric patients, since the most disturbed patients tend to be excluded from participation in studies of decisionmaking abilities.

COMPETENCE-RELATED IMPAIRMENTS IN NEUROBIOLOGICAL RESEARCH

What are the implications of these findings for the recruitment of subjects in research on psychiatric and other neurobiological disorders? Clearly, the available data do not support the conclusions at either extreme of the spectrum. It cannot reasonably be maintained that psychiatric patients as a class are incompetent to offer (or refuse) their informed consent for participation in research. Many persons with psychiatric disorders, perhaps most, retain decisionmaking abilities indistinguishable from those found in the general population. Neither, however, can it be argued that decisional incompetence is simply not a problem among psychiatric patients. Substantial impairment is not uncommon.

How then should policymakers react to current concern that incompetent patients may be consenting to research participation in violation of the principles of informed consent and to their personal detriment? Whatever action is taken should be sensitive to the balance of interests involved. Protection of the rights and well-being of potential subjects is of great importance. So too, however, is the advancement of knowledge regarding disorders affecting the brain. As we guard patients' rights and interests, the incremental burden of additional protections on the conduct of research must be taken into account. One more safeguard can always be added. The key question is whether the additional protection it provides is worth the potential negative impact on the advance of knowledge. Economists would call this a "cost-benefit analysis." In some cases, additional protective measures will meet this test, in others, undoubtedly, not.

Within that framework, I submit that efforts should be made in four areas to address problems related to competence to consent to neurobiological research: more precise formulation of the degree of capacity required for competent consent to research; development of means for identifying persons who may lack the requisite capacities; implementation of approaches to compensate for decision-making impairment among potential research subjects; and design of mechanisms for more easily obtaining substituted consent when that appears to be appropriate. I consider each in turn.

First, with regard to formulating the degree of capacity required for competent consent, it is evident that subgroups of patients with psychiatric disorders, perhaps particularly those with the greatest levels of thought disturbance, perform much worse than control subjects on competence-related measures. Are they therefore incompetent to consent to participation in research? Perhaps surprisingly, it is difficult to say. Although a consensus has developed regarding which abilities are relevant to decisionmaking competence, almost no attention has been paid to the question of where the cut-offs should be. That is, below what level of performance should we say that a person is not competent to give or withhold consent?

To date, resolution of this issue has been left to the decentralized mechanisms we have developed for overseeing research. Institutional review boards (IRBs) may ask investigators to provide assurance that incompetent subjects will be excluded from participation, but I am unaware of any IRB that has specified how to define that population. (My unawareness, of course, does not mean that such events have not occurred, only that, if they have, they are uncommon and poorly publicized.) Investigators themselves, my experience indicates, are often untutored in the nuances of assessing competence, and usually leave recruitment in the hands of research assistants. The latter may be given a general charge regarding exclusion of incompetent subjects, but subsequent monitoring is made difficult by the absence of clear performance standards (Appelbaum and Roth, 1983).

Demanding total comprehension and appreciation, along with high-level reasoning abilities, from psychiatric patient populations is unrealistic because most members of the general population would fail such tests. Comparisons with non-patient groups, suitably matched, might provide a basis for identifying levels of performance that fall so far below the general norm as to justify calling a person "incompetent"—a label that our society, with its emphasis on protecting individual autonomy, does not lightly bestow.

A strictly statistical measure of competence, however, is likely to be unsatisfactory for two reasons. First, despite generally good performance, there may be some functions so critical to a competent consent that we would not feel comfortable allowing a person to make a decision about research participation in their absence. For example, whatever else they understand, should subjects who fail to comprehend that they are being asked to enter a research project ever be considered competent? Similarly, if they lack appreciation of the differences between ordinary treatment and research—for example, randomized assignment to placebos—might that not be considered an insuperable obstacle to a competent consent?

The second reason why a narrow, statistical measure of competence may be inadequate is that we may be inclined to require different levels of capacity depending on the nature of the research project to which consent is being sought. As has been suggested in the context of treatment (Drane, 1984; President's Commission, 1982), it may be reasonable to slide standards or cut-offs (Grisso, 1986) according to the risks that may accrue if an incompetent decision is made. For minimal risk research with no therapeutic benefit, it may be reasonable to allow persons to make their own decisions, even in the face of considerable impairment, knowing that adverse consequences are unlikely to ensue whichever way they decide. As the risks of either participating or not participating rise, however, more demanding levels of capacity may be required.

To date, we have elected not to work these problems out in a uniform manner, with the result that practices vary widely across research projects. Moreover, our inability to say firmly what level of ability is required in any project makes it difficult to supervise investigators, and unfair to criticize them for not following standards that may be formulated only after the fact. Were general standards to be developed, they would have to be flexible enough to allow application in a wide variety of research projects with very different patient populations. Although difficult, this should not be an impossible task. Representatives of the research community, patients and their advocates, ethicists and regulators should be involved, with a broad consensus the goal. If accomplished, it would provide greater assurance to patients and their advocates that subjects likely to be incompetent were not being allowed to consent to research; and it would alleviate investigators' concerns that their subject recruitment practices were subject to unfair *post facto* review.

Once clear criteria exist for determining whether potential subjects are competent to consent to research, one can envision the next step in the process: development of means for identifying persons who may lack the requisite capacities. Now, except in the most extreme cases, investigators are left in something of a quandary about who should be excluded from ordinary consent procedures. The creation of generally agreed-upon criteria would permit the development of screening mechanisms for identifying subjects at risk of incompetence, and perhaps more detailed means of assessment for that high-risk population. Screening might be based on clinical judgment, on measures of psychopathology shown to correlate with significant impairment (recall the finding in the MacArthur Treatment Competence Study that disorders of thought predicted poorer performance) (see also Schachter, Kleinman, Predergast, Remington, and Schertzer, 1994), or on tests aimed at consent-related abilities *per se*. My colleague Tom Grisso and I are now testing a condensed version of our competence assessment instruments for use in clinical treatment settings. It is not far-fetched to imagine similar devices being designed to assist in the assessment of competence to consent to research when subjects at elevated risk for impaired decision making are identified.

Development of policies for identification of potential subjects with impaired capacities should reflect the cost-benefit analysis suggested earlier. Formal screening of the capacities of all potential research subjects would be an enormously costly and time-consuming endeavor. It may be worthwhile, for example, in studies employing interventions of known high risk or uncertain risk/benefit ratios, or recruiting extremely impaired populations who run some risk of harm. In many cases, however, it seems reasonable to rely on the judgment of trained research assistants who are responsible for patient recruitment, with investigators encouraged to monitor their decisions.

A third effort in this area should involve the implementation of approaches to compensate for decisionmaking impairment among potential research subjects. Persons' decisionmaking abilities are not fixed, but highly context dependent. Categorization of a potential subject as impaired does not mean that the person will be unable, under all circumstances, to offer a competent consent for research participation. Indeed, several studies now suggest that modifications of disclosure methods can significantly improve potential subjects' understanding (Grisso and Appelbaum, 1995; Benson, Roth, Appelbaum, Lidz, and Winslade, 1988). Among these methods are disclosure of information in smaller segments, use of video-taped disclosures to augment discussions with researchers, and the employment of independent educators, whose sole job is to teach subjects about the studies to which they are being asked to consent. Other approaches, drawn from the considerable literature on patient education in clinical settings, may be helpful as well. To be sure, the most highly impaired subjects are unlikely to benefit from these techniques, but they may permit potential subjects of marginal competence to make acceptable decisions.

Finally, it is important to acknowledge that some important research questions may never be answerable without the capacity to involve subjects with severe and irremediable decisionmaking impairment in research. Lacking now are generally accepted and easily accessed mechanisms for including such persons in studies. As a result, investigators' practices have varied around the country. Sometimes patients' impairments are ignored, and they are permitted to consent to participation. Familial consent may be accepted in some cases, or such patients simply omitted from the recruitment pool in others. The most rigorous procedure from a legal point of view—declaration of incompetence by a court, and appointment of a guardian with the power to consent to research participation—is rarely pursued because of its costs. Recognizing that definitive resolution of this problem will be difficult because it deals with issues that are largely subject to state law, creative approaches to authorizing surrogate decision makers to act for patients in a manner protective of their interests would be welcome.

REFERENCES

Appelbaum, P.S., and Grisso, T. (1995) The MacArthur Treatment Competence Study, I: mental illness and competence to consent to treatment. *Law and Human Behavior* 19, 105–126.

Appelbaum, P.S., Lidz, C.W., and Meisel, A. (1987) *Informed Consent: Legal Theory and Clinical Practice.* New York: Oxford University Press.

Appelbaum, P.S., and Roth, L.H. (1982) Competency to consent to research: a psychiatric overview. *Archives of General Psychiatry* 39, 951–958.

Appelbaum, P.S., and Roth, L.H. (1983) The structure of informed consent in psychiatric research. *Behavioral Sciences and the Law* 1, 9–19.

Appelbaum, P.S., Roth, L.H., Lidz, C.W., Benson, P., and Winslade, W. (1987) False hopes and best data: consent to research and the therapeutic misconception. *Hastings Center Report* 17, 20–24.

Benson, P.R., Roth, L.H., Appelbaum, P.S., Lidz, C.W., and Winslade, W.J. (1988) Information disclosure, subject understanding, and informed consent in psychiatric research. *Law and Human Behavior* 12, 455–476.

Bonnie, R.J. (1993) The competence of criminal defendants: beyond Dusky and Drope. *University of Miami Law Review* 47, 539–601.

Carpenter, W.T., Bartko, J.J., Carpenter, C.L., and Strauss, J.S. (1976) Another view of schizophrenia subtypes: a report from the International Pilot Study of Schizophrenia. *Archives of General Psychiatry* 33, 508–516.

45 Code of Federal Regulations 46 (1991).

Cutter, M.A.G., and Shelp, E.E. (eds.) (1991) *Competency: A Study of Informal Competency Determinations in Primary Care.* Dordrecht: Kluwer Academic Publishers.

Drane, J.F. (1984) Competency to give an informed consent: a model for making clinical assessments. *Journal of the American Medical Association* 252, 925–927.

Grisso, T. (1986) *Evaluating Competencies: Forensic Assessments and Instruments.* New York: Plenum Press.

Grisso, T., Appelbaum, P.S., Mulvey, E.P., and Fletcher, K. (1995) The MacArthur Treatment Competence Study, II: measures of abilities related to competence to consent to treatment. *Law and Human Behavior* 19, 127–148.

Grisso, T., and Appelbaum, P.S. (1995) The MacArthur Treatment Competence Study, III: abilities of patients to consent to psychiatric and medical treatment. *Law and Human Behavior* 19, 149–174.

McEvoy, J.P., Apperson, L.J., Appelbaum, P.S., Ortlip, P., Brecosky, J., Hammill, K., Geller, J.L., and Roth, L.H. (1989) Insight in schizophrenia: its relationship to acute psychopathology. *Journal of Nervous and Mental Diseases* 177, 43–47.

President's Commission for the Study of Ethical Problems in Medicine and Biomedical and Behavioral Research (1982) *Making Health Care Decisions,* Vol. 1, Washington, D.C.: U.S. Government Printing Office.

Roth, L.H., Lidz, C.W., Meisel, A., Soloff, P.H., Kaufman, K., Spiker, D.G., and Foster, F.G. (1982) Competency to decide about treatment or research: an overview of some empirical data. *International Journal of Law and Psychiatry* 5, 29–50.

Schachter, D., Kleinman, I., Prendergast, P., Remington, G., and Schertzer, S. (1994) The effect of psychopathology and the ability of schizophrenic patients to give informed consent. *Journal of Nervous and Mental Disease* 182, 360–362.

Shamoo, A.E., and Irving, D.N. (1993) Accountability in research using persons with mental illness. *Accountability in Research* 3, 1–17.

Stanley, B., Stanley, M., Lautin, A., Kane, J., and Schwartz, N. (1981) Preliminary findings on psychiatric patients as research participants: a population at risk? *American Journal of Psychiatry* 138, 669–671.

Stanley, B., Stanley, M., Peselow, E., Wolkin, A., Deutsch, S., Platt, L., Speicher, J., Golash, L., and Kaufman, M. (1982) The effects of psychotropic drugs on informed consent. *Psychopharmacology Bulletin* 18, 102–104.

White, P.D., and Denise, S.H. (1991) Medical treatment decisions and competency in the eyes of the law: a brief survey. In M.A.G. Cutter & E.E. Shelp (eds.), *Competency: A Study of Informal Competency Determinations in Primary Care.* Dordrecht: Kluwer Academic Publishers.

Willwerth, J. (1994) Madness in fine print. *Time,* November 7, 1994, 62–63.

The International Covenant on Civil and Political Rights and the Rights of Research Subjects

Eric Rosenthal

Director, Mental Disability Rights International, Washington College of Law, American University, 4400 Massachusetts Avenue, N.W., Washington, D.C. 20016-8084

INTRODUCTION

In June 1992, the United States ratified the International Covenant on Civil and Political Rights (hereinafter the "ICCPR"; see Senate Committee on Foreign Relations, 1992). Drafted by an international committee over a number of years after the Second World War, the ICCPR entered into force in 1976, and it slowly gained recognition as countries ratified it over the next two decades. Today, more than 115 countries representing more than two thirds of the world's population have ratified this international convention (Posner, 1992, p. 1212).

Article 7 of the ICCPR prohibits medical or scientific experimentation on human subjects without their "free consent." Despite the ICCPR's recognition of consent as a fundamental human right, the ICCPR has been largely overshadowed by the Nuremberg Code (1947) and the World Medical Association's Declaration of Helsinki (1965, amended 1989) as an international standard relating to the use of human experimentation in research. In part, this is because the ICCPR's brief reference to experimentation lacks sufficient detail to guide researchers seeking to design a proper experiment, as does the Nuremberg Code or the Declaration of Helsinki (Bassiouni *et al.*, 1981, p. 1662). In addition, as the Nuremberg Code, the Declaration of Helsinki, and subsequent international standards were developed, the ICCPR had not yet gained the status of binding legal authority in many of the countries in which research was conducted.

The international legal landscape has been changed by the widespread ratification of the ICCPR, and the ICCPR is now a binding legal instrument that governs experimentation on human subjects throughout most of the world, including all major industrialized countries. Indeed, the ICCPR is now widely considered the most important international treaty and the cornerstone of international human rights law (Nowak, 1993, p. xxviii). However, the ICCPR's provision on consent

This article will also appear in the journal: *Accountability in Research*, Volume 4, pp. 253–260 (1996).

remains largely unknown by the research community. Particularly in countries that have ratified the ICCPR, it is incumbent upon researchers and regulatory authorities to understand the ICCPR's protections for individuals subject to experimentation.

The use of human subjects in medical experimentation can be crucial to developing important new medical treatments, and the protection of the rights of people subject to such experimentation is complex. Crafting a meaningful definition of consent (including such elements as competence, knowledge, and voluntariness) is difficult, and this must be complimented with mechanisms to ensure oversight and enforcement of those rights. On its face, Article 7 of the ICCPR lacks sufficient detail to guide researchers and protect research subjects in all cases, and it is no substitute for detailed ethical standards, independent oversight boards, and legal enforcement mechanisms. However, the ICCPR is subject to interpretation by the United Nations Human Rights Committee. Most important, the ICCPR establishes that some form of "free consent" is an absolute minimum right of research subjects recognized by international human rights law. As I will describe, the ICCPR creates an obligation on States to make the right to free consent legally enforceable.

The United Nations has recently begun to develop and clarify the exact requirements of the right to free consent as protected by the ICCPR. The relevance of the ICCPR will continue to grow as the United Nations machinery for the ICCPR's interpretation and oversight becomes increasingly developed and respected.

THE ICCPR'S PROTECTION OF RESEARCH SUBJECTS

The ICCPR's provision on human experimentation appears in article 7, which reads in full:

No one shall be subjected to torture or to cruel, inhuman or degrading treatment or punishment. In particular, no one shall be subjected without his free consent to medical or scientific experimentation.

In the context of its broad protections of the spectrum of civil and political rights, the language of the ICCPR is strikingly specific in its reference to experimentation. The ICCPR's drafters were influenced by evidence of atrocities by Nazi physicians being heard at the Nuremberg trial, which was in session on June 18, 1947, when the language on human experimentation was first proposed (Perley, Fluss, Bankowski, and Simon, 1992, p. 153). At the same time, the language of article 7 is absolute and does not limit the protection to cases of intentional abuse on the Nazi scale (McGoldrick, 1991, p. 366). By describing non-consensual experimentation as a "particular" form of torture or inhuman treatment or punishment, the drafters linked the practice to one of the most fundamental prohibitions in international human rights law. As specified in the ICCPR, article 7 is among the few provisions that allows no exceptions and cannot be limited ("derogated") under any circumstances, including cases of national emergency [article 4(2)].

The process of interpreting the ICCPR's requirements has been greatly improved since 1981, when the United Nations Human Rights Committee began issuing its "General Comments" pursuant to its authority under Article 40 of the ICCPR. The General Comments are not binding interpretations of the ICCPR because the

Human Rights Committee is not authorized to issue enforceable decisions in any specific case (McGoldrick, 1991, pp. 92–94). However, the General Comments of the Human Rights Committee have gained widespread recognition as "the most authoritative interpretation of the covenant's provisions" (Nowak, 1993, p. xix).

General Comment 20, adopted by the Human Rights Committee in 1992 provides guidance on the ICCPR's provision on experimentation:

Article 7 expressly prohibits medical or scientific experimentation without the free consent of the person concerned.... The Committee also observes that special protection in regard to such experiments is necessary in the case of persons not capable of giving valid consent, and in particular those under any form of detention or imprisonment. Such persons should not be subjected to any medical or scientific experimentation that may be detrimental to their health.

By calling for the special protection of people "not capable of giving valid consent," General Comment 20 interprets the right to "free consent" in the language of article 7 to include a competence component. The General Comment reads article 7 in a way that is consistent with the Nuremberg Code, adopted shortly after this provision of article 7 was first drafted, which uses the term "voluntary consent" and defines it to require competent and knowing consent (Nuremberg Code, 1947, section 1). With regard to experimentation that "may be detrimental" to the health of the subject, General Comment 20 flatly prohibits any participation in experimentation on a sub-population of detained individuals who are not competent to give consent. This is presumably based on the theory that detention is inherently a coercive environment.

The General Comments' references to people "in any form of detention" clearly includes people involuntarily detained in psychiatric hospitals. General Comment 20(5) states that "article 7 protects, in particular ... patients in teaching and medical institutions".

Unfortunately, the language of General Comment 20 is ambiguous regarding the class of people to be excluded from all experimentation. "Such persons"—people who should not be subject to any experimentation—could refer to all people not capable of giving valid consent, a broad class that "in particular" would include any person in any form of detention. General Comment 20 could also be read more narrowly, however, to exclude only that sub-class of people who are both legally incapable of consent and who are detained.

Under the latter, narrow reading of General Comment 20, exclusion from experimentation would be limited to circumstances in which all of the following three elements are present: (1) a person is under "any form of detention," (2) he or she lacks competence to consent, and (3) the experimentation may be detrimental to his or her health. This reading of General Comment 20 leaves a number of questions unanswered. The ICCPR would require "special protection" for non-detained people who lack competence to consent as well as detained individuals who are competent to consent. The General Comment does not specify what "special protection" for consent would be required.

The Human Rights Committee should clarify the exact requirements of article 7 when it next issues General Comments, and it can provide more immediate guidance by adopting a clear statement on the ICCPR's requirements when it considers the United States record. In a critique of the United States record submitted to the Human Rights Committee by Mental Disability Rights International (MDRI), a non-governmental advocacy organization that I direct, MDRI asked the Human Rights Committee to clarify the language of General Comment 20 (MDRI, 1995, p. 75). If the Human Rights Committee fails to adopt such clarifying language, other sources of international law and practice can be used to help interpret the article 7 requirements. A number of important sources exist, the *travaux préparatoires* (the equivalent of the legislative history of the ICCPR), customary international law (which may include the Nuremberg Code), and United Nations General Assembly Resolutions such as the Principles for the Protection of Persons with Mental Illness (Rosenthal & Rubenstein, 1993, p. 270; Villiger, 1985).

Under a broad or a narrow reading of General Comment 20, article 7 of the ICCPR would prohibit potentially dangerous experimentation on people in detention who lack competence to consent. The Helsinki Declaration contains no such blanket prohibition. Thus, in addition to establishing the right to consent as an absolute minimum required in all cases, the ICCPR creates important new rights for the especially vulnerable class of people who are detained and are not in a position to give truly free consent.

THE ICCPR AS ENFORCEABLE LAW

The ICCPR is a particularly important contribution to the field of human experimentation because, unlike the Nuremberg Code, the Helsinki Declaration, or any other existing standard relating to experimentation, the ICCPR is a human rights convention (a multilateral treaty). Countries that ratify the ICCPR ("States Parties" to the convention) recognize it as international law, and they agree to make the ICCPR's provisions enforceable through their own legal systems. By the terms of the ICCPR, any State Party ratifying the ICCPR "undertakes to respect and to ensure to all individuals within its territory ... the rights recognized in the present Covenant." [Article 2(1)]. This includes adopting "such legislative or other measures" necessary to "give effect to the rights recognized in the present Covenant." [Article 2(2)], and to ensure individuals a right to "an effective remedy" for the violation of these rights through "competent judicial, administrative or legislative authorities" [Article 2(3)]. Within the United States, ratification of the treaty makes the document the legal equivalent of a federal law, the "supreme Law of the land ..." United States Constitution, article VI §2 (see Janis, 1988, p. 72).

One major limitation of the ICCPR is the lack of an international mechanism for enforcement. However, the ICCPR does create a system of international oversight through the United Nations Human Rights Committee (the "Committee"). The ICCPR requires States Parties to report to the Committee on "measures they have adopted which give effect to the rights" recognized in the ICCPR, and the ICCPR empowers the Committee to issue General Comments to clarify the requirements of the ICCPR. (Article 40). The first U.S. report on enforcement of the ICCPR was submitted to the Committee on July 28, 1994, and hearings will be conducted in March of 1995 (Department of State, 1994). These hearings can provide an opportunity to bring public attention to the enforcement of internationally recognized human rights in reporting countries, and they have helped in the past to elicit relevant information about compliance (Shelton, 1986, p. 416) [see postscript to this article]. They require States Parties to go on record about their efforts to enforce the ICCPR, and they provide the Human Rights Committee the needed opportunity to comment on the specific requirements of the ICCPR with regard to domestic laws and practices.

Despite the process of international supervision, enforcement of the ICCPR remains within the hands of States Parties. In some countries, the ICCPR is directly enforceable through the courts, but in most countries implementing legislation is necessary. (Janis, 1988, p. 71) The United States ratified the ICCPR with extensive reservations and limitations, yet even in the United States the ICCPR is an important legal instrument. The United States ratified the ICCPR with a "declaration" that the treaty would be "non self-executing" (Senate Committee on Foreign Relations, 1992, p. 19). This means that the ICCPR requires implementing legislation and does not create a private right of action before the courts. (Stewart, 1993, p. 1202). The United States also ratified the ICCPR with a reservation that article 7 shall provide no more protection than does the United States Constitution (Senate Committee on Foreign Relations, 1992, p. 12). Despite the United States' understanding and reservation, the ICCPR can be used as a guide to the requirements of domestic United States law and established Federal policy (Neier, 1993, p. 1239).[1]

Ironically, United States law provides a more direct recognition of the ICCPR when experimentation is conducted outside the United States than when it is conducted within our borders. Regulations adopted by the Food and Drug Administration (FDA) do not permit new pharmaceuticals to be introduced in the United States that were not tested "in accordance with ethical principles acceptable to the world community" (21 CFR 312.120). The FDA regulations require compliance with the Declaration of Helsinki "or the laws and regulations of the country in which the research was conducted, whichever represents the greater protection of the individual" [21 CFR 312.120(c)].

As described above, there are some areas of practice—such as the use of involuntarily detained people who lack the capacity to consent—in which the ICCPR provides greater protections than does the Helsinki Declaration. In such areas of practice, the FDA regulations require compliance with the interpretation of the ICCPR recognized in the "laws and regulations in which the research was conducted." For research conducted abroad, the United States reservations on the interpretation of article 7 would not be relevant.

The FDA regulation effects a large proportion of all new drugs admitted in the United States. As of 1982, one study found that forty percent of all new drugs admitted in the United States were first tested abroad (Jordan, 1992, p. 501). Worldwide, two thirds of pharmaceuticals newly introduced each year are produced by firms located in the European Community (Jordan, 1992, p. 497). The percentage of pharmaceuticals first tested abroad is likely to grow as the cost of research in the United States continues to increase and pharmaceutical manufacturers can save money by conducting research abroad. The internationalization of pharmaceutical research may also increase as procedures for researching and approving new drugs are harmonized around the world (Jordan, 1992, p. 491).

The need for international regulation and enforcement of the rights of research subjects abroad is particularly important because procedures for informed consent are practically non-existent in some countries (see, e.g. MDRI report on Uruguay, MDRI, 1995, p. 39). In the absence of international protections, researchers may be motivated by their ability to avoid burdensome regulations in developing countries or to gain access to otherwise unavailable research subjects (Gostin, 1991, p. 199).

CONCLUSION

The ICCPR represents an important contribution to the field of human experimentation because it firmly implants the protection of consent within the sphere of international human rights law. As a legal instrument ratified by nations around the world, it is no longer possible to state that informed consent is a particularly Western construct of limited relevance in parts of the world that do not share the same cultural values or concept of the individual, as some researchers have suggested (see e.g. Christakis and Panner, 1991, p. 214). Indeed, the legitimacy of the General Comments derives, in part, from the "geographic, political, and legal balance" of the Human Rights Committee (Shelton, 1986, p. 417) and its success in establishing international consensus and cooperation (Nowak, 1993, p. xix). At the World Conference on Human Rights in Vienna, the international community reaffirmed that international human rights law applies universally to all people, including people with mental disabilities [Vienna Declaration, section II(B)(6)].

An international effort is needed to develop the meaning of article 7 of the ICCPR to provide clearer and more enforceable protections for the subjects of human experimentation. For too long, the international research community and the international human rights community have failed to examine or demand the development of international law in this area. Those who disregard the requirements of the ICCPR lose the opportunity to contribute to its development. The ambiguities that exist in the interpretation of the ICCPR are not a fatal limitation to its relevance but are an opportunity for more international attention and cooperation on a matter of common concern. The United Nations Human Rights Committee has proven to be very open to governmental and non-governmental perspectives, particularly to the views of people who will be directly affected by the law. We must not only enforce international human rights law, we would do well to participate in its future development.

NOTE

1. See Postscript: During the hearings on the United States compliance with the ICCPR in March 1995, the United Nations Human Rights Committee questioned the United States delegation extensively on this point. Among the Committee's positive findings on the U.S. record issued subsequent to the hearing, the "Committee [took] note of the position expressed by the [United States] delegation that, notwithstanding the non-self-executing declaration of the United States, American courts are not prevented from seeking guidance from the Covenant in interpreting American Law." (Human Rights Committee, 1995, paragraph 11).

REFERENCES

Bassiouni, M.C., Baffes, T.G., Evrard, J.T (1981) An appraisal of human experimentation in international law and practice: the need for international regulation of human experimentation, *Journal of Criminal Law and Criminology* 72, 4:1597–1666.

Christakis, N.A. and Panner, M.J. (1991) Existing international ethical questions for human subjects research: some open questions. *Law, Medicine, and Health Care* 19, 3–4:214–221.

Department of State (July 28, 1994). *Report of the United States of America Under the International Covenant on Civil and Political Rights.* Washington D.C.: Department of State.

Gostin, L. (1991) Ethical principles for the conduct of human subject research: population-based research and ethics. *Law, Medicine and Health Care* 19:191–201.

International Covenant on Civil and Political Rights, *opened for signature* Dec. 19, 1966, 999 U.N.T.S. 171, (entered into force Mar. 23, 1976; adopted by the United States Sept. 8, 1992).

Janis, M. (1988) *An Introduction to International Law.* Boston: Little, Brown and Co.

Jordan, D.W. (1992) International regulatory harmonization: a new era in prescription drug approval, *Vanderbilt Journal of Transnational Law* 25:471–507.

Mental Disability Rights International (1995) Discrimination Based on Mental Disability. *The Status of Human Rights in the United States: An Analysis of the Initial U.S. Government Report to the Human Rights Committee of the United Nations Under the International Covenant on Civil and Political Rights.* Washington, D.C.: American Association for the Advancement of Science pp. 75–78.

Mental Disability Rights International (1995). *Human Rights & Mental Health: Uruguay.* Washington, D.C.: Mental Disability Rights International.

McGoldrick, D. (1991) *The Human Rights Committee: Its Role in the Development of the International Covenant on Civil and Political Rights.* Oxford: Clarendon Press.

Neier, A. (1993) Political consequences of the United States ratification of the international covenant on civil and political rights. *DePaul Law Review* 42:1233–1239.

Nowak, M. (1993) *U.N. Covenant on Civil and Political Rights: CCPR Commentary.* Kehl: N.P. Engel.

Perley, S., Fluss, S.S., Bankowski, Z., and Simon, F. (1992) The Nuremberg Code: An International Overview, in George J. Annas and Michael A. Grodin (eds.) *The Nazi Doctors and the Nuremberg Code: Human Rights in Human Experimentation.* New York: Oxford University Press.

Posner, M.H., and Spiro, P.J. (1993) Adding Teeth to the United States Ratification of the Covenant on Civil and Political Rights: The International Human Rights Conformity Act of 1993, *DePaul Law Review* 42:1209–1231.

Rosenthal, E. and Rubenstein, L. (1993) International Human Rights Advocacy under the "Principles for the Protection of Persons with Mental Illness." *International Journal of Law and Psychiatry* 16:257–300.

Senate Committee on Foreign Relations (1992) *International Covenant on Civil and Political Rights.* S. Exec. Rep. No. 23, 102d Cong., 2d Sess, reprinted in *International Legal Materials* 31:645.

Shelton, D. (1986) Supervision Implementation of the Covenants: The First Ten Years of the Human Rights Committee. *Proceedings of the American Society of International Law* 1986:413–418.

Stewart, D.P. (1993) United States Ratification of the Covenant on Civil and Political Rights: The Significance of the Reservations, Understandings, and Declarations. *DePaul Law Review* 42:1183–1207.

United Nations Human Rights Committee (April 6, 1995) Consideration of Reports Submitted by States Parties under Article 40 of the ICCPR: Comments of the Human Rights Committee. U.N. Doc. CCPR/ C/79/Add 50.

United Nations Human Rights Committee (1992) General Comment No. 20 (44) (art. 7), U.N. GAOR, Hum. Rts. Comm., 47th Sess., Supp. No. 40 at 194. U.N. Doc. A/47/40.

United States Code of Federal Regulations: Investigational New Drug Application 21 CFR 312.120(c) (1994).

World Conference on Human Rights, The Vienna Declaration and Programme of Action (adopted June 25, 1993), reprinted in Center for the Study of the Global South and Center for Human Rights and Humanitarian Law, Washington College of Law (1993). *Evaluating the Vienna Declaration: Advancing the Human Rights Agenda.* Washington, D.C.: American University.

Villiger, M. (1985). *Customary International Law and Treaties.* New York: Oxford University Press.

POSTSCRIPT

In March 1995, Mental Disability Rights International (MDRI) submitted a critique of the United States Report to the United Nations Human Rights Committee (MDRI, 1995, p. 75). The critique pointed out the failure of the United States Report to note that U.S. law on human experimentation does not comply with the requirements of the ICCPR. In a meeting organized by the American Association for the

Advancement of Science the day before the Committee's hearing on U.S. practice, Eric Rosenthal, Director of Mental Disability Rights International and Cliff Zucker, Executive Director of Disability Advocates, Inc. met with five members of the Human Rights Committee. Zucker described scientific experiments permissible under New York State Office of Mental Health (OMH) regulations, N.Y.C.R.R. §527.10., including the use of non-therapeutic, potentially dangerous medications on individuals declared not competent to make treatment decisions based on the consent of a self-appointed "close friend." The OMH regulations also permit research performed on patients without their knowledge, research with possible therapeutic effects on legally incompetent adults performed over their objections, non-therapeutic research on children with parental consent, and research on children without parental consent. In *T.D. et al. v New York State Office of Mental Health*, Disability Advocates, Inc. and New York Lawyers for the Public Interest represent six involuntarily institutionalized individuals ruled not competent to make treatment decisions who are challenging the regulations on the grounds that they may be subject to this experimentation. On technical grounds, the New York State Supreme Court, New York County invalidated the OMH regulations (unpublished decision, New York Law Journal, March 6, 1995, p. 27). Cross-appeals are pending.

At the March 1995 UN hearing on U.S. compliance with the ICCPR, members of the Committee questioned the United States representatives about United States practices concerning human experimentation. In its Final Comments on the United States record, the Human Rights Committee noted as one of thirteen "[p]rincipal subjects of concern" about the United States human rights record:

[t]he Committee is concerned that, in some states, non-therapeutic research may be conducted on minors or mentally-ill patients on the basis of surrogate consent, in violation of the provisions of article 7 of the Covenant. (paragraph 21).

As one of its suggestions and recommendations to the United States,

The Committee recommends that further measures be taken to amend any Federal or State regulation which allow, in some States, non-therapeutic research to be conducted on minors or mentally-ill patients on the basis of surrogate consent.

Attorneys representing patients in New York reported to MDRI that they intend to cite the Committee's findings in their appeal to the Appellate Division of the Supreme Court of New York (Interview with Cliff Zucker, Disability Advocates, May 22, 1995).

Neurobiological Research Involving Human Subjects: Perspectives from the Office for Protection from Research Risks

J. Thomas Puglisi and Gary B. Ellis

Office for Protection from Research Risks, 6100 Executive Blvd, Suite 3B01, National Institutes of Health MSC 7507, Rockville, MD 20892-7507

Department of Health and Human Services (HHS) regulations at 45 CFR 46 require that all human subjects research supported by HHS be reviewed and approved by a local Institutional Review Board (IRB). Investigators may not involve human subjects in research without their informed consent, and additional safeguards are required when subjects are likely to be vulnerable to coercion or undue influence. Application of the regulations to neurobiological research is discussed.

Keywords: Human subject protection, Institutional Review Board (IRB), informed consent

The Office for Protection from Research Risks (OPRR), housed administratively at the National Institutes of Health, is charged with implementing the Department of Health and Human Services (HHS) regulations for the protection of human research subjects (Ellis, 1994). These regulations appear at Title 45, Part 46 of the Code of Federal Regulations (45 CFR 46). Subpart A of the HHS regulations constitutes the "Common Rule" or "Federal Policy" for the protection of human subjects, which has also been adopted by the 15 other Federal departments and agencies supporting human subjects research.

The American system of protection for human subjects of research is based on a succession of judgments made by ordinary people in the context of the regulations. The success of the system depends upon well-meaning and thoughtful people examining research protocols and weighing risks and potential benefits on a continuing basis. It involves several levels of protection at which serious efforts are made to protect the rights and welfare of research subjects.

IRB REVIEW

The institutional review board, or IRB, is the cornerstone of this system. No federally supported human subjects research may be initiated, and no ongoing

---

This article will also appear in the journal: *Accountability in Research*, Volume 4, pp. 261–265 (1996).

research may continue, in the absence of IRB approval. Federal agencies cannot provide funds for human subjects research unless an IRB approves the protocol for such studies (45 CFR 46.101).

Under the regulations, the IRB is to be established at the local level (45 CFR 46.103) and must have a minimum of five people, including at least one scientist, one nonscientist, and one person not otherwise affiliated with the institution conducting the research (45 CFR 46.107). IRB review is to be a prospective and continuing review of proposed research by a group of individuals with no formal interest in the research. It is a local review, by individuals who are in the best position to know the resources of the institution, the capabilities and reputations of the investigators and staff, and the prevailing values and ethics of the community and likely subject population.

IRB review assures that: (a) risks are minimized; (b) risks are reasonable in relation to anticipated benefits; (c) selection of subjects is equitable; (d) there is proper informed consent; (d) adequate provisions have been made to protect the privacy of subjects and the confidentiality of data; and (e) the rights and welfare of subjects are maintained in other ways (45 CFR 46.111).

Downstream from the IRB are (a) the executive official of the research site, (b) the scientific review group at the federal funding entity, and (c) the executive official of that funding entity. Each has the authority and the responsibility to review proposed research and express concerns about human subjects issues. Exerting oversight of the whole process is OPRR, and, of course, Congress.

In the second session of the recently closed 103rd Congress, there were at least 11 hearings directly related to human subjects research. This indicates broad and diverse congressional interest, and with each successive hearing, general awareness of, and alertness to, protecting human subjects have grown.

INFORMED CONSENT

The most basic and fundamental element in the system of protections for research subjects involves the interaction between the research volunteer and the research investigator. This is where the informed consent process takes place. Informed consent must be an ongoing, dynamic process, as new information becomes available or is desired. The informed consent document is one component—the *written* component—of the informed consent process.

HHS conducts or supports about $5 billion per year in biomedical human subjects research at some 1,000 institutions receiving HHS funds and many more affiliated performance sites. OPRR responds to allegations of noncompliance from all sources (e.g., research subjects, institutions, media, Congress) and actively pursues investigations. About 150 such investigations have been completed over the past four years, and about 100 are presently ongoing, a relatively small number given the volume of research supported by HHS.

Many of OPRR's compliance oversight activities—many of the problems that OPRR sees—involve some facet of the informed consent process. Although most members of the research community are generally familiar with the concept of informed consent, a thorough and detailed understanding of the specific regulatory requirements for informed consent is essential.

HHS regulations at 45 CFR 46.116 stipulate that informed consent shall be sought only under circumstances that provide the prospective subject, or the subject's legally authorized representative, sufficient opportunity to consider whether or not to participate and that minimize the possibility of coercion or undue influence. The information presented must be in language understandable to the subject or the legally authorized representative.

In addition, the regulations [45 CFR 46.116(a)] specify eight basic, required elements of informed consent. Except in a few limited circumstances [see 45 CFR 46.116(d) and 46.117] the *written* informed consent document must *embody* each of these eight elements:

1. A statement that the study involves research, an explanation of the purposes of the research and the expected duration of the subject's participation, a description of the procedures to be followed, and identification of any procedures which are experimental. Note that *all* procedures to be followed in the research must be described, not just those which are experimental. This is a common misunderstanding.
2. A description of any reasonably foreseeable risks or discomforts to the subjects. Again note the regulatory language: the requirement is for a description of *any reasonably foreseeable* risks, not just the most common risks, or the risks an investigator considers most likely, or the least objectionable risks, but any reasonably foreseeable risks.
3. A description of any benefits to the subject or to others which may reasonably be expected from the research.
4. A disclosure of appropriate alternative procedures or courses of treatment that might be advantageous to the subject. This is an important element that is often omitted entirely or included only to an incomplete or superficial degree. The disclosure of appropriate alternatives must be complete enough for the subject to understand the nature of the alternatives and make a reasonable decision about research participation.
5. A statement describing the extent, if any, to which confidentiality of records identifying the subject will be maintained.
6. For research involving more than minimal risk, an explanation as to whether any compensation and an explanation as to whether any medical treatments are available if injury occurs and, if so, what they consist of, or where further information may be obtained.
7. An explanation of whom to contact for answers to pertinent questions about the research and research subjects' rights, and whom to contact in the event of a research-related injury to the subject. Each of these three areas must be explicitly stated and addressed. The principal investigator or other members of the research team are appropriate sources of information about the research itself. OPRR strongly recommends that the informed consent document also provide contact persons who are not members of the research team to answer questions about subjects' rights and research-related injury.
8. A statement that participation is voluntary, refusal to participate will involve

no penalty or loss of benefits to which the subject is otherwise entitled, and the subject may discontinue participation at any time without penalty or loss of benefits to which the subject is otherwise entitled. Each of these three conditions must be stated explicitly.

Again, HHS regulations generally require that the *written* informed consent document *embody* each of the eight elements described above (45 CFR 46.117). This is an important concept. Oral communication of these elements to subjects, or written communication outside the informed consent document, while important, are not sufficient under the regulations. The eight required elements of informed consent must be included in the written informed consent document.

ADDITIONAL PROTECTIONS FOR VULNERABLE SUBJECTS

Attention to the welfare of research subjects and to the integrity of the informed consent process is particularly important when subjects are likely to be vulnerable to coercion or undue influence, including persons with neurobiological disorders. The HHS regulations provide extra protection for vulnerable subjects in several ways.

If an IRB regularly reviews research that involves a vulnerable category of subjects, consideration must be given to including as IRB members one or more individuals who are knowledgeable about, and experienced in working with, the vulnerable subjects [45 CFR 46.107(a)]. When some or all of the subjects are likely to be vulnerable to coercion or undue influence, IRBs must see that additional safeguards are included in the study protocol [45 CFR 46.111(b)].

What kinds of additional safeguards might be employed for subjects of neurobiological research? OPRR has identified several which seem reasonable and recommends that IRBs seriously consider their appropriateness when reviewing neurobiological research:

1. Where the volume of neurobiological research (or other research) warrants it, OPRR recommends that IRBs engage one or more subject representatives as members who can assist the IRB in the review of issues related to the rights and welfare of subjects with neurobiological disorders. Where the volume neurological research is not large, OPRR recommends that IRBs invite the assistance of subject representatives as nonvoting consultants [see 45 CFR 46.107(f)].

2. OPRR strongly recommends that informed consent documents make clear when the clinicians delivering treatment are also members of the research team. Subjects sometimes have difficulty understanding the distinctions characterizing clinical treatment versus research, especially when they are being asked to participate in treatment-oriented research. Clarification of the role of clinician-researchers helps subjects understand that they are being asked to participate in research.

3. IRBs themselves have the authority under the HHS regulations to observe the informed consent process directly [45 CFR 46.109(e)] and to monitor the collected data to ensure the safety of subjects [45 CFR 46.111(a)(6)]. IRBs should

consider exercising these options on a regular basis, especially where vulnerable subjects are involved.

4. In certain instances, it may be appropriate for an independent consent auditor, monitor, or research subject advocate to be available to subjects or even to be present during the informed consent process. IRBs should consider whether and under what circumstances such independent monitors are warranted, and how often they should report to the IRB.

5. Individual, independent data monitors are also appropriate in some circumstances. Again, IRBs should consider whether and under what circumstances such independent monitors are warranted, and how often they should report to the IRB.

6. Where the nature or volume of neurobiological or other research warrants it, an additional layer of review that is sometimes employed is an independent data and safety monitoring board, or DSMB. The DSMB is appointed to oversee and to evaluate the research investigation. At periodic intervals during the course of an experiment, the DSMB reviews the accumulated data and makes recommendations on the continuation, or modification, of the study. For example, when a study is stopped prematurely because of a toxic effect or because a strong positive effect was seen, and it would be unethical to continue with some subjects not receiving the beneficial test article, it is likely due to the intervention of a DSMB.

This list of safeguards is certainly not exhaustive. Undoubtedly, IRBs, investigators, and subject advocates will identify additional safeguards, as they consider the specific needs of their own local subject populations in specific research contexts. Indeed, such safeguards must be tailored to meet the specific needs of subjects in their particular research settings. Our challenge is to continue efforts on all levels of the human subject protection system to ensure the best possible protection for subjects of neurobiological research, and for all human subjects.

REFERENCES

Ellis, G.B. (1994) Profile: Office for Protection from Research Risks (OPRR). *Politics and the Life Sciences*, 13, 271–273.
United States Code of Federal Regulations. Title 45—Public Welfare, Part 46—Protection of Human Subjects. Department of Health and Human Services.

Problems in Interpreting Active Control Equivalence Trials

Robert Temple

Office of Drug Evaluation, Center for Drug Evaluation and Research, Food and Drug Administration, 5600 Fishers Lane, Rockville, Maryland 20857

INTRODUCTION

I will discuss today the problems associated with interpreting active control equivalence trials, i.e., studies in which the effectiveness of a new drug is to be proved by showing it is as good as a known effective drug. These problems have long been recognized (Lasagna, 1979; Temple, 1982, 1983; Freston, 1986; Prien, 1988) and the need for placebo controls appreciated, but the importance of understanding the limitations of equivalence trials has recently been heightened by a publication by Rothman and Michels in the *New England Journal of Medicine* (Rothman, 1994). The authors argued that placebo controlled trials are invariably unethical when an effective therapy exists, no matter what the condition being studied, whether it is a serious illness or one that is no threat to the patient's health, such as tension headache, back pain, the common cold, or daytime anxiety, no matter how the study is designed and no matter how well the patient is informed. They based this conclusion on a section of the Declaration of Helsinki that says "in any medical study, every patient, including those in the control group, if any, should be assured of the best proven diagnostic and therapeutic method." That phrase was added to the Declaration in Tokyo in 1975.

ETHICS OF PLACEBO-CONTROLLED TRIALS

There are two principal problems with Rothman and Michels' interpretation of the Declaration. First, taken literally, as it would seem the literal or "fundamentalist view" they put forth requires, the statement bars active control trials, the kind of study preferred by Rothman and Michels, as well as placebo-controlled trials, because the person being randomized to the *new* (test) therapy is not getting the "best proven diagnostic or therapeutic method." That person is getting something a commercial sponsor hopes might work, but many new drugs don't prove effective when tested and sometimes they prove more toxic than available treatments.

This article will also appear in the journal: *Accountability in Research*, Volume 4, pp. 267–275 (1996).

On ethical grounds, reading the Declaration literally, therefore, active control trials seem as much barred as placebo controlled trials. Rothman and Michels seem to sense this problem, although they do not acknowledge it explicitly. In response to the question "suppose withholding standard therapy won't really do any harm, but would only lead to discomfort," they responded (Rothman, 1994) that it doesn't matter, because even in that case, use of a placebo violates the ethical principal that "every patient, including those in a control group, should receive *either the best available treatment or a new treatment thought to be as good or better*" (emphasis added). The last part of that statement, referring to new treatment thought to be as good or better, is not a citation of the Declaration of Helsinki but is newly created by Rothman and Michels, perhaps because without so modifying the Declaration, it would seem to bar virtually *all* studies of new agents, at least when any effective therapy is available, a conclusion the authors apparently found uncomfortable. But if Rothman and Michels can "interpret" the Declaration, and not take it literally, surely others can also do so.

The second problem with Rothman and Michels' interpretation is that it makes no ethical sense. What ethical principle *requires* that a headache be treated immediately, that anxiety be suppressed at once, that insomnia be eradicated the night it is identified, or that baldness be covered over at the first opportunity? There are many conditions in which an informed person might reasonably agree, and ethically be asked, to defer treatment and to be randomized to drug or placebo, with the assurance that no harm would result. It is very difficult to imagine that the 1975 revision of the Declaration of Helsinki meant what Rothman and Michels say it meant, and that so radical a proposition, the barring of all active or placebo-controlled studies, would have been incorporated without any discussion or explanation. The passage can in fact be read differently, as saying the physician must not abandon the person in the trial or let that person come to harm, a reminder of the investigator's responsibility to the subject as a patient. Of course it would not be ethical to randomize patients to placebo when there is therapy known to affect survival or irreversible morbidity.

I need to make just a few more points before I get to the matter of problems in interpreting active control equivalent trials. Dr. Rothman indicates his belief that the unethical use of placebos he perceives is going on in secret, that nobody really knows the extent to which placebos are being used, that the use is "clandestine." He is incorrect. The subject has been discussed extensively. FDA's published clinical guidelines and international guidelines frequently call for use of placebos in trials of a wide range of drugs. I have discussed the problems of active control trials in print on several occasions (Temple, 1982, 1983) and wrote recently (Temple, 1994) about some ways of making placebo controlled trials more comfortable to investigators and patients. There are countless published reports of placebo-controlled trials. In a recent review of fluoxetine in the *New England Journal of Medicine* (Gram, 1994), for example, it is easy to identify at least 10 placebo controlled trials. There have also been published discussions of this issue by representatives of the American College of Gastroenterology (Freston, 1986). In 1991 FDA reported (Glassner, 1991) a meta-analysis of patients in placebo-controlled angina trials to see whether the placebo-treated patients fared badly compared to patients receiving active drugs. We found that the placebo-treated patients were at least slightly better off when all kinds of problems, including adverse reactions, were measured, and were approximately equal to treated patients with respect to serious morbid events, such as heart attacks and death. That report, in the *Journal of the American Medical Association*, conspicuously displayed the existence of many placebo-controlled studies in angina.

Rothman and Michels attribute the wide use of placebo-controls to FDA preferences and insistence. This is, in a sense, correct, but needs to be understood clearly. FDA does not have a "preference" for placebos in the sense that one might like chocolate better than vanilla. The Food, Drug, and Cosmetic Act says that FDA may approve a drug only if it is shown effective in adequate and well controlled studies. We accept a variety of kinds of control groups as constituting adequate controls, but the choice has to be appropriate, so that experts can conclude from the results that the drug studied will have the effect it is supposed to have. Thus, even though historical controls are cited in our regulations as one acceptable kind of study design, they are obviously not acceptable all the time, but only where the serious consequences of the disease are frequent and reasonably predictable. Active control equivalence designs similarly may be difficult to interpret and are often inadequate. If they are not persuasive in a particular case, we cannot accept them as evidence of effectiveness.

Finally, it goes without saying that the need for something (e.g., a placebo-controlled trial) does not make it ethical. There are some trials that people might like to do that simply cannot be done because they would expose people to unacceptable risks by denying them established therapy. FDA is as conscious of those situations as anybody and we would not countenance a trial in that situation. We have, however, generally felt that where participation in a trial would involve only discomfort, an informed patient could agree to participate.

PROBLEMS IN INTERPRETING ACTIVE CONTROL EQUIVALENCE TRIALS

Let me now turn to my main subject, which is active control equivalence trials, i.e., trials in which the intent is to show *no difference* between treatments. If the active control trial is designed to show that the new therapy is *better*, that is not a problem; dose response studies are also not a problem because they show differences between groups or a dose-response slope.

Trials in which the intent is to show equivalence have a fundamental problem that must be addressed and resolved if such trials are to be interpretable. There is, in the analysis of these trials, an inherent, but often unstated and unrecognized, assumption, namely, that the active control was effective in that particular study, i.e., that it had an effect that could have been measured had there been a placebo group. Put another way, it must be assured that the study had "assay sensitivity and could distinguish effective from ineffective treatments." If it did not, then a finding of no difference between treatments would obviously be meaningless. But the assumption is not testable in the data collected if there is no placebo group, and it is often incorrect, as I will show. Many unequivocally effective drugs are not effective every time they are studied, and the reason for this is not usually apparent; i.e., it isn't just that the study was too small. What this means is that an active control equivalence study is really very much like a historically controlled trial. Information about what would have happened to an untreated group has to come from somewhere else because the study does not provide it. Sometimes this information can be obtained; sometimes it cannot. But if it cannot, if the study design and study population are not ones in which you can assuredly tell inactive from active therapy, no amount of precision, no hugeness of sample size, no degree of care in conducting the study will enable that study to be meaningful if it fails to show a difference between treatments.

There are two related but somewhat different approaches to validating an equivalence trial, i.e., to supporting the probability that the study has assay sensitivity. One is to show, by reference to past similar studies of similar populations, that the response in an untreated group is well-defined and is clearly different from what was observed; this is essentially what one does in supporting any historical control. The second is to show that, even if the response in an untreated control is variable, the active control drug is essentially always superior to the untreated control by a well-defined amount. The opportunity to use this latter approach as support for the validity of the study is the principal advantage, with respect to showing effectiveness, that the active control has over a pure historical control.

Let me now illustrate how difficult it is to distinguish drug and placebo in some specific situations and thereby show how little meaning should be attached to showing of equivalence in these cases. Table I shows all six studies in the marketing application for an antidepressant called nomifensine that used a three-group randomized study design: the new drug group (nomifensine, an effective but toxic antidepressant), imipramine, a standard antidepressant, and placebo. Table I does not show the placebo results. The studies are all in depressed people with Hamilton depression scale scores in the 25–30 range. In each trial the four week results showed a sizable treatment effect, measured as improvement from baseline, but none of the trials showed a statistically significant difference between nomifensine

TABLE I   Results (4 week adjusted endpoint Ham-D total scores) of 6 trials comparing nomifensine, imipramine, and placebo showing only the new drug vs. imipramine comparison

Study	Item	Common Baseline	NOM	IMI	"p" two tail	Power to detect 30% difference
R301	HAM-D	23.9	13.4	12.8	0.78	0.40
	(n)		33	33		
G305	HAM-D	26.0	13.0	13.4	0.86	0.45
	(n)		39	30		
C311 (1)	HAM-D	28.1	19.4	20.3	0.81	0.18
	(n)		11	11		
V311 (2)	HAM-D	29.6	7.3	9.5	0.63	0.09
	(n)		7	8		
F313	HAM-D	37.6	21.9	21.9	1.0	0.26
	(n)		7	8		
K317	HAM-D	26.1	11.2	10.8	0.85	0.33
	(n)		37	32		

and; imipramine. This is not a matter of trends that don't reach statistical significance, the numbers are virtually identical. If one could believe that active control trials could ever persuasively show equivalence in depression, these trials must surely have done it. The only reservation one would have is that the studies are not very big, but at the time the studies were done, a 30 patient per treatment trial was typical for depression, and many such trials have distinguished active drug from placebo. In any case there are six trials and they don't show any difference at all, even if one were to pool ("metaanalyze") them.

In Table II are the same trials, now showing the placebo group results as well. In five of the six trials there is no difference at all between either active drug and placebo; again, this is not a matter of trends that don't quite reach statistical significance. These are thus almost all trials that are incapable of distinguishing active from inactive therapy. The one exception is study V311(2), a trial that would, rightly, have been dismissed as useless for an equivalence trial because it only had seven patients per group, obviously too small to be meaningful. The trials do not seem to have failed because they were too small. Rather, although one cannot prove this, the successful and unsuccessful trials seem to have encountered qualitatively different populations, one of which could, and five which could not, respond to antidepressants. Of course, if you don't have a placebo group, you can't tell whether a given trial is one that *is* an effective means of distinguishing active from inactive therapy, or *is not*. That is, you can't tell whether you have got a valid assay without an internal measure of assay sensitivity.

This is not an unusual example. For a recently approved antidepressant, nefazadone, 3 of 7 studies of adequate size and using adequate doses of nefazadone showed superiority to placebo while 4 of 6 trials with imipramine did so. In general, about one-third to one-half of adequate-sized antidepressant trials seem to show a significant difference between active drug and placebo. The same problem arises in many therapeutic areas. Among the four studies on which we based

TABLE II   Results (4 week adjusted endpoint Ham-D total scores)
of 6 trials comparing nomifensine, imipramine,
and placebo showing all comparisons

Study	Item	NOM	IMI	PBO	Baseline HAM-D adjusted
R301	HAM-D	13.4	12.8	14.8	23.9
	(n)	33	33	36	
G305	HAM-D	13.0	13.4	13.9	26.0
	(n)	39	30	36	
C311 (1)	HAM-D	19.4	20.3	18.9	28.1
	(n)	11	11	13	
V311 (2)	HAM-D	7.3	9.5	23.5	29.6
	(n)	7	8	7	
F313	HAM-D	21.9	21.9	22	37.6
	(n)	7	8	8	
K317	HAM-D	11.2	10.8	10.5	26.1
	(n)	37	32	36	

*IMI, NEW vs PBO, "p" less than 0.001

approval of cimetidine for treatment of duodenal ulcer (2 two week studies, a four week study and a six week study), the four and six week studies couldn't distinguish drug from placebo statistically, although there were favorable trends. The two week studies showed significant differences. Ulcers heal by themselves to a considerable degree, with 50–60% healing rates by 4–6 weeks. Turning to another example, Merck carried out two large studies of enalapril in patients with symptomatic heart failure, one in Europe, and one in the U.S., of similar design and size. The European study easily showed that all symptoms and signs of heart failure and exercise ability were improved. The U.S. study showed no significant effect on any measurement. We do not know the reason for this; there was nothing wrong with the U.S. study that we could see. It seemed as well designed and conducted as the European study and there is no doubt that enalapril is effective in heart failure. Again, there was no way to look at the domestic study and know that it would be unable to distinguish active and inactive treatments.

Now, having described the fundamental problem of active controlled equivalence trials, which is that effective drugs don't always show themselves to be effective in seemingly well designed studies of adequate size, I have to tell you that problem is even worse than I have indicated so far. The examples I have been telling you about occurred in a setting where everybody's incentive was to minimize variance and thus maximize the study's ability to show a difference from placebo. That is, everyone was on the best possible clinical trial behavior, making sure the patients really have the disease, trying to exclude people whose measurements vary too much, standardizing measurement techniques. They were, in other words, trying to minimize beta error. But if the objective of the trial is to show *no difference*, there is no incentive to minimize beta error. Indeed, it is not too cynical to say that there is some reason to want to *maximize* beta error. Remember, *sloppiness obscures differences*, so when "no difference" is the goal, there is no incentive not to be sloppy.

Note also that not all kinds of sloppiness show up as increased variance, which would widen the confidence interval for the result and perhaps prevent a showing of equivalence. Suppose, for example, that the patients entered do not really have the disease in question, but do have symptoms that are likely to get better spontaneously. This is a disaster in a study trying to show a difference from placebo but is an "ideal population" for the purpose of showing no difference between two therapies. Table III shows the efforts made in typical hypertension study to allow the study to show a drug effect if there is one. Great effort goes into making sure people actually have high blood pressure, so therapy is stopped for a few weeks until blood pressure is increased and then is stable. There are attempts to measure

TABLE III   Efforts to decrease variance and beta error

Example of Hypertension
1. Be sure they have hypertension; stop therapy, make sure it stays up over a few weeks.
2. Measurement precision: time of day, standard operating procedures for technique, avoiding patients who vary too much.
3. Encourage compliance, record other drugs, don't mix up treatments.

blood pressure at the same time of day, using standard operating procedures, often with training sessions. Investigators avoid patients who vary too much, encourage compliance, record other drugs, don't mix up treatments, all so that a difference from placebo can be detected. There is no incentive to make those efforts if the objective is to show no difference.

The last problem with active control trials, a surmountable one, is that a good deal of discussion must occur before you can say what constitutes "statistically significant similarity." There really isn't any such thing as statistically significant similarity, of course. All you can really show is that a difference of an agreed upon size can be excluded with an agreed upon degree of confidence. When you start trying to do that, and want to be, e.g., 95% sure that the new drug is at least 75% as good as the standard drug, you end up with some very large studies. That may not be a fatal problem, but people have to recognize its existence. More fundamentally, until you can say *with assurance* what the size of the effect of the standard drug is vs. placebo, you can't even begin to carry out the calculation of how much of the standard drug's effect is lost.

Consider a depression study in which the change from baseline for the standard drug in the HamD depression scale is 12 units *(24 baseline to 12 on treatment) and for the new drug is 10 units (24 to 14). If the whole change is due to drug, the effect of the new drug is 10/12 of the old, probably a trivial difference and one might, in an adequately sized study, be able to conclude with 95% confidence that the new drug has at least 50% of the effect of the standard. Suppose, though, that the placebo response is typically 8 units in a depression trial. Then, subtracting this from the 10–12 point changes, the actual drug effects are 2 vs. 4. The ability to conclude anything about the relative efficacy of the two drugs from such a study is in doubt; a very large study would be needed to know that 50% of the standard drug's effect was preserved. In reality, without a placebo group, we can't know whether the placebo response is 0, 8 or *more*, so there is no way to determine what fraction of the standard drug's effect is obtained by the new drug.

STUDY DESIGNS THAT LIMIT OR SHORTEN PLACEBO EXPOSURE

There are some study designs to consider that do not involve use of a placebo, yet are still rigorous. One such design is a dose response study. If the study shows a positive slope, that constitutes unequivocal evidence that the therapy has an effect. But it is essential to show a positive slope. If all doses have the same effect, the study is often uninformative, for the reasons described above, unless a placebo group is included too.

Another study design is the "add on" design. In such a study people who don't respond adequately to a standard treatment are randomized to receive either the standard therapy with the new treatment or the standard therapy with added placebo. In this design, everybody receives standard therapy. Such studies can be successful when testing a drug with a mechanism of action different from the standard and they are widely used. The add-on is the only design that can be used in long-term heart failure studies today, for example, as standard therapy is life-saving and cannot be denied. It is also a useful study design for cancer trials and is widely used for anti-epileptic drugs. It is not useful, however, for study of a new drug of the same class as the standard because the new drug in that case would not be expected to add to the other therapy.

Even when a placebo control is needed, there are study designs that minimize the duration of exposure to placebo. I have written some about this in a recent publication (Temple, 1994). One method is to build "early escape" into the design, removing the patient from the study if symptoms or signs reach a defined level or fail to respond to a defined extent, and counting failure rate as an effectiveness end point. In this way, no patient receives a therapy that is ineffective more than briefly. Patients can, of course, always leave a study if they wish; the difference here is that departures for lack of effectiveness are carefully assessed and anticipated.

Another design that uses the same principle is the randomized withdrawal study. In this design responders to a drug in either an open or blinded study are randomized to continued therapy with the drug or to placebo substitution. During the trial return of signs or symptoms or ability to continue on trial are study endpoints. Again, in this design, no patient has to spend a long time on a therapy that doesn't work.

CONCLUSION

The desire to treat patients with effective therapies is understandable, but active control equivalence trials are uninformative in many situations and in those cases cannot provide evidence that a new therapy is effective. This does not appear to be a matter of study size or power, but reflects real differences among populations in their ability to discriminate between active and inactive treatments. Where standard therapy is known to prevent death or irreversible morbidity in a population this therapy cannot ethically be denied to that population, and even a comparative trial would need special attention because the patients receiving the new drug would not be getting the standard drug. But where no permanent harm will come to patients from deferral or foregoing of therapy, fully informed patients can participate in placebo-controlled trials.

The Declaration of Helsinki, in demanding "best proven therapeutic method" for patients in a study, including those in a control group, surely cannot have meant to eliminate essentially all controlled trials of new agents, both placebo-controlled and active control trials, without ever addressing the issue explicitly. FDA's role in implementing the Food, Drug, and Cosmetic Act requires that we approve only drugs shown effective by adequate and well-controlled studies, and we cannot responsibly approve such drugs as antidepressants on the basis of facially inadequate active control studies. This is not, however, an FDA matter alone. There is no community interest in the marketing of antidepressants, agents for heart failure or angina, or drugs for ulcer disease if those drugs are not effective.

REFERENCES

Freston, J. (1986) Dose-ranging in clinical trials: rationale and proposed use with placebo or active controls. *American Journal of Gastroenterology* 81(5):307–311.

Glassner, S.P., Clark, P.I., Lipicky, R.J., Hubbard, J.M., and Yusuf, S. (1991) Exposing patients with chronic, stable, exertional angina to placebo periods in drug trials. *Journal of the American Medical Association* 265(12):1550–1554.

Gram, L.F. (1994) Fluoxetine. *New England Journal of Medicine* 331(20):1354–1361.

Lasagna, L. (1979) Editorial: placebos and controlled trials under attack. *British Journal of Clinical Pharmacology* 15(6):373–374.

Prien, R.F. (1988) Methods and models for placebo use in pharmacotherapeutic trials. *Psychopharmacology Bulletin* 24(1):4–8.

Rothman, K.J., and Michels, K.B. (1994) The continued unethical use of placebo controls. *New England Journal of Medicine* 331(6):394–398.

Temple, R. (1982) Government viewpoint of clinical trials. *Drug Information Journal* Jan/June:10–17.

Temple, R. (1983) Difficulties in evaluating positive control trials. *Proceedings of the American Statistical Association*, Biopharmaceutical Section, 1–7.

Temple, R. (1994) Special study designs: early escape, enrichment, studies in non-responders. *Communications in Statistics* 23(2):499–531.

Sounding Board—The Continuing Unethical Use of Placebo Controls

K. J. Rothman[†] and K. B. Michels[‡]

[†]Boston University School of Public Health, Boston, MA 02118
[‡]Harvard University School of Public Health, Boston, MA 02115

Is it ethical to use a placebo? The answer to this question will depend, I suggest, upon whether there is already available an orthodox treatment of proved or accepted value. If there is such an orthodox treatment the question will hardly arise, for the doctor will wish to know whether a new treatment is more, or less, effective than the old, not that it is more effective than nothing.

—A. Bradford Hill[1]

Unaccountably, in these times of raised ethical consciousness, placebo treatments are still commonly used in medical research in circumstances in which their use is unethical. We refer not to the deceptive use of placebo, but to studies in which patients are informed that they may receive a placebo and then give their consent. Even so, such studies are unethical if patients are assigned a placebo instead of a therapy effective in treating their condition. Here we examine why this ethical breach persists and suggest ways to reduce it.

THE ETHICS OF PLACEBO CONTROLS

The Nuremberg Code, "the cornerstone of modern human experimentation ethics,"[2] was formulated shortly after World War II in response to Nazi atrocities. The World Health Organization adopted a version of the code in 1964 as the Declaration of Helsinki.[3] The declaration elevates concern for the health and rights of individual patients in a study over concern for society, for future patients, or for science. "In any medical study," it asserts "every patient—including those of a control group, if any—should be assured of the best proven diagnostic and therapeutic method."[4] This statement effectively proscribes the use of placebo as control when a "proven" therapeutic method exists. The declaration also directs that a study that violates its precepts should not be accepted for publication.

Nevertheless, studies that breach this provision of the Declaration of Helsinki are still commonly conducted, with the full knowledge of regulatory agencies and institutional review boards. Although some are published in peer-reviewed medi-

Reprinted by permission of *The New England Journal of Medicine*, Vol. 331, 394–398 (1994), Massachusetts Medical Society.

cal journals, the declaration notwithstanding, many trials that are conducted in order to gain regulatory approval for new drugs or devices never reach libraries. Thus, there is no straightforward way to estimate how many trials are undertaken that involve the unethical use of placebos.

Below are a few examples from among those that have actually been published. Some of these examples might be challenged by specialists in the disciplines involved, who might argue that the use of placebo was justifiable in the case under discussion. In the aggregate, however, the examples indicate that patients in trials are often denied "best proven" treatments.

Ivermectin Trial

In 1985 a group of investigators reported the efficacy of ivermectin to treat onchocerciasis, or river blindness.[5] The investigators assigned some of the study participants to placebo when, according to the investigators themselves, the drug diethylcarbamazine had been "the standard therapy ... for over three decades." The study participants were illiterate Liberian seamen, some of whom indicated their "informed consent" by thumbprint.

Rheumatoid Arthritis Trials

In recent years there have been numerous placebo-controlled trials of secondary treatments for rheumatoid arthritis. In many of these trials,[6-9] some enduring for years, all the patients were assigned to receive a primary therapy, such as a non-steroidal antiinflammatory agent, and were then randomly assigned to receive either a new secondary treatment or a placebo in addition. New placebo trials of secondary treatments for arthritis continue to be proposed and conducted, even though many such trials have shown various secondary treatments to be more effective than placebo.[10] Participants who receive placebo in these studies are at risk for serious and irreversible degenerative changes that can, to some extent, be prevented.

Antidepressant-Drug Trials

A 1992 report of a randomized trial of treatment for major depression began with the statement "Effective antidepressant compounds have been available for over 30 years."[11] Nevertheless, the investigators in that study assigned half the seriously depressed patients in the trial to receive placebo and the other half to receive paroxetine. Placebo controls are commonplace in trials of antidepressant drugs, despite the availability of therapies whose success is acknowledged.[12-18]

Ondansetron Trials

Considerable advances have been made in controlling chemotherapy-induced emesis in recent years.[19] Several drugs are available for use singly or in combination; they include metoclopramide, phenothiazines, substituted benzamides, corticosteroids, and benzodiazepines.[20-22] Nevertheless, when a new agent, ondansetron, was tested, it was compared with placebo in several trials.[23-25] (This use of placebo was criticized in an editorial accompanying the published report of one of the trials.[26])

Trials of Drugs for Congestive Heart Failure

Angiotensin-converting—enzyme inhibitors are accepted as a standard treatment for congestive heart failure.[27] Although a number of these drugs have been approved, new ones, as well as other drugs for congestive heart failure, are commonly evaluated against placebo.[28-30]

Antihypertensive-Drug Trials

Trials of new drugs for mild-to-moderate hypertension typically use placebo controls, despite the established efficacy of many agents in treating mild-to-moderate hypertension.[31-33] For example, in the introduction to a "dose-ranging" study of the calcium antagonist verapamil,[33] verapamil was described as "an effective antihypertensive drug, which is dose dependent, superior to placebo, comparable to or more effective than propranolol, and comparable with nifedipine." Despite these assertions, the investigators assigned some patients in the study to receive placebo.

PLACEBO CONTROLS AND DRUG APPROVAL

In the United States, many drug studies are conducted to meet the requirements of the Food and Drug Administration (FDA) so that the drug can be marketed. The Code of Federal Regulations under which the FDA operates is ambiguous about the acceptability of placebo controls. In one place it suggests that they should be avoided: "The test drug is compared with known effective therapy; for example, where the condition treated is such that administration of placebo or no treatment would be contrary to the interest of the patient."[34] The regulations go on, however, to suggest including both placebo controls and active-treatment controls in a study: "An active treatment study may include additional treatment groups, however, such as a placebo control...."[34]

In practice, FDA officials consider placebo controls the "gold standard." Agency guidelines specify the study designs required to obtain approval for new drugs. Placebo controls are, in effect, required for disorders of moderate severity and pain, even when an alternative treatment is available. For example, in its "Guidelines for the Clinical Evaluation of Anti-Inflammatory and Antirheumatic Drugs,"[35] the FDA demands the inclusion of a placebo group when new-drug applications are submitted for fixed-dose combinations of nonsteroidal antiinflammatory drugs (NSAIDs) with codeine: "The combination must be shown to be superior to each component and the NSAID must be superior to placebo in order for the study to be persuasive." For the clinical evaluation of disease-modifying antirheumatic drugs (DMARDs), placebo controls also appear necessary: "In order to develop the body of information necessary for approval of a DMARD, studies using the following different control groups should generally be conducted: Comparison of the drug with a placebo...."

In at least one instance, the FDA refused to approve a new drug, a beta-blocker for use in angina pectoris, even though the application showed that the new drug had an effect similar to that of propranolol, an already approved drug. The application was rejected because the drug had not been tested against placebo,[36] even though a placebo-controlled trial would have violated the Declaration of Helsinki.

IS THERE A SCIENTIFIC RATIONALE FOR PLACEBOS?

The FDA is not alone in pushing for placebo controls. For example, a recent textbook on clinical drug trials advocates using them because "if a new drug has only been compared to an active control (without a placebo-controlled trial), this is not a convincing proof of efficacy (even if equivalence can be demonstrated)."[37] Without justification, such statements confer on placebo control a stature that ranks it with double blinding and randomization as a hallmark of good science.

The randomized, controlled trial is well recognized as the most desirable type of study in which to evaluate a new treatment. This recognition acknowledges the essential role of comparison and the importance of randomization in enhancing the comparability of two or more treatment groups. Using a placebo for comparison controls for the psychological effects of receiving some treatment and also permits blinding. No scientific principle, however, requires the comparison in a trial to involve a placebo instead of, or in addition to, an active treatment. Why, then, are placebo controls considered important? Three arguments have been advanced, none of which withstands scrutiny.

Establishing a Reference Point

By allowing the investigator to determine whether a new treatment is better than nothing (beyond the psychological benefits of treatment), a placebo control offers a clear benchmark. After all, even if a new treatment is worse than an existing one, it may still be "effective" in that it is better than no treatment. On the other hand, as Hill pointed out in 1963, the essential medical question at issue is how the new treatment compares with the old one, not whether the new treatment is better than nothing.[1]

Avoiding Difficult Decisions about Comparison Treatments

Determining whether one treatment is better than another is not always a straightforward matter. Beyond the question of efficacy, one can and should take into account unintended effects, interactions, costs, routes of administration, and other factors. Thus, it may appear simplistic to demand that the best proven treatment be chosen as the standard for comparison, if "best proven" refers only to efficacy. For some patients there may be advantages to a treatment that is inferior to a current standard with regard to efficacy but better with respect to cost or quality of life. For example, the adverse effects of some accepted treatments might offset the thera-peutic benefits for some patients sufficiently that a placebo control would be ethically justified. This reasoning involves a complex decision that should be defended in submitted research proposals and published reports. It is not justifi-able, however, to assign placebo controls simply to avoid the complex decision of which treatment should be used as a standard. Investigators are ethically obliged to make such decisions.

Bolstering Statistical Significance

One FDA scientist contends that placebo-controlled trials are superior to studies using an active treatment as the control because it is much easier to demonstrate a statistically significant effect in the former case.[36] The FDA relies heavily on statistical significance in judging the efficacy of new drugs.[36] Despite its popularity, however, this tool is not a good one for measuring efficacy.[38-42] The significance of an association depends on two characteristics—the strength of the association and its statistical variability. A weak effect can be "significant" if there is little statistical variability in its measurement, whereas a strong effect may not be "significant" if there is substantial variability in its measurement. Of the two characteristics, only the strength of the effect should be fundamental to the decision about approval of the drug. Ideally, statistical variability should be reduced nearly to zero when the magnitude of a drug effect is assessed, so that random error does not influence the assessment.

Unfortunately, the main way to reduce statistical variability is to conduct large studies, which are expensive. Statistical significance, on the other hand, can be obtained even in small studies, if the effect estimate is strong enough. When a placebo control is used instead of an effective treatment, the effect of a new drug appears large and may be statistically significant even in a small study. The scientific benefit, however, is illusory. Because the study is small, the measurement of the effect is subject to considerable statistical error. Thus, the actual size of the effect, even when a new drug is compared with placebo, remains obscure, and the study does not address the question of the effectiveness of the new treatment as compared with currently accepted treatments.

The small placebo-controlled studies fostered by the FDA benefit drug com-panies, which can more easily obtain approval of an inferior drug by comparing it with placebo than they can be testing it against a serious competitor. Smaller studies are also cheaper. Unfortunately, the costs saved by the drug company are borne by patients, who receive placebos instead of effective treatments, and by the public at large, which is supplied with a drug of undetermined efficacy.

There is no sound scientific basis for these arguments on behalf of placebo controls. Furthermore, regardless of any apparent merit these arguments have, scientific considerations should not take precedence over ethical ones, even if the use of active controls requires more difficult decisions about study design, more costly studies, and more complicated analyses.

ETHICAL COUNTERARGUMENTS

Two ethical arguments are sometimes advanced to justify the use of placebos when effective therapies exist. First, one can argue that withholding an accepted treatment may not lead to serious harm. For example, treating pain or nausea with a placebo may cause no long-term adverse effects, and the patient can call attention to any treatment failure or even choose to drop out of the study. Nevertheless, although withholding an accepted treatment may occasionally seem innocuous, allowing investigators to do so runs counter to the ethical principle that every patient, including those in a control group, should receive either the best available treatment or a new treatment thought to be as good or better. Instead, it concedes to individual investigators and to institutional review boards the right to determine how much discomfort or temporary disability patients should endure for the purpose of research. Ethical codes in medical experimentation have been developed expressly to shield patients from such vulnerability.

The second justification offered is that of informed consent. This argument says that if patients are fully informed about the risks of entering a trial and still agree to participate, there is no reason to prevent them from doing so. The ethical burden is passed directly to the patients. Informed consent is always desirable, but investigators should not put patients in a position in which their health and well-being could be compromised, even if the patients agree. There are several reasons. Despite the best efforts to inform patients, they will rarely if ever be as well informed about their treatment options as their physicians.[43] Moreover, even informed patients may not be disinterested enough to decide rationally whether it is tolerable to be deprived of an accepted treatment. Finally, patients are given the choice of participating in a trial or not, but they are given no choice about which treatments will be studied. It may be more desirable to a patient to be a part of the trial than to decline to participate, but it might have been preferable to be in a different trial that did not have a placebo arm.

RECOMMENDATIONS

Placebo is likely to continue to be used in place of an effective control until all parties to such studies are held strictly accountable for the ethical conduct of the research. We recognize that in some situations an accepted treatment may not be better than placebo for a given indication and that arguments can be made to justify the use of placebo instead of an existing treatment. The burden of justification, however, should fall not on critics but on those responsible for the research, including investigators, regulatory agencies, research sponsors, institutional review boards, and journal editors. All these parties should adhere to the precept that patients ought not to face unnecessary pain or disease on account of a medical experiment, and they should question the ethical legitimacy of using placebos in any experiment. Investigators should be routinely required by regulatory agencies, institutional review boards, and funding agencies to justify in writing the use of placebos in any study that uses them. This explanation should be part of all proposals, protocols, and published papers. Editors should be vigilant about questioning the use of placebos in experiments involving humans; regardless of assertions authors make about institutional review, editors should always require authors to justify in their manuscripts any use of placebo controls.

The change needed most is the enforcement of ethical guidelines at regulatory agencies, such as the FDA, which review research that may never be published. The FDA should conduct an ethical review of every study submitted to it. Any study proposing to use placebos in place of effective treatments without making a persuasive ethical justification should be disapproved. Studies involving unethical use of placebos should be ignored in the drug-approval process. Above all, scientific imperatives should never be weighed against established ethical canons.

ACKNOWLEDGMENTS

We are indebted to Alexander Walker, David Felson, Philip Cole, Howard Brody, Tom Beauchamp, Hans-Olov Adami, Ezekiel Emmanuel, Nancy Dreyer, Charles R. McCarthy, Luis García Rodríguez, Cristina Cann, Eleanor Druckman, Troyen Brennan, Arthur Caplan, Matthew Gillman, Jerry Avorn, Harris Pastides, and Howard Frazier for useful criticisms and suggestions.

REFERENCES

1. Hill, A.B., Medical ethics and controlled trials. *British Medical Journal* 1963;1:1043–1049.
2. Grodin, M.A., Historical origins of the Nuremberg Code. In: Annas, G.J., and Grodin, M.A., eds. *The Nazi Doctors and the Nuremberg Code: Human Rights in Human Experimentation*. New York: Oxford University Press, 1992:121–144.
3. Appendix 3. In: Annas, G.J., and Grodin, M.A., eds. *The Nazi Doctors and the Nuremberg Code: Human Rights in Human Experimentation*. New York: Oxford University Press, 1992:331–342.
4. Declaration of Helsinki IV, World Medical Association, 41st World Medical Assembly, Hong Kong, September 1989. In: Annas, G.J., and Grodin, M.A., eds. *The Nazi Doctors and the Nuremberg Code: Human Rights in Human Experimentation*. New York: Oxford University Press, 1992:339–342.
5. Greene, B.M., Taylor, H.R., and Cupp, E.W., et al., Comparisons of ivermectin and diethylcarbamazine in the treatment of onchocerciasis. *New England Journal of Medicine* 1985;313:133–138.
6. Tugwell, P., Bombardier, C., Gent, M., et al., Low-dose cyclosporin versus placebo in patients with rheumatoid arthritis. *Lancet* 1990;335:1051–1055.
7. Johnsen, V., Borg, G., Trang, L.E., Berg, E., Brodin, U., Auranofin (SK&F) in early rheumatoid arthritis: results from a 24-month double-blind, placebo-controlled study: effect on clinical and biochemical assessments. *Scandinavian Journal of Rheumatology* 1989;18:251–260.
8. Williams, H.J., Ward, J.R., Dahl, S.L., et al., A controlled trial comparing sulfasalazine, gold sodium thiomalate, and placebo in rheumatoid arthritis. *Arthritis and Rheumatism* 1988;31:702–713.
9. Trentham, D.E., Dynesius-Trentham, R.A., Orav, E.J., et al., Effects of oral administration of type II collagen on rheumatoid arthritis. *Science* 1993;261:1727–1730.
10. Felson, D.T., Anderson, J.J., and Meenan, R.F. The comparative efficacy and toxicity of second-line drugs in rheumatoid arthritis: results of two metaanalyses. *Arthritis and Rheumatism* 1990;33:1449–1461.
11. Rickels, K., Amsterdam, J., Clary, C., Fox, I., Schweizer, E., Weise, C., The efficacy and safety of

paroxetine compared with placebo in outpatients with major depression. *Journal of Clinical Psychiatry* 1992;53:Suppl:30–32.

12. Kiev, A., A double-blind, placebo-controlled study of paroxetine in depressed outpatients. *Journal of Clinical Psychiatry* 1992;53:Suppl:27–29.

13. Smith, W.T., Glaudin, V., A placebo-controlled trial of paroxetine in the treatment of major depression. *Journal of Clinical Psychiatry* 1991;53:Suppl:36–39.

14. Fabre, L.F., Buspirone in the management of major depression: a placebo-controlled comparison. *Journal of Clinical Psychiatry* 1990;51:Suppl:55–61.

15. Amsterdam, J.D., Dunner, D.L., Fabre, L.F., Kiev, A., Rush, A.J., and Goodman, L.I., Double-blind placebo-controlled, fixed dose trial of minaprine in patietns with major depression. *Pharmacopsychiatry* 1989;22:137–143.

16. Silverstone, T., Moclobemide—placebo controlled trials. *International Clinical Psychopharmacology* 1993;7:133–136.

17. Fabre, L.F., Double-blind multi-center study comparing the safety and efficacy of sertratine with placebo in major depression. Presented at the Fifth World Congress of Biological Psychiatry, Florence, Italy, June 1991, abstract.

18. Bowden, C.L., Brugger, A.M., Swann, A.C., *et al.*, Efficacy of divalproex vs lithium and placebo in the treatment of mania. *Journal of the American Medical Association* 1994;271:918–924.

19. Gralla, R.J., Tyson, L.B., Kris, M.G., and Clark, R.A., The management of chemotherapy-induced nausea and vomiting. *Medical Clinics of North America* 1987;71:289–301.

20. Cuningham, D., Evans, C., Gazet, J.-C., *et al.*, Comparison of antiemetic efficacy of domperidone, metoclopramide, and dexamethasone in patients receiving outpatient chemotherapy regimens. *British Medical Journal* 1987;295:250.

21. Cox, R., Newman, C.E., and Leyland, M.J., Metoclopramide in the reduction of nausea and vomiting associated with combined chemotherapy. *Cancer Chemotherary and Pharmacology* 1982;8:133–135.

22. Edge, S.B., Funkhouser, W.K., Berman, A., *et al.* High-dose oral and intravenous metoclopramide in doxorubicin/cyclophosphamide-induced emesis: a randomized double-blind study. *American Journal of Clinical Oncology* 1987;10:257–263.

23. Beck, T.M., Ciociola, A.A., Jones, S.E., *et al.*, Efficacy of oral ondansetron in the prevention of emesis in outpatients receiving cyclophosphamide-based chemotherapy. *Annals of Internal Medicine* 1993;118:407–413.

24. Gandara, D.R., Harvey, W.H., Monaghan, G.G., *et al.*, The delayed-emesis syndrome from cisplatin: phase III evaluation of ondansetron versus placebo. *Seminars in Oncology* 1992;19:Suppl:67–71.

25. Cubeddu, L.X., Hoffmann, I.S., Fuenmayor, N.T., and Finn, A.L., Efficacy of ondansetron (GR 38032F) and the role of serotonin in cisplatin-induced nausea and vomiting. *New England Journal of Medicine* 1990;322:810–816.

26. Citron, M.L., Placebos and principles: a trial of ondansetron. *Annals of Internal Medicine* 1993;118: 470–471.

27. Braunwald, E., ACE inhibitors—a cornerstone of the treatment of heart failure. *New England Journal of Medicine* 1991;325:351–353.

28. Kelbæk, H., Agner, E., Wroblewski, H., Vasehus Madsen, P., and Marving, J., Angiotensin converting enzyme inhibition at rest and during exercise in congestive heart fialure. *European Heart Journal* 1993;14:692–695.

29. Packer, M., Narahara, K.A., Elkayam, U., *et al.*, Double-blind, placebo-controlled study of the efficacy of flosequinan in patients with chronic heart failure. *Journal of the American College of Cardiology* 1993;22:65–72.

30. Cowley, A.J., and McEntegart, D.J. Placebo-controlled trial of flosequinan in moderate heart failure: the possible importance of aetiology and method of analysis in the interpretation of the results of heart failure trials. *International Journal of Cardiology* 1993;38:167–175.

31. Svetkey, L.P., Brobyn, R., Deedwania, P., Graham, R.M., Morganroth, J., Klotman, P.E., Double-blind comparison of doxazosin, nadolol, and placebo in patients with mild-to-moderate hypertension. *Current Therapy and Research* 1988;43:969–978.

32. Torvik, D., and Madsbu, H.P., Multicentre 12-week double-blind comparison of doxazosin, prazosin and placebo in patients with mild to moderate essential hypertension. *British Journal of Clinical Pharmacology* 1986;21:Suppl 1:69S–75S.

33. Carr, A.A., Bottini, P.B., Prisant, L.M., *et al.*, Once-daily verapamil in the treatment of mild-to-

moderate hypertension: a double-blind placebo-controlled dose-ranging study. *Journal of Clinical Pharmacology* 1991;31:144–150.

34. Code of Federal Regulations, Food and Drugs, 21. Parts 300 to 499. Revised as of April 1, 1993. § 314.126. Washington, D.C.: Government Printing Office, 1987.
35. Guidelines for the clinical evaluation of anti-inflammatory and antirheumatic drugs. Washington, D.C.: Department of Health and Human Services, 1988.
36. Temple, R., Government viewpoint of clinical trials. *Drug Information Journal* 1982; January/June: 10–17.
37. Spriet, A., Dupin-Spriet, T., and Simon, P., Choice of the comparator: placebo or active drug? In: *Methodology of Clinical Drug Trials*. 2nd ed. New York: Karger, 1994.
38. Salsburg, D.S., The religion of statistics as practiced in medical journals. *American Statistics* 1985;39:220–223.
39. Gardner, M.J., and Altman, D.G., Confidence intervals rather than P values: estimation rather than hypothesis testing. *British Medical Journal* 1986;292:746–750.
40. Rothman, K.J., Significance questing. *Annals of Internal Medicine* 1986;105:445–447.
41. Walker, A.M., Reporting the results of epidemiologic studies. *American Journal of Public Health* 1986;76:556–558.
42. Savitz, D.A., Is statistical significance testing useful in interpreting data? *Reproductive Toxicology* 1993;7:95–100.
43. Cassileth, B.R., Zupkis, R.V., Sutton-Smith, K., and March V., Informed consent—why are its goals imperfectly realized? *New England Journal of Medicine* 1980;302:896–900.

Reforming the IRB Process: Towards New Guidelines for Quality and Accountability in Protecting Human Subjects

Jean Campbell, Ph.D.

Research Assistant Professor in Psychiatry, University of Missouri School of Medicine-Columbia, Missouri Institute of Mental Health, 5247 Fyler Avenue, St. Louis, Missouri 63139-1361

Degradation, anger, and sadness are often voiced by people with mental illness when they speak about their experiences as research subjects. These experiences stand as a robust critique of biomedical and services research. Rather than a shared reality, research represents two quite separate worlds, the meanings of one being significantly different from the other. Separated by role and function from the person and his/her suffering, the researcher monitors functionality, recidivism, and symptom control. However, people judge benefit and harm by the impact on the quality of their lives. This inherent tension has vast implications for the production and distribution of scientific knowledge, and for the development of the cultural reference points that underlie every aspect of the personal and social relationships in the research process. Ultimately, the social relations of science have the potential to alter the symbolic processes where reality is produced, maintained, repaired and transformed. The point of departure is visionary and comes into existence through what one knows, does, and values as a human being.

The growing tide of health consumerism is one of the most compelling forces in the future of psychiatric research. It is based on the assumption that persons who seek health services are consumers just as are persons who seek other types of services. The doctor is perceived to be the purveyor of a service, and the patient is viewed as the buyer. The consumer listens to the thoughts of the provider, but ultimately makes his or her own decisions. Consumerism implies that values derived from principles of good medical care must be interpreted and operationalized through reference to the patients personal health care values and desires.

Although consumerism has periodically emerged as a force in American society, its current application in health care draws its roots from two marketplace trends: consumer rights protections with its concerns regarding manufacturing and product safety, and Total Quality Management (TQM) with its focus on customer satisfaction. The former is grounded in a profound distrust by individuals of the actions and motives of providers of products and services, the latter promotes collaboration between provider and consumer and seeks to answer such questions as what do customers prefer. Both trends emphasize the need for information that accommodates consumer rights and interests.

However, while there is general acceptance of medical consumerism, psychiatric consumers must overcome tremendous personal and political barriers to gain control over and responsibility for their own lives in the medical marketplace. Persons with psychiatric disabilities are embraced by the mass media as secular versions of the devil. With madness the scapegoat for violence and uncontrollable mayhem in modern society, mental health consumers are seen as the quintessential Other—people who represent the subterranean depths of humanity and whose differentness makes them not really human at all. This is compounded by the assumption that the individual is morally at fault. The diagnosis of mental illness permeates all aspects of the lives of mental health consumers by establishing authority over their ideological, material, and emotional resources. It strips away a persons objective power and will to power as he/she encounters the social world and participates in the reproduction of daily life.

Because of pervasive cultural stereotypes of people with mental illness as violent and incompetent, as well as the profound poverty of most, it is not surprising that mental health systems have been slow to change traditional paternalistic medical relationships to ones that empower the customers of psychiatric services and products. The problem of enabling psychiatric consumers to participate in research and to serve on review boards is that the perspectives of the mental health consumer movement are discredited by large sectors of its potential audience. In general, the role of the mental health research professional is to direct and prescribe; the role of the subject is to cooperate. While many ex-patients are angry at treatment they have received or witnessed as research subjects in the mental health system, it is presumed that they do not know what is in their own best interests.

People with mental illness and their families have fought with limited success to redress perceived injustices in the research of mental health treatments and services by organizing advocacy groups. The National Alliance for the Mentally Ill (NAMI) has worked for most of the past decade to improve service delivery in the public mental health system. Although often at odds with mental health consumer leadership on issues such as involuntary commitment and psychiatric medications, this conference points to human subjects' rights protections as one important area where families and consumers can work together.

Mental health consumers have also built a national self-help movement with a growing number now participating in research and evaluation, and taking leadership roles in policy and administration of mental health services. Consumers have led efforts to determine housing preferences, to define outcome measures, and to develop partnership models with mental health professionals. Some professionals and policymakers have responded to consumer demands by redesigning professional roles and creating opportunities for people who receive services to provide input and perspective. The proliferation of Offices of Consumer Affairs in state mental health agencies, state-wide consumer conferences, consumer-directed social centers across the nation, model consumer case management programs, and support for consumer research and policy professionals are indicators of a vibrant mental health consumer empowerment movement.

It is not surprising that scientists have been unable to cope with growing demands for changes in the research process, particularly those coming from people with mental illness and their families. As with science in general, psychiatric research is bound by the past and inscribed with power, bias, and stereotypy. Reform in research methods, protocols, and human subjects protections has historically been based on the assumption that the expert knows best about the operating principles of research. Further, scientists believe they know values and preferences of their subjects; they are disinterested and objective without biases of their own; and, they choose for their subjects what the subject would choose for themselves if they had the same knowledge. This seduction by authority inevitably influences future behavior to mimic current practices, and has led to the creation of protocols and protections without knowledge of how research subjects would address these issues.

Therefore, one of the most important forces for reform of human subjects protections is dialogue and collaborative work regarding ethical issues in psychiatric research. By listening respectfully and treating research subjects with dignity, dialogue has the power to bridge the differences between scientist and subject and to generate new knowledge and understanding. Instead of resisting criticism, researchers should welcome mental health consumers and their families to the process, saying, Gee, how can we improve? The first step is for each of us to interrogate the a priori assumptions of the research enterprise. What are the values, sensibilities, biases, and stereotypes that inform the work of researchers?

The next step is to look at the context in which the IRB and human subjects' protections operate, not just examine the regulations themselves. The gritty work of churning over knowledge is much like any other labor process. In the production of research, dehumanization naturally occurs: one forgets that the objects of inquiry are human beings. As relationships are routinized, you lose touch with the relatedness of one person to another whether you are serving burgers at McDonalds or interviewing people with schizophrenia. Therefore, you need the conscience of others. One way is to adopt a person-centered language in psychiatric research rather than continuing to use terms such as schizophrenic that implicitly dehumanizes people with mental illness. A people first construction challenges the social relations of domination in the review and conduct of psychiatric research and could help keep our consciousness alive to peoples' humanity and worth.

A third step is to strive for excellence and not just focus on minimal compliance to regulations. We need to rigorously apply the protections that already exist; and we need to look at the applications of protections to determine if they are effective. We also need to create an environment in which whistle-blowing is really supported. One approach to fostering excellence, is to develop a gold standard for human subjects' protections, and to find and support those people that have gone beyond the regulations to develop superior types of protocols. I believe that people who have a mental illness and are researchers should be sought out and supported as the jewels of a IRB reform process. In some cases, efforts to enhance and strengthen protections would restrict the conduct of research. However, this has the potential to force psychiatric research to develop better protocols and methodologies. Consider the positive impact of the curtailment of animal testing in cosmetic research that prompted creativity rather than suppressing such endeavors.

Finally, we focus much of our attention on ethical dilemmas in neurobiological research (especially experimental psychotropic drug research). Yet, the services researchers and professionals who conduct survey research or design and direct quality assurance and management information systems also face important challenges in their investigations and data collection efforts involving mental health consumers. There is the assumption that these forms of data collection pose minimal risk to participants. However, subjects report that there has been considerable abuse and greater risks are involved than most researchers realize. Ethical issues are rarely considered in this field and the research and data community has yet to achieve consensus about standards and policy recommendations. Such trends as the growth of survey and services research in the community and in institutions; increasingly rigorous evaluation of services demonstrations; the ascendance of randomized controlled trials in services research; the development of management information systems used for patient tracking, managed care and quality assurance programs; and the expanding influence of consumers, family members and consumer advocates in the research and data collection arena have clouded traditional human subjects' protections and compromised boundaries of appropriate ethical authority and protocol. There are exceedingly complicated questions in the following areas: appropriate use and language of informed consent forms and protocols; confidentiality and privacy of subject identified data; professional conflicts of interest; limited or bias evaluations of risks and benefits to subjects; consumer participation in research and data system designs, implementation, and monitoring; provisions for adequate human subjects review and oversight in community settings not covered by IRBs; the ethics of randomization and sample recruitment; use of coercion in subject recruitment and retention; and, gender and ethnic parity in research samples. These are the gray zones of IRB responsibility. IRBs do not appreciate ethical issues in survey research or management information systems, although they compromise client confidentiality and privacy as well as the ability of participants to consent to data collection. In addition, in research demonstration projects where variables such as choice prove to have significant therapeutic value, the risk of recidivism should be considered with the same caution as in neurobiological interventions.

Since feedback is essential for a reiterative, continuous quality improvement process, the promotion of alliances in human subjects' protections is essential. It is especially important to engage mental health system stakeholders in dialogue related to ethical concepts and current protocols, provide training, and begin to develop recommendations from the field because of evolving complexities and ambiguities in research and data collection. Since most current policies and protocols were developed without the insight and perspective of the consumers themselves, researchers have to be open and welcome input as critical to the excellence of their own work. Ironically, the existence of IRBs is often used as a shield against reform to counter the concerns of research subjects and their families.

One model that redresses the asymmetrical power relationships between researcher and subject is the emerging multi-stakeholder design for research collaborations. In the 1992 amendments reauthorizing the Rehabilitation Act of 1973, Congress also indicated a strong desire for the beneficiaries and constituencies of the National Institute on Disability and Rehabilitation (NIDRR) to participate in the research and implementation process. The Constituency Oriented Research and Dissemination initiative (CORD) was drafted to provide a mechanism for all research stakeholders to actively participate in identifying research needs, setting priorities, RFP development, the application preparation process, peer review, making awards, conducting projects, dissemination and utilization of results, and conducting evaluations. As noted in the CORD policy statement, The acronym CORD is apt, in that beneficiaries and constituency participation in the NIDRR research and dissemination process is the cord that ties together the producers, disseminators, and users of research outcomes to achieve a common goal— improving the lives of individuals with disabilities." CORD is a blueprint for generating research inputs by promoting the opportunity for each individual to share and utilize his or her skills, background, and experiences so that common objectives can be defined and achieved in partnership with people with disabilities. Certainly, the protections of human subjects should not solely be delegated to those with special research knowledge or expertise, and IRB membership could be expanded to include other research stakeholders.

With the initiative and support of some progressive scientists, the research community is slowly moving towards more accepting attitudes and reciprocal relationships. No set of regulations, no matter how carefully drawn or enforced, can really protect human subjects without a profound change in how scientists relate to the people they research. We need only to remember that there were protections in place in Germany prior to World War II. Still scientists brutalized people without compunction, seeing them as less than human and expendable for scientific progress. It is clear that neither psychiatric research nor the methodological and communication technologies that facilitate its activities are simply mechanical, electronic, or intellectual tools and protocols that serve the needs of individuals and groups within society. Since reform in protecting research subjects is limited by the attitudes and social relationships of those that research, manage, and deliver services, it therefore follows that the source of new knowledge to guide the generation of human subjects' protections may lie in the incongruities between the perceptual and experiential framework of the experts and those that receive services.

In conclusion, as we develop models to involve consumers and their families in developing new guidelines for quality and accountability in protecting human subjects in psychiatric research, we must understand that not everyone is going to be a team player. In fact, some people are going to get very angry, and others get uncomfortable as we collaborate. This tension is inextricably linked to the movement to common ground. On the other hand, to amend Dr. Shamoo's statement on the conference brochure, our hopes are that we can reach a consensus as how to conduct research ethically *in collaboration* with human subjects with neurobiological disorders.

Comments of Laurie M. Flynn

Laurie M. Flynn

Executive Director, National Alliance for the Mentally Ill, 2101 Wilson Boulevard, Suite #302, Arlington, Virginia 22201

I would very much like to thank Dr. Adil Shamoo and the organizing committee for putting together this very important conference. I am especially appreciative to Dr. Shamoo for the opportunity to offer some reflection on these discussions from a family and consumer perspective.

As many of you know, the National Alliance for the Mentally Ill (NAMI) has been a vigorous supporter of psychiatric research since our founding fifteen years ago. I am very proud of our effective advocacy, which has resulted in a tripling of the research appropriations for the National Institute of Mental Health over the past decade. In the current climate of cutbacks in Congress, our advocacy must, and will, continue.

Clinical research, especially, has been strongly advocated by NAMI, and we have sought a greater focus in that research on those who are most severely ill, because they are our family members. A national study, conducted by Johns Hopkins University in 1991, indicated that nearly 65 percent of NAMI members identify schizophrenia as the psychiatric disorder confronting their relatives. Clinical research on severe mental illnesses is our only hope for improvements in treatment and is a key source of new and more effective medications.

Let me preface the rest of my remarks by stating a personal belief. I do not believe that there is a widespread pattern of abuse of human rights on human subjects in psychiatric research today. I do believe improvements in our research settings and systems are needed. This conference offers a unique opportunity to identify those improvements and work together to implement them. Even though I believe the problems we face are not major ones, I want to emphasize that nothing—no study, no research, no scientific advance—is more important than vigorously protecting the human rights of subjects of research, especially those who are most disabled and vulnerable.

As we have heard from many speakers, informed consent must be a process, not an administrative event. Clearly, much effort at improvements in the informed consent process is crucial. I believe the model utilized for patients with Alzheimer's disease can be an especially useful guide. I want to thank Dr. Trey Sunderland for offering his very instructive presentation.

The comments were delivered after dinner at 7 PM, January 7, 1995.

I believe we must proceed carefully to develop clearer criteria to assess competency. We must develop stronger safeguards for those who are seriously impaired. In some situations, consent auditors may be useful in strengthening the informed consent process, I believe NAMI members are a potentially valuable resource in this effort.

As we have heard, institutional review boards (IRBs) vary considerably across the nation. IRBs are an impressive mechanism, but certainly their function can be strengthened. Again, I believe more family and consumer participation in the local IRBs will be an important component in the protection of psychiatric research subjects. We must mandate a voice for vulnerable populations in the IRB process.

Additional training focused on serious mental illness and its impact on individuals and families is another way to improve the work of the IRB. IRB members need a clear understanding of the complex ethical issues of risk and benefit involved in many studies, especially those which involve placebo controls.

As a community, we must work together to minimize the exposure of patients to harm by developing designs that limit the numbers of individuals and the length of their exposure to placebo.

IRBs must vigorously monitor and follow-up on all approved projects, especially those involving moderate to high risk, using impaired subjects, such as those with a diagnosis of schizophrenia, manic-depressive illness, or Alzheimer's disease. Such monitoring will be especially important where large numbers of institutionalized human subjects are involved in the research design.

We must explore an expanded role for substituted judgment, designated power of attorney, and creative use of advanced directives.

One of the most controversial issues among families and consumers today is the use of placebo in medication trials. Many NAMI members have come to feel a placebo is not acceptable because we view schizophrenia and other psychotic disorders and affective illnesses as life-threatening diseases. In fact, NAMI three years ago asked such designation from the Food and Drug Administration for schizophrenia and manic-depressive disorder. Clearly, this is a subject which needs a candid and comprehensive dialogue involving clinicians, researchers, administrators, and most importantly, families and consumers.

As a field, we need to avoid oversimplifying the issues. To do so will create a continuing stream of damaging, sensationalized media coverage and could lead to the creation of undue barriers and excessive protectionism.

If we believe that informed consent is not an event, but a process, we have a responsibility to educate and inform not only individual consumers, but families and consumers in organized settings. Advocacy groups like the National Alliance, the National Depressive and Manic-Depressive Association, and others, can be helpful communicators.

It is vital that we recognize, even as we identify psychiatric patients as vulnerable, that most consumers are fully capable much at the time. They must be directly engaged and involved in decisions relating to research protocols for which they volunteer. We also know that many individuals with a psychiatric illness will need special protection in order to participate in today's research protocol. All families and consumers need ongoing information sharing. One of the strongest messages we have received from our own NAMI consumers is their frustration at not hearing more about the purposes and the results of research in which they have participated.

I believe the Federal Office for Protection from Research Risks (OPRR) must have a more active oversight role. It is important that OPRR have the ability to monitor research in progress and the funding for regular site visits, especially in long-term studies utilizing high risk design with vulnerable populations.

As an advocate, I believe in the power of creative tension. Research on schizophrenia, manic-depressive illness, and other neurobiological disorders is vital and must go forward. We must frame solutions that are sized to the actual scope of the problem, as we can together define it. And we must continue the dialogue of honesty and respect among all stakeholders. Perhaps most important, we must remember the shared value on which our enterprise must be based. Our goal is to alleviate human suffering while respecting human rights. This conference has been an extraordinary opportunity to begin that critical work.

Thank you.

Enhancing the Climate of Trust in Clinical Psychiatric Research

Frederick K. Goodwin, M.D., and Suzanne W. Hadley, Ph.D.

The Center on Neuroscience, Medical Progress & Society, George Washington University Medical Center, Department of Psychiatry, 2300 I Street N.W., Room 514, Washington, DC 20037

Given the varied backgrounds of the participants of this meeting, it is important that we identify some basic premises associated with specific strategies that might enhance the climate of trust in clinical psychiatric research. We address these issues primarily from the perspective of the clinician and clinical investigator—two roles and identities—which, although distinct, overlap substantially. Our premises, hopefully shared by most of you, are as follows:

1. Incompletely or unsuccessfully treated mental illness, especially severe mental illness, is an enormous personal tragedy burdening patients, families and society as much as any severe chronic illness in medicine and, indeed, more than most.

2. Biomedical and behavioral research has been central to the interests of the mentally ill for a number of interrelated reasons. One of these is, of course, the development of new understanding of the illnesses and new treatments which are, at least in part, based on research with patients. One need hardly mention the example of the new antipsychotic agents such as clozaril, which have revolutionized the outcome for a group of very treatment resistant schizophrenic patients.

Another "output" of research, often overlooked, has been its important contribution to the destigmatization of mental illness and the mentally ill, and the legitimization of the professions that serve them. Studies of prejudices against the mentally ill make it clear that such attitudes are more likely to flourish in an environment of ignorance and, thus, the best antidote is enhanced knowledge, especially about biomedical aspects of mental illness. It is precisely the kind of knowledge produced by the biologically based investigations of mental illness that has reinforced the message that mental illness is an integral part of the health spectrum and is comparable to "physical" illnesses in every important respect.

Even more specific than this general destigmatization/legitimization has been the field's ability recently to use solid research data on diagnosis and treatment of mental illness to argue persuasively for equity in the reimbursement sphere. Indeed, patient and family groups have long understood their debt to research and

309

have paid it back many times over in their sustained, vigorous and sophisticated support for the National Institute of Mental Health (NIMH) research budget over many years.

This might be the appropriate juncture to take note of the increasing sense of partnership that has developed at the national level between the research community, and consumer organizations such as the National Alliance for the Mentally Ill (NAMI) and the National Depressive and Manic-Depressive Association. A concrete example of this partnership is reflected in the NIMH Knowledge Exchange program, a program that one of us (F.G.) initiated. This program formally brings together researchers and consumers to consider important research questions jointly. It is the Institute's expectation that the research community will learn from the patient perspective and vice-versa and that both the selection and conduct of research will be enhanced. Another example of this increasing partnership is the arrangement by which representatives from NAMI review the NIMH research portfolio and provide advice regarding its content and balance. Finally, the new partnership is being expressed in the specific involvement of patient and family members in National Institutes of Health conferences dealing with the protection of human research subjects. The NIMH is justifiably proud that it is the first Institute to include consumers in formal deliberations about human subject protection.

3.   Ignorance is our common enemy. A corollary of this basic assumption is that anything which slows the pace of biomedical research on mental illness has significant ethical consequences. This is simply another way of stating the obvious— most human endeavors of consequential value are not risk-free. Indeed, risk characterizes both "sides" of this issue, that is, the risk associated with slowing research by "protections" that are overly cumbersome versus the risk to patients that are inherent in certain research procedures. It is important that we not tacitly assume a goal of zero or nearly zero risk for research. To do so would be putting clinical research in a special class, with risk standards that would be virtually impossible to meet in most other endeavors. We must continuously remind ourselves that most worthwhile human endeavors from medical care to getting from one place to another, involve risk, and often these everyday activities incur more risk than what one encounters in psychiatric research. We must watch for the unspoken assumptions that our goal for psychiatric research should be zero risk. It cannot be.

4.   Ultimately to understand mental illness one must study actual patients with the disease. And, with respect to studies of treatment, they simply cannot be limited to those that can be expected to benefit the individual patient/subject. If that were so, it means that we already know the answer (i.e., that the experimental treatment is likely to work) and, therefore, not worth doing in the first place. Also patients with mental illness must be free to volunteer for research which is not intended to evaluate a possible benefit at all. Examples here include spinal taps for biochemical studies, brain imaging involving radiation exposure and "challenge" studies using a pharmacological "probe." Without such studies an important bridge between laboratory science (including animal studies) and the patient could not be crossed. On the other hand, it is reasonable to expect that direct clinical research of more than minimal risk will, in the aggregate contribute to the well being of a particular class of patients in the future.

5.   As identified in both the Helsinki Declaration and the Nuremberg Code, the ultimate and essential foundation of patient protection is clarity about the investigator's *personal moral responsibility*, a point already emphasized by Loren Roth (1988). Therefore, anything which dilutes this responsibility or seems to allow investigators to skirt the obligation to continually exercise and strengthen their own personal moral compass has the potential to be harmful to the patient's interest. One instance of this is overalliance on formal patient consent as a way of blunting questions about whether the risk benefit ratio associated with a particular study is, in fact, morally acceptable to the investigator and his or her peers. The primary responsibility for the assessment of risk rests with the investigator, not with the patient, no matter how well informed. In a word, written informed consent (or documented ascent) is necessary but not sufficient. As noted by others at this meeting, true consent is an ongoing process. The limited number of studies that do exist in the field suggest that, ultimately, one critical factor determining a patient's consent or refusal is the level of trust in the investigator, not the specific detail in the consent process. Indeed, when improperly handled, the consent process can become a barrier between patient and investigator, ultimately, eroding trust.

6.   Across the spectrum of our society, people from all walks of life are expressing increasing concern about two unhealthy and interrelated trends in our contemporary culture. The first, which seems to have started a generation ago, perhaps arising out of the post-Vietnam, post-Watergate era is the steady institutionalization of mistrust, with the emergence of a veritable mistrust industry as reflected in new bureaucratic structures for "oversight." This trend is also evident in the proliferation, some would say explosion, of regulations and a cynical mistrusting tone in our media.

The second worrisome trend is the ever increasing tendency to look to the law to solve problems that are essentially ethical dilemmas about reconciling competing goods. Noting the more than threefold increase in the number of lawyers per capita in the U.S. over the last twenty-five years, an increasing number of legal scholars are sounding the alarm about the extension of the principles and machinery of the law into arenas for which its processes, especially the adversarial process, are not particularly well-suited.

The legalistic tendency to treat every *confluence* of interest as a *conflict* of interest helps balkanize us into competing factions or "special interest groups;" this balkanization facilitates simplistic "all or nothing" thinking, freeing us from the more demanding moral task of using our own best judgment in balancing competing values. We would submit that the topic of this meeting represents one of those areas that should not find its solutions by turning primarily to our legal infrastructure. Certainly the law is relevant to these issues and legal experts must have a place at the table, but none of us—clinicians, researchers, consumers—should abdicate our responsibility by "turning over" these tough problems in the hope that they will be resolved in our legal system.

7.   Adversarial processes, which in the abstract can seem benign, may have outcomes that are hardly benign when operated by human beings. For example, if people whose orientation and reflexes derive essentially from the adversarial process become patient advocates charged with negotiating research on behalf of patients, a situation is created in which the advocates may measure their success by

the number of patients who refuse to participate in research. The same might be said for consent auditors. The only antidote to this adversarial splitting between the researcher and the "protector" (with the patient caught in the middle) is to insist that those involved in the process have a strong and believable personal stake in both the research and the welfare of the individual research subjects. Obviously, two groups which come to mind that do have or should have a stake in both are members of the patient's family and the professionals who are treating the patient. Along with the patient they share the confluence of two important interests: new knowledge about the illness and the well-being of the individual patient. It is the complex process of dealing with the confluences of many different interests within an individual that constitutes moral judgment. To transplant the adversarial system from the criminal law, with its either/or judgment of guilty or innocent to the subtle and complex arena of patient participation in research represents an abdication by stakeholders—patients, family members, clinicians and researchers.

A more subtle issue is how should consent be obtained when a patient has impaired capacity to consent, sometimes associated with a severe mental illness. Again, to recognize the confluence of interests involved, the investigator should share this responsibility with another stakeholder who also has an interest in both the research and the patient's well being. Further, when a patient's competence is at issue, who should evaluate it? What's critical here is that this evaluation should be independent of the investigator, it should be systematic, and it should involve the clinicians who know the patient, as long as those clinicians have an understanding of and appreciation for the importance of the research in question.

8.   We each owe it to ourselves and to each other to acknowledge that personal motives are mixed with more noble altruistic motives in everyone involved in this process. It seems to us that the bulk of the concern about personal motives contaminating altruism has focused on investigators, not on those who see themselves as protectors of research subjects. So let us remind ourselves that we all can and probably do have personal agendas admixed with altruism-earning public esteem, praise from colleagues, the competitive gratification of winning in the extreme called self-aggrandizement, etc.

9.   The category of behavioral research known as decision science has yielded insights that suggest that human systems which imply or assume mistrust have negative impacts on attitude and behavior. We are not talking simply about the erosion of the doctor/patient relationship, but rather the general tendency for all of us to resent systems or structures whose premise is that we are not to be trusted. Depending on the individual, this might contribute to negative behaviors ("acting out") or the strict adherence to the letter of the rules while not really engaging one's conscience in moral choices.

10.   History tells us that when guidelines meet bureaucracies they almost inevitably get cast in stone as rules and regulations and the flexibility that was implicit during the development of the guidelines becomes ossified. Critical to the maintenance of flexibility is the independence of the local institutional review board in determining how the general principles outlined by government are actually applied. Indeed, it is at the local level that consumers can have their greatest impact either on the IRB or the actual research setting itself. An example of the latter is when informed consent is incorporated into a group process; in this way patients can help each other with these complex issues.

11. It is axiomatic that the protection of human subjects issue has special valence with respect to the mentally ill, given that some segment of this population, however small, will have impaired capacity to consent. But we might consider other factors that could add to the extra attention to the ethics of research with the mentally ill, as is exemplified in the recent media coverage of the UCLA case. One possibility, to our knowledge not yet mentioned, is stigma. In addition to the mentally ill, the professionals who study them are also stigmatized and the public may be more disposed to believe the worst.

Another reason why mental illness research stands out in the ethical arena is simply that we've paid more attention to these issues, which is not surprising given our professional orientation. Thus, for example, in the McArthur Treatment Competence Study—a study with implications for research consent—Appelbaum and Grisso found that while schizophrenic subjects showed significantly poorer understanding of treatment-related information than did angina patients, depressed patients' understanding did not differ from that of angina patients. Schizophrenic and depressed patients both showed poorer appreciation of illness and likely benefits from treatment, as well as poorer rational manipulation of information, as compared with angina patients. However, the majority of schizophrenic and depressed patients performed in the same range as their matched, non-ill community controls on all competence variables—including understanding of treatment information, appreciation of illness/likely benefits from treatment, and rational manipulation of information—leading Appelbaum to observe that "... the glass can be seen as either half-empty or half-full" (Appelbaum, 1996, p. 247; and in this volume).

Obviously, it would be inappropriate and dangerous if our long-standing attention to this complex issue reinforced the automatic assumption of many that mental illness research should be segregated from the mainstream, with a special set of procedures and regulations.

In closing, let us suggest that this whole area needs more study. How big a problem is abusive clinical research? If it is indeed rare, then we must insist that the machinery established to protect against a rare event not end up doing more harm than good. What is the actual impact of clinical research on clinical care? In the one study we know which addressed this (Kocsis *et al.*, 1981), there was a trend toward better clinical outcomes for patients on the research ward when compared with similar patients on a purely clinical ward, which may be attributable to differences in the treatments given to patients on the two wards.

And what about the flip side of that—is research actually being inhibited? Anecdotes suggest that it is, but we need more solid information. What about randomization? What about informed consent? Although there are ongoing studies of these issues, important questions remain. Among patients who have been research subjects, what are their actual attitudes about their participation? There are some data indicating that a very high percentage of research subjects would volunteer again; broader surveys are needed.

In addressing each of these questions we must consider what is known about clinical research in general. How do we compare? Just how unique is psychiatric research? Finally, an area that clearly has special relevance to psychiatric research, what is the impact of research participation on patients' self-esteem? This is an especially important question for the mentally ill because of the shame that is uniquely associated with these illnesses. As one of our research patients said many years ago, "Doctor, this is the first time that anything good has come out of these terrible episodes."

This conference and the confluence of interests represented here could provide an initial rough template for a research agenda. If that is an outcome of this meeting, then in our view, all of the effort has been worthwhile.

REFERENCES

Appelbaum, P.S. (1996) Patients' competence to consent to neurobiological research. *Accountability in Research* 4:241–251.

Benson, P.R., Roth, L.H., Appelbaum, P.S., Lidz, C.W., and Winslade, W. J. (1988) Information disclosure, subject understanding, and informed consent in psychiatric research. *Law and Human Behavior* 12:455–476.

Kocsis, J.H., Frances, A., Kalman, T.P., and Shear, M.K. (1981) The effect of psychobiological research on treatment outcome. *Archives of General Psychiatry* 38:511–515.

Science and Ethics—
The Search for a Balance

Herb Pardes

Vice-President and Dean, College of Physicians and Surgeons, Columbia University, 630 West 168th Street, New York, New York 10032

I'm very pleased that this conference is taking place, and I say that pointedly because I think the stakes are substantial. The most important thing in my mind is that conversation be encouraged and that there be a real exchange. If you'll pardon a personal moment, about a year-and-a-half ago my own child was threatened by a life-threatening illness. It sometimes takes that kind of experience to realize how horrifying it is to be a parent watching your child suffer from an illness and realize also the enormous damage and distress it does to one's life. Fortunately, that episode came out okay, but in the process the stress is dramatic. I believe that psychiatric providers, researchers, patients and families are all interested, concerned, dramatically affected by psychiatric illness and therefore should form a natural and national partnership.

The seriousness of the illnesses we are dealing with we all know. Whether it's patients on very potent medications, whether it's people who can go on to commit suicide in the course of a psychiatric illness, whether it's patients suffering from symptoms or families suffering secondarily from the effects on the patients, the effects are dramatic. And we are all happy to have seen progress in psychiatric research.

I remember when I first came into this field too many decades ago, that there was very little that the psychiatric community had to offer therapeutically. I believe that research and improvement in clinical treatment availability is of value to a large number of people. But still there's a very long way to go. Theoretically there should be no stronger alliance than that between families and scientists, with both using their aggressive energies to gain control over psychiatric illness, and relieve much of the pain patients and their families experience.

There have been recent unfortunate tensions in that relationship centered around the nature of certain research approaches. We in the academic community must be sensitive to the citizen community, and assume that the reasonable rights of patients are observed. On the other hand, families and patients must help us

---

These are edited transcripts of Dr. Pardes' remarks delivered Monday morning, January 9, 1995 during the conference in Baltimore.

weigh the alternatives of the benefits of certain research approaches versus the risks. we do not want to condemn patients and people to a level of treatment destined to remain unimproved because of an absence of good clinical research. The possibility of using less drugs, whether in dosage or in numbers of administration, is attractive, and that's particularly true given the fact that these drugs are potent drugs and create side effects of substantial consequence. On the other hand, the clinician's responsibility is first and foremost the best interest of his or her patient.

Given that increased knowledge regarding the use of treatments is a goal for the larger population, the interest of the individual and the larger group can come into conflict. Informed consent is one of the ways of addressing this conflict, but it must be conducted with genuine sensitivity and concern that the patients understand. In an eloquent article by Sam and Deborah Hellman in the *New England Journal of Medicine*, these authors refer to moral theory for an assertion that human beings, by virtue of their unique capacity for rational thought, are bearers of dignity. As such, they ought not be treated merely as a means to an end, rather they must always be treated as ends unto themselves. Furthermore, the doctor/patient relationship requires that doctors see patients as bearers of rights who cannot merely be used for the greater goal of humanity. And the Hellmans argue that such rights cannot be waived or abrogated. They are inalienable.

In the same issue of the *New England Journal of Medicine*, Eugene Passamani notes the utility of clinical trials as a preferred means of evaluating an ever-increasing flow of innovative diagnostic and therapeutic maneuvers. He further notes the mechanisms society has created, some of them at least, to insure that individual patient interests are attended to should they choose to participate in a clinical trial.

First, again, is informed consent and Passamani stresses that this informed consent should involve explicitly informing a potential participant of the goals of the research, its potential benefits and risks, showing alternatives to participation and also emphasizing the right to withdraw from the trial at any time. He adds that a patient must always be aware that he or she is part of an experiment.

Further, Passamani calls attention to the clinical considerations for the patient. By this he refers to the fact that on the basis of available data that a community of competent physicians would be content to have their patients pursue any of the treatment strategies being tested in a randomized trial since none has been clearly established as preferable.

He also states, the trial must be designed as a critical test of therapeutic alternatives being assessed, and it must have a good chance of settling an open question. I would add the institutional review board as yet another mechanism for protecting patients' individual rights.

Some would argue that a patient is deserving of the benefits of the clinician's best judgment and at times that may be valuable information. One often has to go with clinical judgment. Sometimes that can compete with data from randomized trials. I believe that patients should get the best a doctor can provide which, from my vantage point, would mean both information from data derived from clinical trials but also from doctors' best clinical judgments.

Still, patients and their families have to realize that many pet hunches have been discarded over the years as a result of systemic clinical research. Many of us in this field have heard all kinds of claims for treatments and conviction by certain physicians and providers that X treatment or intervention is superior. And yet until one puts that to the critical test of systemic research, one can't be sure. For Passamani the obligation to the patients involved in the assessment of new drugs is the best that science and ethics can deliver. He states that today, for most unproven treatments, that means a properly performed randomized clinical trial.

There is a clear difference of emphasis in Hellman's and Passamani's papers. Perhaps that's why the *New England Journal* decided to put them both in the same edition. They represent contending positions, articulated by observers of the science/ethics interface. Yet the positions of the major participants, in my opinion, should be reconcilable. Perhaps that's really the direction that this family of concerned people here and around the country should take.

We are all interested, I would assume, in finding and delivering the best clinical care. Patients and families must be involved. They should help clinicians and researchers find rapprochement between concerns focused on the individual and concerns focused on the larger group. My sense of this is that these are changing situations. It would be productive to bring people from the array of constituencies together in such a fashion as to lay out a way of going forward that would be sensitive to the multiple interests and insure optimal continuation of research designed to produce better treatments.

Now I urge this, not just to utter a romantic wish, though I plead guilty to being a romanticist. But I believe our collective community has a very great outside threat with which to contend, and that is the threat of constricting resources for health and mental health treatments, and for research. Without appropriate reimbursement, patients will be shunted aside and treated shabbily, if at all. With diminished research monies, with decreased resources for the institutions that support research, new medications, new social and psychosocial treatments, will not be developed or refined.

I know there was acrimony that arose in connection with this conference. I purposely reinstated myself at the behest of Vera Hassner, who called me, to support the idea of researchers, clinicians, patients and families joining hands to solve their differences and turn their attention to protecting the programs that bring clinical care and psychiatric research to patients and families, while simultaneously protecting the rights of individuals. If we need resolutions, let the resolutions be to reassert an alliance supportive of patients and of the development of better treatments. The largest enemy, unfortunately, is the very large part of our country who today, regrettably, seem oblivious and at times even hostile to their co-citizens' needs.

I want to stop at this point. I made my statement relatively brief. My understanding is that this conference has previously used a discussion format, so I'm happy to stop and either hear other peoples' comments, respond to questions or stand aside and let the next presenter present. Thank you very much.

Audience: Dr. Pardes, thank you very much for that very articulate presentation. It was outstanding. Your comment about the threat to our collective well-being and diminishing resources, could you also comment as to your perception as to the effects of managed care arriving over the horizon.

Pardes: I think managed care is inevitable and yet worrisome. I think that the need to get control of the cost of health care was compelling. Unfortunately, no congressional, no executive action, I should say no federal action was taken. Managed care is one way by which costs are being addressed. My concern is the extent to which the need to address the bottom line starts to take dominance over the needs of individual people. And I think there is yet another version of managed care which may make that worry even more severe. That is capitation. In capitation one turns over resources to the provider community. The concern is as follows: that the patient walking into a doctor's office will be seeking care, and the doctor will have in mind the fact that the more tests, the more consultations and the like that he or she provides, the more he or she is putting himself at financial risk. He may even be going into a deficit which he has to handle. So one can get patients' needs and the financial bottom line conflicting right within the doctor's office. I thought there were a number of things in the bills that the congress considered which might have moderated some of that concern. The thing that troubles me particularly is that the resources that are being saved are being converted into bottom line profits for private companies and are not being reinvested in either better clinical care, teaching, research or the like. That's a very worrisome development.

Audience: Are you concerned at all that apparently, other than perhaps Dr. Sharfstein, no other representatives are here from the managed-care operation at this conference?

Pardes: I don't know the conference well enough to know, I can't comment on who should be here. What seems to me worth thinking about is to create a situation in which all the primary constituents sit together. If that includes the managed-care companies, that's fine. I think the first order constituents are patients, families, doctors, and researchers, and then there are many people who conduct the business of health care who belong as well.

Audience: I'd like to make a quick comment in relation to this, and that is, that the growth of managed care in this country has been extraordinary in health and mental health and this has been, I think, a broad experiment in which there has been no informed consent. And what I mean by that is that suddenly the payers came to a consensus that they had to cut costs and they did it through a process of very aggressive utilization review, utilization management, and second guessing clinicians' judgments, cutting off funds and this was done to peoples' insurance policies and nobody signed an informed consent or even participated in any informed consent process. I, you know, that's one of the interesting aspects of this, there was an Institute of Medicine study on the impact of managed care and they basically called it in terms of what's the outcome of it, they have no idea, but it's been a very broad social experiment that has occurred in this country. I do think that it is, there are, opportunities to take a look at what the impact has been, but at this point, but already there's been a lot of harm.

Pardes: If we are going to successfully contain health care costs in this country, the success of that effort must rely very, heavily on research. What we are dealing with in many cases is choice between alternatives of care, issues of cost effectiveness without data to drive some of those decisions. The likelihood is that they will be made on a cost basis alone. We have to work hard to insure there is enough of a database. I think unfortunately, the origin of this is that the business community has decided that they cannot tolerate the cost of health care, so a momentum has been developed which I see as irresistible. Here's where the role of federal government is important, and obviously up for grabs right now. The public is saying they want to decrease federal government involvement. Yet sometimes the federal government is critical to assert a sense of social justice and social equality. The marketplace is not preoccupied with social equality.

Audience: I'm Clare Griffin Francell, Director of Development for NAMI Curriculum and Training Network. I want to speak to you about developing relationships. You talked about interdeveloping relationships between psychiatry, researchers and families and consumers. We have had some experience back and forth at a national level at national conferences and at conferences like this, but do you have any suggestions about moving your peers in the psychiatric community towards some reality in states in exchanges of information, etc.? At state NAMI conventions we often ask researchers or psychiatrists to come and present at our meetings. We never receive the same invitations to present at their annual continuing education meetings. We never do. And when we ask sometimes for exhibit space at their continuing education conferences, it's denied. I think that if we are really going to go beyond lip service in relationship building, then I think that we have to ask each group, whether it's families, consumers or psychiatry, to go back to their grass roots opportunity and please make more opportunities for us. The other piece that families increasingly are aggravated and irritated about is that we have built a very good support system for families in difficulty, yet the referrals that come to us from either private and sometimes public psychiatrists are very infrequent. Yet we could help your profession by being extended, we are the care givers, so I'd like to ask if you have any thoughts on it or if you can use your influence to really put a relationship into practical, pragmatic terms.

Pardes: Well, if you will be patient with my recalling a bit of history. I was Director of National Institute of Mental Health I think in 1979 or 1980 when I was invited to a conference of families who had relatives with psychiatric illness in Madison, Wisconsin. Before that there was no organized citizen group other than the Mental Health Association. So when we talked about constituencies in the days when Steve Sharfstein and I were working on creating a new Mental Health Systems Act, the people there were professionals, state mental health commissioners, community mental health center directors and the Mental Health Association. At that meeting I was asked in a small session by about 10–12 family members whether I thought it was a good idea to start a national association, and I said yes. I said I'd be happy to help and we did, we tried to help by securing resources. NAMI grew up from that, and I feel I was at the birth of the NAMI organization. I don't have to be convinced about the wisdom of citizen and NAMI participation in a variety of educational programs. The mental health awareness program we do in the congress every year always includes a speaker from NAMI as well as research speakers and the American Psychiatric Association. At Columbia what we've done is develop an annual conference in which consumers and a variety of people play quite a role. We've tried to get away from this notion of having hierarchical divisions. I purposely made the point, and it's not peculiar to me, that I'm just a father and a parent who also has kids who get ill. We should break that notion of some differentiation between the two of us. My feeling is that NAMI can be helpful and an important contributor to educational conferences of whatever sort. It's my impression, and Harold Pincus might speak to it, that the APA has tried to include more in the way of citizen and consumer participation. As a matter of fact, Laurie Flynn now represents NAMI and sits at the board of trustee meetings of the APA with regularity. That doesn't mean that we are where we should be, but I would applaud your direction. I will certainly, try to use whatever influence I have to encourage that even further. I think it should be diffused not only at the federal level but throughout the states.

Audience: Thank you, and I do encourage you to move beyond the national scene in partnerships to the local.

Pardes: I heard you. Thank you.

Audience: I hate to get back to managed care, but I think that the issue of managed care really does point to something that's very important for this conference and that is that in an era of limited resources and limited health care resources allocation decisions are going to be made based upon scientific data and what kind of interventions either work or can be documented to work, and it's not just a situation where it's the health system looking for money, it's going to be different areas of health care that are going to be pitted against each other, and traditionally, I think mental health has been seen as a soft science, and to the extent that mental health can produce good research and show itself to be on as sound a therapeutic and scientific basis, it can only be of benefit to everybody.

Pardes: I'm sorry, you didn't introduce yourself.

Audience: I'm sorry, my name is Jay Zucker. I'm from New York State Office of Mental Health.

Pardes: Good to see you. You're right! There's going to be competition between the various parts of health and mental health has traditionally been a loser in those competitions. On the other hand, despite justified criticisms of the overall mental health movement, there can be some collective feeling of satisfaction in the fact that there's been substantial movement from 30–40 years ago. I was at the time a resident in Kings County Hospital and we were inundated with people coming in in enormous numbers and usually staying for a short while. We had very little in the way of medications. The acute phase of the psychotic process would run its course and people might go on to state hospital if it didn't or they might be released if it did. And so one had the sense of people coming in and out of a system with not a lot happening therapeutically. The good news is that psychiatrists and other mental health providers have something valuable to offer. Having said that, your point about the fact that we have to fight for resources to sustain the research effort in order to make sure that that movement continues, is, something I couldn't applaud loudly enough.

But I think the point goes even beyond that, because everybody in this room recognizes that there have been tremendous strides made. Even in the general public there is more of an awareness. But still there is not parity, even if you look at the Clinton health plan, it probably represented an increased awareness, but on the other hand, mental health was treated separately from general health care with limits and caps placed on it. The popular perception, I think someone referred to it yesterday, is Woody Allen's psychiatry. The popular perception is that people sit down on the coach and bitch and moan, and many conclude it's something that is not worthy of resources. Unless some consensus evolves so that there is a united front we have a big political challenge. That doesn't mean people have to agree on everything, but there needs to be at least the perception that the mental health community which includes the four groups that you mentioned before, is united on the idea that this is an area where we need care, we need treatment and it's just as valid as bypass surgery, and just as valid as chemotherapy for cancer. Unless that becomes a popular perception, then in an era of limited resources money is going to go elsewhere. That means that treatment and care is going to go unfunded or poorly funded, and advances in therapy will be pursued for the other diseases that people believe are real diseases and have a scientific basis.

Audience: I appreciated your statements. My name is Jean Campbell and yesterday I did a presentation that also talked about the issue of relationships and the multi-stake holder approach and collaboration, and people came up to me afterwards and applauded my idealism, not my romanticism, and I often say that without that, I mean, why go forward, I mean you have to, why is it necessarily that the baseline has to be a pessimistic cynical approach and I think much more in this field particularly in research, which is so proactive, is that you have to have that concept of hope and ideals. I don't think you need to separate that. One of the interesting things that I think is important to talk about too, that you didn't mention, is some of the relationships inside the club of research often what I find, I'm a consumer survivor as well as a researcher, but often in conferences I have to distinguish that I'm not just a consumer, that a consumer isn't a career choice, that I'm a researcher who has been informed by a lot of life experiences, one of the most profound one is my experience with psychiatric illness. However, in my professional roles, what I find is a tremendous amount of discrimination both towards myself and towards my field of work, and where we have so many young researchers that have, and a lot of this is due to the new treatments, who are now approaching research as a place where they can get an authenticity and make a contribution is that they are not supported, neither is their work, and then also within the professional communities they are not welcomed in the same way. There really is a struggle. Instead of saying please come in, we've been looking for you, you really have unique perspective both with professional skills and with this perspective to really move the field forward. There is the opposite. The opposite type of reaction and also, I think about all of the researchers that are already in our academic institutions who hide their psychiatric diagnosis and can't even speak up and make these contributions, and it's very sorrowful when you think about there is within the club itself a natural alliance that could talk about these things.

Pardes: I just want to make sure I understood what you said. You said that people who do psychiatric research, who coincidently suffered from a psychiatric illness themselves are treated in a second-class fashion and intimidated by virtue of the stigma within the very system itself.

Audience: Absolutely.

Pardes: I'm not entirely surprised by what you say, though disappointed perhaps. What's been impressive to me however is picking up on something you said, I think the number of people who are doing psychiatric research who've had some brush with psychiatric illness themselves or in their families, is more than anybody ever knows. If one could ever get the information out, it would be impressive. And there have been some very profound examples of that. I don't know how many of you may have been at the Lasker Award when Mogan Schou received the award for the development of lithium. Patty Duke was there, and told about her life before lithium and how much it had been changed by virtue of lithium. She said that since Schou had helped save her life, she was going to work to help save other people's lives through supporting research of that sort.

Schou, the researcher, broke into tears at the podium, and described the fact that he had gotten into the field because of manic-depressive illness in his family. What was poignant was that while he was there and upset, Patty Duke was comforting him.

It's interesting that you should raise a question about scientists and their own bouts with psychiatric problems. Coincidentally a few days ago an academic who is doing psychiatric research came to talk to me about the fact that his work had been compromised since he had had a couple of bouts of major depression. He wanted me to know that to understand that.

I don't pretend to know what's going on all over the field, but my impression is that this is a topic which has not been given very much attention and deserves it. You've sensitized me and all of us in the community should try to kind of make stronger points of it.

Stigma is a funny thing. It's not easy to just talk it away. Several things that are important have diminished stigma. One has been the development of knowledge and the ability to treat better. I agree with the previous speaker, that there's a long way to go, but the fact that one has something to offer seems to have some real effect on the intensity of stigma and second, the rise of the citizen movement in the last 15 years has been perhaps one of the most important things that's ever happened in this field. We did not have an alliance before because there weren't people able to get out and speak up. I believe that's getting better and will continue to get better.

Related to the rise of the citizen movement is the view in research that we have to have a certain humility that we only have partial truth. It isn't just a matter of doing the right thing in terms of rights, but it's also in doing good science that we need to have multiple perspectives at the table in order for our field to move forward.

Audience: Amen.

Audience: Just a quick comment on what was brought up by the gentleman from New York, Sam Keith and others, when Sam was at NIMH, prepared some very nice material and very compelling slides to compare the effectiveness of treatments in psychiatry with many other areas of medicine. I think anyone who's discussed this in public forums will find that information very useful. Someone here may know how to access those slides and that material and if they did, it would be useful to say because they are very, very helpful.

Pardes: I think what I would do is ask Paul Sirovatka, if he's still there or Harold Pincus.

(comment in background—not able to hear well)

I agree with Dr. Carpenter, that is very valuable information to have because what it shows is that the given effectiveness of a psychiatric treatment compared, for example, to other treatments like treatments for heart disease, is impressive. That's not designed to say that there aren't large numbers of people for whom we don't have good enough treatments, but it's designed to help the argument to a person outside the field who's making budgetary decisions that you're helping a field which can produce help. That's important.

(comment in background)

Pardes: Just a comment again, the issue of patient outcome, perhaps that's one positive that may come out of the requirements of managed care, which is to look at patients' outcome. Good patient outcome after all is precisely what we want. The APA is now coming out with clinical practice guidelines for the various conditions. While people argue whether guidelines become set in stone, the point is until there were no guidelines and so how to evaluate outcome was difficult.

Audience: I'm Bob Levine from Yale. Hi, Herb. On the points of building relationships, I think one thing that has not been discussed yet is that the economic crunch related to managed care is having a major effect on the academic institution, as the managed-care corporations buy up all of our affiliates and dictate who can and cannot be referred for consultation, and as they have an affect on relationships within the physician or health care provider patient relationship dictating that you must see a patient one patient every ten or 7½ minutes and so on, the same thing is happening in the university. As our sources of funding are beginning to dry up or get curtailed by things of this sort, the faculty are told you must spend more and more of your time in money producing efforts and in the clinical departments, that often translates into seeing patients. We get no credit for seeing patients who don't have insurance. We are called upon to be private philanthropists to see patients who don't have any insurance coverage, so there's pressure on us to do what we all believe is the wrong thing, and to set poor examples for the medical students. I bring this up now, because as I hear several articulate speakers calling for increased relationships between the communities of consumers and communities of researchers and health care practitioners, this is a powerful force that's standing in the way of that, as academic values are more and more being interpreted in terms of what is their impact on the bottom line. We're not talking about avarice we're talking about in many cases institutional survival. There are some institutions that simply will not survive this unless they can discover new ways to get financing for their institutional operations. Thank you.

Pardes: I want to first comment on what Vera said and then I want to go back to what Bob said. As you point out, managed care is a mixed blessing. But, one of the values is going to be a greater attention to outcome. Services research is going to be necessary, because provider institutions are going to have to take a look at what they do and try to demonstrate that to the community.

Now I want to just pick up on what Bob said because it's the subject of a whole other talk, but to those of us who care about seeing a steady sustaining of research to enable us to really make the progress that can be made in the field, his comment about institutional survival should be underlined. I have never seen a period of greater concern in the academic biomedical research community than there is today. And the reason for that is several fold. To expand a little bit on what Dr. Levine said, number 1, is the fact that the monies that are trickling into the bottom line of managed-care companies are being squeezed out of clinical care and often that means out of academic revenues. Number 2, academic institutions do a lot of unreimbursed care. The teaching hospitals in this country do over 50 percent of the total indigent care and don't get paid for it. There are some mechanisms to modify this, but they're under threat right now all over. And third, the research institutions, be they Yale, Hopkins, Duke, Stanford or Columbia, could be in mortal danger if this current government makes an excessive cut in indirect costs, which is what basically supports institutions doing research. And it's something that all of us who are concerned about this field worry about.

It seems unthinkable that the biomedical research effort in the United States, which is probably better than anything else even close to it around the world, could be under any threat. But it is. And this collective set of pressures to which Bob introduced us, if you will, is very worrisome.

Audience: I'd like to add another underscore to what Dr. Levine and what you are just saying and from a public policy perspective. I think what has happened is that the physicians, the hospitals and the patients, particularly I think that through managed care the physicians and patients are being screwed out of the system. And I think that if we are talking about a war, I think it's time that we also join forces in fighting what's happening that is allowing the insurance industry to become a health provider rather than the physicians in the hospitals, and the people who are making the money are the board of directors of the managed-care systems.

Pardes: I totally agree. You said it well enough, I applaud what you say. Let me close by saying I want to thank the organizers of the conference for inviting me. I want to thank Vera for calling me to make sure that I came. It is critical that the alliance of patients, families and scientists be sustained to insure progress in the field.

NAMI'S Standards for Protection of Individuals with Severe Mental Illnesses who Participate as Human Subjects in Research

Adil E. Shamoo,[†‡] J. Rock Johnson,[‡] Ron Honberg,[‡] and Laurie Flynn[‡]

[†]*Center for Biomedical Ethics, and the Department of Biochemistry, School of Medicine, University of Maryland at Baltimore, 108 N. Greene St., Baltimore, Maryland 21201*
[‡]*National Alliance for the Mentally Ill, 200 N. Glebe Rd., Arlington, Virginia 22203*

In response to inquiries from one of its members, the National Alliance for the Mentally Ill (NAMI) began to focus attention in 1991 on informed consent and other issues related to the participating of individuals with mental illness as human subjects in research. In 1993, Adil Shamoo, Ph.D., a bioethicist, father of a child with mental illness, and current member of the NAMI Board of Directors, published a paper with Diane Irving, Ph.D. in *Accountability in Research* entitled "Accountability in Research Using Persons with Mental Illness."

During this same year, at NAMI's initiative, a series of meetings were held between NAMI and the National Institute of Mental Health (NIMH), involving discussions between leading scientists, family members and consumers. These meetings facilitated greater understanding of the complex issues involved in the use of human subjects in research in psychiatry. An important symposium was held on this subject at the 1993 NAMI Convention in Miami Beach, Florida, which included presentations by Dr. Shamoo; David Shore, M.D., NIMH; Fred Frese, Ph.D., a psychologist and leading consumer advocate; and John Petrila, J.D., Director of the Law and Mental Health program at the University of South Florida.

As an outgrowth of the NAMI-NIMH meetings and the symposium at the NAMI Convention, NAMI resolved to undertake the development of policies governing protections of individuals with severe mental illnesses who participate as human subjects in research. These policies were drafted by Dr. Shamoo, in collaboration with J. Rock Johnson, a consumer, NAMI Board Member and attorney; Laurie Flynn, NAMI's Executive Director; and Ron Honberg, NAMI's Legal Counsel. Extensive input was received from the scientific community, from the NAMI Consumer Council and from other members of NAMI. In early January, 1995, the conference on which these proceedings is the product was held in Baltimore, Maryland.

The conference lent greater understanding and appreciation to the discussions of these issues. After nearly one year of consideration and revisions, the NAMI Board of Directors adopted the attached policies at its February, 1995 meeting.

The NAMI policies seek to achieve a balance between protection of the rights of individuals and their families who participate in research as human subjects and the requirements of the research community. The policies set forth standards pertaining to many aspects of the research process, including informed consent procedures, the important role of families and consumers on institutional review boards (IRBs), procedures for substitute consent when individuals are incompetent to provide informed consent, the responsibility of researchers to link individuals who terminate participation in research protocols with necessary and appropriate aftercare, and mechanisms for ensuring that individuals who participate in protocols involving assessment of new medications have continuing access to those medications.

NAMI urges the psychiatric research community to take all appropriate steps, in accordance with these standards, to facilitate maximum protections for vulnerable individuals who participate as human subjects. In the future, we intend to consider additional policies concerning important issues regarding research on human subjects which have not been addressed in these policies. As occurred in the development of these policies, extensive input will be sought from families, consumers and members of the scientific community. As always, we appreciate the vitally important role of science in discovering important new ways for treating and perhaps ultimately curing devastating diseases of the brain such as schizophrenia and major affective disorders.

POLICIES ON STRENGTHENED STANDARDS FOR PROTECTION OF INDIVIDUALS WITH SEVERE MENTAL ILLNESSES WHO PARTICIPATE AS HUMAN SUBJECTS IN RESEARCH (ADOPTED BY THE NAMI BOARD OF DIRECTORS, 2/4/95)

1. NAMI accepts the critical necessity for research using human subjects, acknowledges the important contribution of persons who become human subjects, and affirms that all such research should be conducted in accordance with the highest medical, ethical and scientific standards.

2. National standards to govern voluntary consent, comprehensive exchange of information, and related protections of persons with cognitive impairments who become research subjects must be developed, in which the interests of persons who become human subjects, families and other caregivers are included.

3(A). Participants in research and their involved family members must be fully and continuously informed, orally and in writing, about all aspects of the research throughout the process. Research investigators must provide information in a clear, accessible manner to ensure that participants and their involved families fully understand the nature, risks and benefits of the research.

3(B). The consent protocol must provide information which is clear and understandable on an individual basis for each participant and their family members. The consent protocol must provide information on the purposes and scale of the research, what is hoped to be learned and prospects for success, potential benefits and potential risks to the individual (including options for treatment other than participation in research, since research is not the same as treatment). The consent protocol should also contain information concerning the function of the institutional review board (IRB), the identity, the address, and telephone number of the IRB administrator and other information, as appropriate.

3(C). Whenever consent is given by someone other than the research participant, the participant and involved family members must receive information on the same basis as the person actually giving consent.

4. Research participants should be carefully evaluated before and throughout the research for their capacity to comprehend information and their capacity to consent to continued participation in the research. The determination of competence shall be made by someone other than the principal investigator or others involved in the research. Except for research protocols approved by the Institutional Review Board (IRB) as minimal risk, whenever it is determined that the subject is not able to continue to provide consent, consent to continue participation in the research shall be sought from families or others legally entrusted to act in the participant's best interests.

5. Institutional review boards which regularly review research proposals on severe mental illnesses must include consumers and family members who have direct and personal experience with severe mental illness.

6. Members of IRBs approving research on individuals with severe mental illness must receive specialized training about mental illness and other cognitive impairments and the needs of individuals who experience these disorders. Persons with severe mental illness and members of their families must be integrally involved in the development, provision and evaluation of this training.

7. Without penalty, a research participant is free to withdraw consent at any time, with or without a stated reason. Any time a participant terminates participation, regardless of reason, investigators will make every effort to ensure that linkgages to appropriate services occurs, with followup to assist that participant to establish contact with appropriate service providers and/or care-givers. If a participant disappears or terminates their continued consent, the investigator shall contact his/her family or others designated to receive notification and information.

8. When participation by an individual in a research protocol is completed, participants and/or their families are entitled to be informed of results as soon as this information is available, to have the opportunity to receive feedback concerning their individual participation in the protocol, to critique the protocol, and to provide input concerning possible additional research.

9. All participants in research protocols involving the assessment of new medications will be provided with opportunities by the investigator for a trial on the medication being studied, so long as other research on the new medication has demonstrated potential safety and efficacy.

10. All individuals who have benefitted from the administration of experimental medications in research will be provided continual access to the medication by the investigator without cost until a source of third party payment is found.

Ethics in Neurobiological Research with Human Subjects—Final Reflections

Jay Katz

Yale Law School, 127 Wall Street, New Haven, Connecticut 06511

We owe a great debt to Adil Shamoo for convening this conference and for not caving in to pressures to cancel it. Now, with our three-day meeting being almost over, it is clear that the conference did not serve, as some had feared, as a platform to indict research or to malign physician-investigators. Instead, our gathering allowed us to listen to many different voices; voices that challenged us, voices that expressed frustration, resentment, disbelief and righteous indignation. That this happened here bears testimony to the significance of creating fora, all too rare in medicine, which are designed to invite, rather than to stifle, controversy. The conference's ultimate impact, however, will not be realized unless we take the time, once we have returned home, to reflect on what has transpired here. The different voices we have heard should jar long-held beliefs. Only then may we be able to consider whether reconciliation of our disparate views is possible, at least to a greater extent than we had envisioned before coming to Baltimore.

All of us agreed that research with human beings is important and necessary. The question was not *whether* human research should be conducted or not; the question was *how* human research should be conducted. From the perspective of the history of human experimentation, throughout which many argued either against any human experimentation or against the necessity for its regulation, the shift from whether to how is quite a remarkable achievement in at least two ways: (1) It recognizes that therapy without being informed by research makes the former such an uncertain enterprise that every therapeutic intervention becomes guesswork, subject to the vagaries of undisciplined observations. (2) It draws attention to the fact that research, however promising in relieving human suffering, cannot be pursued with unfettered license, that it must be constrained by whatever answers we give to the "how."

Central to the question of how—and on this question there was disagreement among us—is the crucial problem of the ways in which the invitation to participate in research should be extended. Before turning to that issue, I would like to suggest that for human research the vexing issue, contrary voices notwithstanding, is not the *physical* harm that research may inflict on subjects of research but the *dignitary*

This article will also appear in the journal: *Accountability in Research*, Volume 4, pp. 277–283 (1996).

harm inflicted whenever subjects' consent is manipulated by incomplete, evasive, misleading and deceptive disclosures. My reasons for being less concerned about physical harm are many: (1) The possibility of physical harm is an inevitable byproduct of research. After all, research is a voyage into the unknown and physical harm therefore may always be a consequence, however much moderated by preliminary data on toxicity and efficacy accumulated from all pertinent sources. (2) Subjects may be willing to suffer physical harm not only for altruistic reasons to advance the frontiers of knowledge but also for self-serving reasons to moderate the deleterious effects of standard therapies. (3) Physician-investigators have always taken great care, some egregious examples to the contrary notwithstanding, to find ways of minimizing physical harm. The ancient Hippocratic commandment—"above all do no harm"—is deeply embedded in physicians' psyche so that they strive to avoid all unnecessary physical suffering. (4) Institutional review board (IRB) scrutiny of research protocols assures that physical harm to research subjects is kept to a minimum, aided by the presence on IRBs of knowledgeable research scientists who can identify avoidable physical risks to research subjects by appropriate modifications in the protocols submitted. Indeed, that aspect of review has been one of the significant contributions that IRBs have made to the greater protection of subjects of research. (5) Finally, and more generally, while some subjects are of course inevitably harmed, the point, frequently made, must be granted that as a group they may be harmed physically less, or at least not more, than are patients with similar diseases in therapeutic settings. This comparative observation has led physician-investigators to complain that it is unjustified to burden them with obligation more onerous than those imposed on physicians in the practice of medicine. While that argument raises profound concerns about therapeutic practices, it should not distract attention from another ethical issue: That in therapy physicians are expected to attend solely to the welfare of the individual patient before them, while in research patient-subjects also serve as means for the ends of science and therefore their individual needs have to yield to the dictates of the research question and protocol. This crucial difference has far-reaching implications for the ethical conduct of research.

The most pressing problem for finding answers to the question of how research should be conducted resides in becoming clearer about the ways in which the invitation to participate must be extended. (To be sure, there are other issues; for example, that the project is important enough, and based on solid methodological grounds, to warrant asking human beings to join investigators in their endeavors. Consent is a necessary, but not sufficient, justification for using human beings as subjects of research.) At the conference we spent considerable time exploring this issue and we disagreed. On reflection I believe that we disagreed less on the proposition that subjects must be informed or give consent and more on the nature and quality of disclosure and consent that we consider adequate for human experimentation to proceed. I want to address our controversy about the adequacy of current informed consent practices in order to illustrate my misgivings about the ways in which the invitation is being proffered. From time to time I shall expand on this problem by addressing another question, particularly important for research with mentally impaired patient-subjects: Should IRBs be authorized to permit research to proceed whenever consent is in doubt or when for good reasons disclosures are not fully made, since doing so conflicts with fundamental democratic values of individual autonomy and self-determination?

Some commentators noted that in research "an exquisite degree of cooperation" exists between investigator, subject, and often their families, and that therefore informed consent is less of a problem than it has been made out to be. But such cooperation, I believe, cannot be presumed. If it exists at all, it must still be grounded in a prior explicit mutual understanding of the competing interests that bring the parties together; for the posited cooperation can be based, for example, on subjects' expectations as to what an investigator—whom they view more as a physician than an investigator—will and will not do, expectations which the investigator, constricted by the dictates of the research protocol, cannot fulfill. Thus, cooperation can result from false assumptions that deserve clarification.

Other commentators argued that it is not informed consent but the moral integrity of the investigator which best protects subjects of research. Such sentiments echo Henry Beecher's who believed that while informed consent was important, "the more reliable safeguard [is] provided by the presence of an intelligent, informed, conscientious, compassionate, responsible investigator" (Beecher, 1966). But how should, or can, "responsible investigators" reconcile their responsibilities to advancing science with those to their patient-subjects, to their careers, to their staff, to the institutions who employ them, or to the institutions who provide grant money and expect results as quickly as possible? In prior writings I have called attention to the current reality which places physician-investigators at the mercy of the institutions in which they work and the private and public grant agencies which support their research. Medical research since World War II has become a research-industrial complex. Academic institutions rely on the revenues which accrues from the assessment of indirect costs to the providers of grants. Research proposals have to be generated and completed at a rapid rate to assure future grant support. Thus, investigators are under considerable pressure to recruit subjects as quickly as possible to support the institutions' buildings, laboratories, staff and salaries. With respect to career advancement and future grant support, physician-investigators are thus the victims of an institutional system (their own institutions and the National Institutes of Health) which penalizes them if in fulfillment of their ethical disclosure obligations toward patient-subjects, the pace of research would be slowed" (Katz, 1993). With all those competing moral obligations, investigators' lot is not an easy one.

It is much too facile to assume that the morality of the research enterprise can be safeguarded, as some argued at the conference, by getting rid of "the few rotten eggs." Viewing it solely as a rotten egg problem is dangerous because it diverts attention from the endemic moral tension inherent in the conduct of research that defy easy resolution: Advancing knowledge for the benefit of mankind, on the one hand, and protecting the inviolability of subjects of research, on the other (Katz, 1972). To resolve, or at least contain this tension so that competing moral values will not be unduly undermined requires constant scrutiny and vigilance.

Let me turn now to informed consent which I consider a cornerstone on which the morality of the research enterprise must rest. Crucial to informed consent are the ways in which the invitation to participate in research is extended. I would like to suggest, and for purposes of further debate I shall put it most starkly, that to obtain a morally valid consent, the physician-investigator must make, among others, the following disclosures: (1) that the subjects are not only patients and, to the extent to which they are patients, that their therapeutic interests, even if not incidental, will be subordinated to scientific interests; (2) that it is problematic and indeterminate whether their welfare will be better served by placing their medical fate in the hands of a physician rather than an investigator; (3) that in opting for the care of a physician they may be better or worse off and for such and such reasons; (4) that clinical research will allow doctors to penetrate the mysteries of medicine's uncertainties about which treatments are best, dangerous, or ineffective; (5) that clinical research may possibly be in the patient's immediate best interest, perhaps promise benefits in the future, or provide no benefit, particularly if the patient is assigned to a control (placebo) arm of a study; (6) that research is governed by a research protocol and a research question and, therefore, his or her interests and needs have to yield to the claims of science; and (7) that physician-investigators will respect whatever decision the subject ultimately makes.

The disclosure obligations which I have set forth emphasize the need to explain to patient-subjects how participation in research differs from how they would ordinarily be treated or would expect to be treated. Thus, the first task in extending the invitation is to be absolutely clear about the research dimension of the invitation, its implications and possible consequences. Such disclosures do not require patient-subjects to understand the esoteric knowledge of medicine and science. Indeed, at present, subjects are overwhelmed with unnecessary scientific information that clarifies little and serves more the purpose of obscuring the crucial information that they need to know: (1) The risks, benefits, alternatives, and uncertainties which patient-subjects face whenever they agree to participate in clinical research; (2) The impact of their participation, known and conjectured, on the quality of their remaining lives which at times is severely compromised by the administration of toxic agents in the face of desperate medical conditions. Investigators have an obligation not to hide behind esoteric scientific information but to translate it into language which is not only understandable but also relevant to patient-subjects' lives and interests.

An observation frequently made at the conference was that "informed consent is not an event but a process." It is not quite clear to me what the commentators had in mind in putting it that way. A number seemed to refer to the informed consent form; others to the importance of the conversation between investigators and subjects. Yet, the latter is a process which in today's world is all too briefly and perfunctorily conducted, and for allegedly "good" reasons: That honest interactions between patient-subjects takes too long, may lead to too many refusals, endangering the scientific reliability of the project by selection bias or worse, may make it impossible to find enough subjects to conduct the project. If that were to happen, progress in science can only suffer. Therefore circumspection in what one reveals is crucial, otherwise one's moral obligations to future generations are subverted. Yet, the morality of human experimentation confronts here another stark choice: whether to opt for a slower rate of progress, with its grievous implications for the conquest of disease, or to opt for progress as an uncompromising commitment, with its grievous implication for safeguarding the dignity of self-determining individuals.

This brings me to the informed consent form as a written document that could testify to the respect accorded to choice-making individuals. Like many others, I too have been a critic of these forms as currently written. They are not only incomprehensible but, for example, as I have already suggested, also include too much distracting technical information of little consequence to the decisions patient-subjects must make. At present, as Alan Meisel cogently observed, "[C]onsent forms play no [meaningful] role in the informed consent process. Where used they [merely] memorialize that in fact the informed consent process transpired" (Meisel, 1982). Indeed, to return once more to the comment often made during the conference that "informed consent is a process rather than an event"—and in putting it that way downgrading the significance of the informed consent form—makes it as Meisel suggested, the meaningless scrap of paper which it has become. It can, however, be more than that if it is viewed as the *final* event at the *end* of the process of conversing with patient-subjects. Then it "memorializes" in simple language what subjects have understood about the research dimension of their participation; i.e., how research differs from therapy, how their interests, but with their consent, have to be compromised by the dictates of the protocol, how quality of life will be affected if the research cannot live up to its promise for a better and longer life. All this can be set forth with candor if it is done in the spirit of allowing subjects to make the best choice they can make, however "rationally" or "irrationally."

I have gone to some length in giving my version of what informed consent is all about, for I want to make another point as well: IRBs do not have the authority to modify the informed consent form in the ways I would like to see them modified. IRBs are compelled by the federal regulations to include in the informed consent form information that is useless and obscures rather than clarifies. For, ever since 1974 when the federal regulations were enacted, investigators and IRBs had to comply with the requirements for informed consent by formalistically adhering to the stated criteria, setting them forth in the protocols and in the informed consent documents signed by subjects. As a consequence the mass of required information obscures the vital information a subject needs to have in order to arrive at a considered judgment about participation. Second, the criteria which would give subjects the best understanding of what they are agreeing to are often not clearly specified. Many members of IRBs with whom I talked have voiced concerns over the flawed nature of the informed consent process. They are aware of protocols and consent forms in which the risks of participation are minimized; in which the unlikelihood of anything of benefit accruing to individual patient-subjects is obscured. I believe that the informed consent form could become, what it is not now, a meaningful and important document attesting to the fact that there has been a meeting of the minds between investigator and subject.

What I have set forth here, i.e., the need to fundamentally revise the informed consent form, is only an example of a larger issue: The need for a radically different approach to the regulation of human experimentation. I made this point already twenty-two years ago when I chaired the subcommittee of the Tuskegee Syphilis Study Ad Hoc Advisory Panel, charged with the task of recommending changes in the then existing research policies and procedures (1973). We proposed that Congress "establish a permanent body [we called it the National Human Investigation Board] with the authority to *regulate* at least all Federally supported research involving human subjects." The Board, we suggested, should be independent of the Department of Health, Education and Welfare, for we did not believe that "the agency which both conducts [research] and supports much of the research that is carried on elsewhere is in a position to carry out disinterestedly the functions we [had] in mind." Most importantly we recommended that the Board must not only *promulgate* research policies but also *administer* and *review* the human experimentation process. Only constant interpretation and review by a body that is not advisory but whose decisions count by virtue of the authority invested in them, we argued, can strike a balance between protecting both the claims of science and society's commitment to the inviolability of subjects of research. I have recently published a brief article on that topic (Katz, 1995). Here I only want to proffer one other observation: The Federal Regulations essentially codified criteria for informed consent in the conduct of research but, with limited and vague additional instructions, left it to local IRBs to review—"approve, require modification in, or disapprove"—research proposals submitted by their local institutions. Thus, IRBs are forced to make decisions that are compromised by the limited time and resources available to them; by the pressures of their institutional colleagues to approve their protocols as quickly and unquestioningly as possible; and by their lack of expertise to consider in any depth the complex legal, ethical, and societal problems that human experimentation poses. The latter is particularly true for research with mentally impaired patient-subjects. For IRBs to authorize physician-investigators to proceed without, or with compromised, consent is tantamount to conscripting citizen-patients for participation in research. Such practices, unless sanctioned by the State, is deeply offensive to democratic values. It surely cannot be left to the discretion of IRBs as it is at present. In short, IRBs are left at sea, given a mandate that they cannot responsibly carry out without recourse to an overarching body that would structure and guide their work.

The problems inherent in research with human subjects—advancing science and protecting subjects of research—are immense. Society can no longer afford to leave the balancing of individual rights against scientific progress to the low-visibility decision-making of IRBs with regulations that are porous and invite abuse. The recent revelations about the radiation experiments conducted by governmental agencies and the medical profession once again confront us with the human and societal costs of too relentless a pursuit of knowledge at the expense of other moral values. If this is a price worth paying, society should be forced to make these difficult moral choices in bright sunlight and through a regulatory process that constantly strives to articulate, confront, and delimit the costs.

Finally, I want to express the hope that we shall meet often in Baltimore and elsewhere and air our frustrations, disagreements, and expectations. If we then listen to one another, as we have done during the past three days, we may in time learn how to reconcile our moral obligations to medicine, science and society as well as to ourselves and our patient-subjects. As Hippocrates reminded us millennia ago: "Life is short, the Art long, Opportunity fleeting, Experiment treacherous, Judgment difficult" (Hippocrates, trans. James, 1967).

REFERENCES

Beecher, H.K. (1966) Ethics in clinical research. *New England Journal of Medicine* 274:1354.

Hippocrates. *Aphorisms,* trans. W. James. (1967) Cambridge: Harvard University Press.

Katz, J. (1972) *Experimentation with Human Beings*. New York: Russell Sage Foundation.

Katz, J. (1993) Human experimentation and human rights. *Saint Louis University Law Review* 38:7.

Katz, J. (1995). Do we need another advisory commission on human experimentation? *Hastings Center Report* Jan.–Feb., 25:29.

Meisel, A. (1982) Comments to T.M. Grundner, more on making consent forms more readable. *Institutional Review Boards* 4:9.

Tuskegee Syphilis Study Ad Hoc Advisory Panel. (1973) Final Report.